Principles of Statistical Genomics

Shizhong Xu

Principles of Statistical Genomics

 Springer

Shizhong Xu
Department of Botany and Plant Sciences
University of California
900 University Avenue
Riverside, California
USA

ISBN 978-0-387-70806-5   ISBN 978-0-387-70807-2 (eBook)
DOI 10.1007/978-0-387-70807-2
Springer New York Heidelberg Dordrecht London

Library of Congress Control Number: 2012942325

Printed on acid-free paper

Springer is part of Springer Science+Business Media (www.springer.com)

Preface

Statistical genomics is a new interdisciplinary area of science, including statistics, genetics, computer science, genomics, and bioinformatics. The rapid advances in these areas have dramatically changed the amount and type of information available for characterization of genes. In many genomic applications, existing methods coupled with new computational technology have successfully directed the exploration of high-dimensional data. What remains to be accomplished is the successful statistical modeling of genomic data to support hypothesis-driven biological research. This will ultimately lead to the exploitation of the predictive wealth that much of the current and impending genomic data have the potential to offer. Statistical development will continue to significantly amplify and focus the molecular advances of the last decades toward general improvements in agriculture and human health.

Using advanced statistical technology to study the behavior of one or a few Mendelian loci defines the field of statistical genetics. For complex traits, such as grain yield in crops and cancers in human, one or two loci are rarely sufficient to explain majority of the trait variation. People then study the behavior of all genes influencing a trait without distinguishing the effects of individual genes, creating a field called quantitative genetics. Taking advantage of saturated DNA markers generated with advanced molecular technology, we are now able to localize individual genes on the genome that affect a complex trait, which leads to this new field of statistical genomics or quantitative genomics. In statistical genomics, we emphasize the notion of whole genome analysis and evaluate the joint effect of the entire genome on a quantitative trait.

Any genome study requires a sample of individuals from a target population and genomic data collected from this sample. Genomic data include (a) genotypes of molecular markers, (b) microarray gene expressions, and (c) phenotypes of clinic or economic traits measured from all individuals of the sample. Any particular genomic study may involve all the three types of data or any two of them. With the advanced biotechnology, molecular marker data will soon be replaced by whole genome sequences. In the narrow sense, phenotypic data are not genomic data but the ultimate purpose of genomic data analysis is to dissect these traits and understand the genetic architectures of these traits. Therefore, phenotypic data are essential in genomic data analysis. This is why phenotypic data are included as genomic data. When a study involves phenotypes and marker genotypes, it is called QTL mapping where QTL stands for quantitative trait loci. A study involving phenotypes and microarray gene expressions is called differential expression (DE) analysis. If a study involves marker genotypes and microarray gene expressions, it is called expression quantitative trait locus (eQTL) analysis. The purpose of QTL mapping is to find the genome locations, the sizes, and other properties of QTL through associations of marker genotypes with the variation of a quantitative trait. In DE analysis, the phenotype of interest is usually binary such as case (represented by one) and control (represented by zero). The primary interest of DE is to find genes that express differently in case and control. The purpose of eQTL mapping is to find regulation pathways of the genes. Transcripts mapped to the same locus of the genome are considered in the same regulation pathway.

Many statistical models, methodologies, and computing algorithms are involved in the textbook. Major statistical models include the linear model (LM), the generalized linear model (GLM), the linear mixed model (LMM), and the generalized linear mixed model (GLMM). In a few places, the hidden Markov model (HMM) is required to infer the unobserved genotypes of QTL given observed marker genotypes. Another important model is the Gaussian mixture model for cluster analysis. Commonly used statistical methods include the least squares (LS) estimation, the maximum likelihood (ML) estimation, the Bayesian estimation implemented via the Markov chain Monte Carlo (MCMC) algorithm, and the Bayesian method via the maximum a posteriori (MAP) estimation. Optimization technologies include the Newton–Raphson algorithm, the Fisher scoring algorithm, and the expectation–maximization (EM) algorithm. For the Newton–Raphson algorithm, if the first- and second-order partial derivatives of the target function with respect to the parameters are easy to derive, an explicit form of the iteration equation will be given. Otherwise, numerical evaluations of the partial derivatives are calculated using some powerful numerical differentiation subroutines. In genomic data analysis, the number of parameters is often very large, updating all parameters simultaneously can be prohibitive. In this case, a coordinate descent approach may be taken, in which one parameter is updated at a time conditional on current values of all other parameters. This approach can improve the robustness of the optimization algorithm and save much computer memory but at the cost of computing time and risk of trapping to a local solution of parameters.

This book was compiled from a collection of lecture notes for the statistical genomics course (BPSC234) offered to UCR graduate students by the author. Approximately half of the material was collected from studies published by the author and his research team. A small proportion of the remaining half consists of some unpublished works conducted by the author. Much of the remaining half of the book represents a collection of the most updated statistical genomic methods published in various journals for the last couple of decades. The topics selected purely reflect the author's choices for the course according to the level of understanding of the target students. The book is not an introduction to statistical genomics because statistical genomic is a diversified area including many different topics, and this book only covers a proportion of the topics. However, the statistical technologies chosen represent the core of statistical genomics. Understanding the principles of these technologies, students will easily extend the methods to other analyses of genomic data generated from different experimental designs. Although the book narrowly focuses on a few topics, each topic introduced is provided with the derivation of the method or at least a direction leading to the derivation. Statistical genomics is a multidisciplinary area with a rapid development. Writing a comprehensive book in such an area is like shooting a moving target. For example, during the time between the completion of the first draft and the publication of this book, new technologies and methodologies may have already been developed. Therefore, the book can only focus on the principles of statistical genomics. Most recently developed methods may not be covered, for which the author owes an apology to those researchers whose works are relevant but not cited in the book.

The book consists of three parts. Part I contains Chaps. 1–4 and covers topics related to linkage map construction for DNA markers. Part II consists of Chaps. 5–16 and is the main part of the book. These chapters cover topics related to genetic mapping for quantitative trait loci using various designs of experiments. Part III (Chaps. 17–25) covers topics related to microarray gene expression data analysis. This book intends to be used as a textbook for graduate students in statistical genomics, but it can be used by researchers as a reference book. For advanced readers, they can choose to read any particular chapters as they desire. However, for junior researchers and graduate students, it is better to study from the beginning and not to escape any chapters because some of the methods introduced in early chapters will be used later in the book and they will only be referenced.

Former and current postdocs and graduate students in the lab all contributed to the material published by the UCR quantitative genetics team. Postdocs who contributed to the material relevant to this book include Damian Gessler, Chongqing Xie, Shaoqi Rao, Nengjun Yi, Claus Vogl, Chenwu Xu, Yuan-Ming Zhang, Lide Han, Zhiqiu Hu, and Fuping Zhao. Graduate students involved in the research include Lang Luo, Yun Lu, Hui Wang, Yi Qu, Zhenyu Jia, Xin Chen, Xiaohong Che, and Haimao Zhan. Without their hard work, the author would not have been able to publish this book. Their contributions are highly appreciated. In the main text, I choose to use the first person plural pronoun "we" instead of "I" for the very reason that the book material was mainly contributed by my research team. In the UCR quantitative genetics team, Nengjun Yi made the most contribution to the material included in the book and thus he deserves a special acknowledgement. A special appreciation goes to the three current members of the UCR quantitative genetics team, Zhiqiu Hu (postdoc), Haimao Zhan (student), and Xiaohong Che (student), for their help in drawing the figures, checking the accuracy of equations, and correcting errors occurred in an early draft of the book.

Riverside, California, USA   Shizhong Xu

Contents

Part I
Genetic Linkage Map

Chapter 1
Map Functions

Genes are physically located on chromosomes. Because chromosomes are string-like structures, genes carried by chromosomes are arranged linearly (Morgan 1928). In other words, a particular location of a chromosome can only be occupied by one and only one gene. Figure 1.1 shows a marker map of the barley genome (Hayes et al. 1993), where the names of markers are given in the right-hand side of the chromosomes and the marker positions measured in centiMorgan (cM) are given in the left-hand side. In genetics, the terms gene and locus are often used interchangeably, but a more precise description of a chromosome location is the locus. Genetic loci carried by the same chromosome are physically linked, and thus, they tend to cosegregate. These loci are said to be in the same linkage group. The distribution of loci among chromosomes and the order of loci within chromosomes are called genetic map. Using observed genotypes of these loci to infer the genetic map is called genetic mapping or linkage analysis.

1.1 Physical Map and Genetic Map

The size of a genome is determined by the number of chromosomes and the size of each chromosome. In general, there are two types of measurements for the size of a chromosome. One measurement is the number of base pairs (bp) of the DNA molecules carried by the chromosome. The order and relative localization of all the loci in a genome with distances between consecutive loci measured by the numbers of base pairs is called the physical map. Unfortunately, one must sequence the entire chromosome to determine the size of the chromosome and sequence the entire genome to determine the genome size. During meiosis, pairs of duplicated homologous chromosomes unite in synapsis, and then nonsister chromatids exchange segments (genetic material) during crossing-over, which produces the recombinant gametes. This phenomenon is called crossover. The number of crossovers between two loci depends on the physical distance of the loci. More distant loci tend to have more crossovers between them. The relationship between

S. Xu, *Principles of Statistical Genomics*, DOI 10.1007/978-0-387-70807-2_1,
© Springer Science+Business Media, LLC 2013

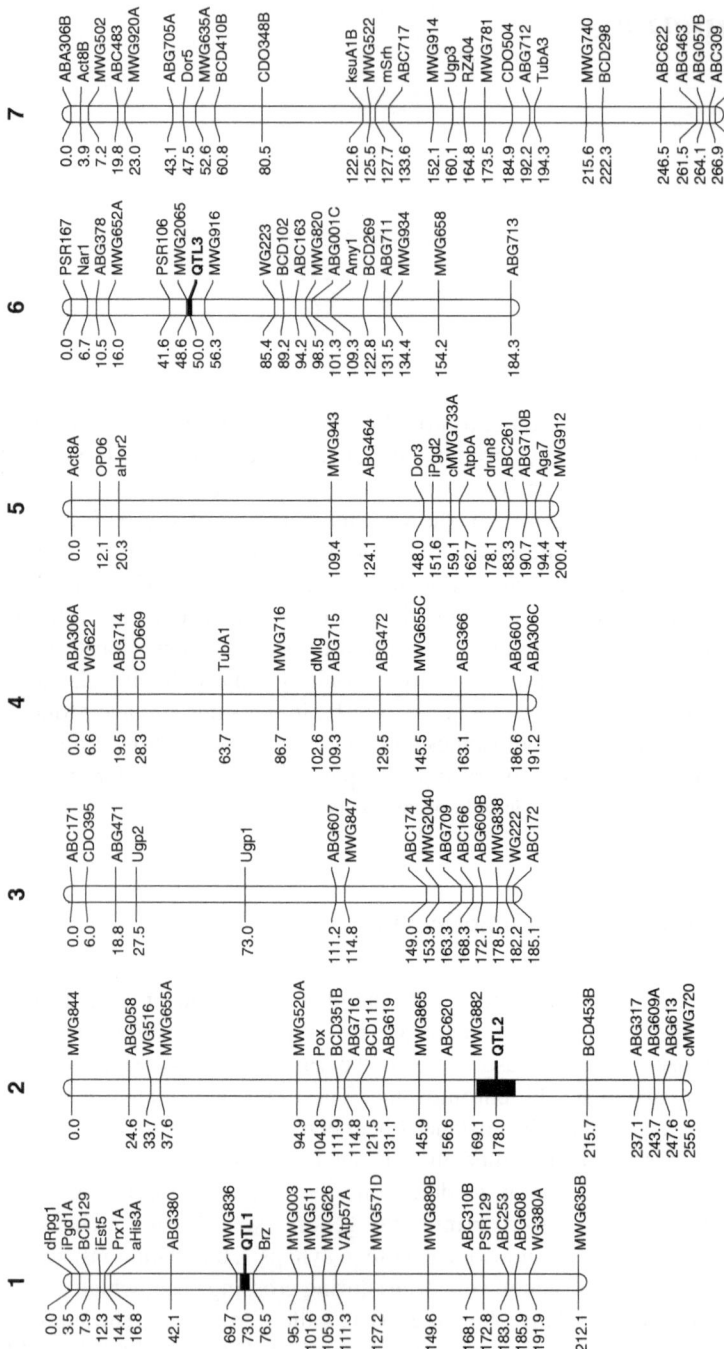

Fig. 1.1 Marker map of the barley genome (seven chromosomes). The names of markers are given in the right-hand side of the chromosomes, and the marker positions measured in centiMorgan (cM) are given in the left-hand side of the chromosomes

the number of crossovers and the number of base pairs may be described by a linear function. The slope of the linear relationship varies from one organism to another and even varies across different chromosome regions within a species (Civardi et al. 1994; Chen et al. 2002). Nevertheless, under the assumption of a constant slope across the genome, one can measure the size of a chromosome by the number of crossovers and then infer the number of base pairs from the number of crossovers. Roughly, 1 cM on a chromosome encompasses 1 megabase (Mb) (1 Mb $= 10^6$ bp) of DNA (Chen et al. 2002). The order and relative positioning of all linked loci in a genome with distances between consecutive loci measured by the numbers of crossovers is called the genetic map, due to the fact that crossover is a genetic phenomenon. Although we can observe crossovers with some special cytogenetic technology, counting the total number of crossovers of the entire genome is still impractical. If a crossover has occurred at a point between two loci during meiosis, the gamete formed will carry a paternal allele for one locus and a maternal allele for the other locus. This mosaic gamete is called the recombined gamete or simply recombinant. Crossover, however, is a random event, and it happens with a certain probability during meiosis. If no crossover has happened, the gamete will carry paternal alleles or maternal alleles for both loci. This type of gamete is called the parental gamete. If two loci are far apart, crossover between the two loci may happen twice or multiple times during meiosis. Only an odd number of crossovers will generate recombinants. The proportion of recombinants in a gametic pool is called the recombination fraction or recombination frequency. This fraction depends on the number of crossovers, although not in a linear fashion. In genetic linkage study, people often use recombination fraction in place of the number of crossovers to measure the distances between loci. The order and relative localization of linked loci with distances between consecutive loci measured by the recombination fractions is also called the genetic map. Therefore, we have two different measurements for the genetic distance between two loci: the number of crossovers and the recombination fraction.

1.2   Derivation of Map Functions

As mentioned earlier, there are two measurements of map distance between two linked loci, the average number of crossovers (x) and the recombination fraction (r). The number of crossovers itself is a random variable in the sense that it varies from one meiosis to another. The unit of map distance measured this way is Morgan (M), named in honor of geneticist Thomas Hunt Morgan. One Morgan is defined as the length of a chromosome segment bracketed by two loci that produces, on average, one crossover per meiosis. In other words, if we can observe many meioses in a genetic experiment for the segment under investigation, some meioses may produce no crossover, some may produce one crossover, and others may produce more than one crossover. But, *on average*, a segment of 1 M produces one crossover. We often use 1/100 Morgan as the unit of map distance, called one centiMorgan (cM). Note that the unit Morgan is rarely used today.

Genetic map distance measured by Morgan or centiMorgan is additive. Assume that three loci are ordered as A, B, and C with x_{AB} and x_{BC} representing the map distance between A and B and the distance between B and C, respectively, then the distance between A and C is $x_{AC} = x_{AB} + x_{BC}$. Because of this property, we call the map distance measured in Morgan or centiMorgan the additive distance, as apposed to the distance measured in recombination fraction. Recombination fraction between two loci is defined as the ratio of the number of recombined gametes to the total number of gametes produced. For example, assume that we sample 100 gametes for a chromosome segment bracketed by loci A and C in a genetic experiment; if 12 gametes are recombinants and the remaining 88 gametes are parental gametes, then the recombination fraction between A and C is $r_{BC} = 0.12$. Recombination fraction is not additive in the sense that $r_{AC} \neq r_{AB} + r_{BC}$. This is because only odd-numbered crossovers can generate a recombinant.

If two loci overlap, no crossover will be expected between the two loci, and both x_{AC} and r_{AC} will be zero. On the other hand, if the two loci are far away (almost unlinked), an infinite number of crossovers will be expected. When the segment is sufficiently long, odd- and even-numbered crossovers will be roughly equal, leading to a recombination fraction of $\frac{1}{2}$. Therefore, recombination fraction ranges between 0 and $\frac{1}{2}$. The relationship between r_{AC} and x_{AC} can be described by a nonlinear function, called the map function. In linkage analysis, we usually estimate the recombination frequency between loci and then convert the frequency into the additive distance and report the additive map distance. Several functional relationships have been proposed (Zhao and Speed 1996; Liu 1998), but the most commonly used functions are the Haldane map function (Haldane 1919) and its extension, called the Kosambi map function (Kosambi 1943).

Consider three ordered loci A–B–C. If the probability of crossover between A and B and that between B and C are independent, then the probability of double crossovers between A and C should take the product of r_{AB} and r_{BC}. The recombination frequency between A and C should be

$$r_{AC} = r_{AB}(1 - r_{BC}) + r_{BC}(1 - r_{AB}) = r_{AB} + r_{BC} - 2r_{AB}r_{BC}. \qquad (1.1)$$

The independence of crossovers between two nearby intervals is called no interference. However, it is often observed that if the two intervals under consideration are too close, the crossover of one interval may completely prevent the occurrence of the crossover of the other interval. This phenomenon is called complete interference, in which case the recombination fraction between A and C is described by

$$r_{AC} = r_{AB} + r_{BC}. \qquad (1.2)$$

This relationship was given by Morgan and Bridges (1916). Under the assumption of complete interference, the recombination frequencies are additive, like the additive map distance measured by the average number of crossovers. When loci A and C are very closely linked, the recombination fractions are approximately additive, even if interference is not complete. In practice, an intermediate level of interference is most likely to occur. Therefore, the recombination fraction between A and C can be described by

$$r_{AC} = r_{AB} + r_{BC} - 2cr_{AB}r_{BC}. \tag{1.3}$$

where $0 \leq c \leq 1$ is called the coefficient of coincidence whose complement $i = 1 - c$ is called the coefficient of interference. Equations (1.1) and (1.2) are special cases of (1.3), where $c = 1$ for (1.1) and $c = 0$ for (1.2).

We now develop a general map function to describe the relationship between the recombination frequency r and the additive map distance x measured in Morgan. Let the map function be $r = r(x)$, indicating that r is a function of x. The above equation can be rewritten as

$$\underbrace{r(x_{AB} + x_{BC})}_{r_{AC}} = \underbrace{r(x_{AB})}_{r_{AB}} + \underbrace{r(x_{BC})}_{r_{BC}} - 2c\,\underbrace{r(x_{AB})}_{r_{AB}}\,\underbrace{r(x_{BC})}_{r_{BC}}. \tag{1.4}$$

Let $x = x_{AB}$ be the map distance between loci A and B and $\Delta x = x_{BC}$ be the increment of the map distance (extension from locus B to locus C). We can then rewrite the above equation as

$$\underbrace{r(x + \Delta x)}_{r_{AC}} = \underbrace{r(x)}_{r_{AB}} + \underbrace{r(\Delta x)}_{r_{BC}} - 2c\,\underbrace{r(x)}_{r_{AB}}\,\underbrace{r(\Delta x)}_{r_{BC}}, \tag{1.5}$$

which can be rearranged into

$$r(x + \Delta x) - r(x) = r(\Delta x) - 2cr(x)r(\Delta x). \tag{1.6}$$

Dividing both sides of the equation by Δx, we have

$$\frac{r(x + \Delta x) - r(x)}{\Delta x} = \frac{r(\Delta x) - 2cr(x)r(\Delta x)}{\Delta x}. \tag{1.7}$$

Recall that when Δx is small, $r(\Delta x) \approx \Delta x$. Therefore, we can take the following limit:

$$\lim_{\Delta x \to 0} \frac{r(x + \Delta x) - r(x)}{\Delta x} = \lim_{\Delta x \to 0} \frac{r(\Delta x)}{\Delta x} - 2c \lim_{\Delta x \to 0} \frac{r(x)r(\Delta x)}{\Delta x}. \tag{1.8}$$

Because $\lim_{\Delta x \to 0} \frac{r(x+\Delta x)-r(x)}{\Delta x} = \frac{d}{dx}r(x)$ and $\lim_{\Delta x \to 0} \frac{r(\Delta x)}{\Delta x} = 1$, we have

$$\frac{d}{dx}r(x) = 1 - 2cr(x). \tag{1.9}$$

This is a differential equation, which can be easily solved if c is specified. This differential equation was derived by Haldane (1919) and is the theoretical basis from which the following specific map functions are derived.

1.3   Haldane Map Function

If c is treated as a constant, we can solve the above differential equation for function $r(x)$. We then obtain

$$x = -\frac{1}{2c} \ln[1 - 2cr(x)]. \tag{1.10}$$

Solving for $r(x)$, we get

$$r(x) = \frac{1}{2c}(1 - e^{-2cx}). \tag{1.11}$$

When $c = 1$ is assumed, i.e., there is no interference, we have obtained the following well-known Haldane map function (Haldane 1919):

$$r(x) = \frac{1}{2}(1 - e^{-2x}). \tag{1.12}$$

Note that when using the Haldane map function, the genetic distance x should be measured in Morgan, not in centiMorgan.

1.4   Kosambi Map Function

Kosambi (1943) expressed the coefficient of incidence as a function of the recombination frequency, i.e., $c = 2r(x)$, based on the notion that $c = 0$ when the two intervals are very close (complete interference) and $c = 1$ when the two intervals are virtually not linked (no interference). Substituting $c = 2r(x)$ into (1.9), we have

$$\frac{d}{dx}r(x) = 1 - 4r^2(x). \tag{1.13}$$

Solving this differential equation for $r(x)$, we have

$$x = \frac{1}{4} \ln \frac{1 + 2r(x)}{1 - 2r(x)}. \tag{1.14}$$

A rearrangement on (1.14) leads to the famous Kosambi map function (Kosambi 1943),

$$r(x) = \frac{1}{2}\left(\frac{e^{4x} - 1}{e^{4x} + 1}\right). \tag{1.15}$$

Fig. 1.2 Comparison of the Haldane and Kosambi map functions

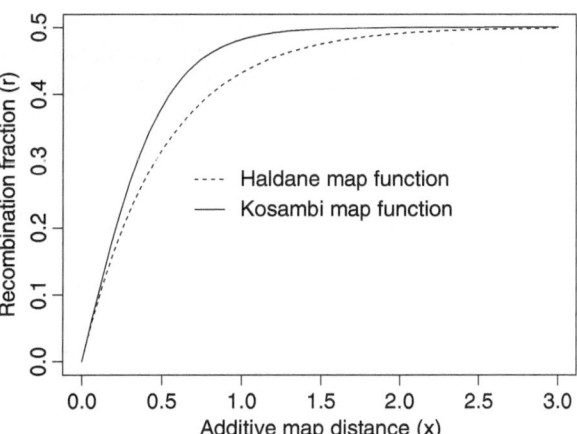

The main difference between Haldane and Kosambi map functions is the assumption of interference between two consecutive intervals. Haldane map function assumes no interference, whereas Kosambi map function allows interference to occur. This can be reflected by the following two equations under the two map functions. Under the Haldane map function,

$$r_{AC} = r_{AB} + r_{BC} - 2r_{AB}r_{BC}, \qquad (1.16)$$

whereas under the Kosambi map function,

$$r_{AC} = \frac{r_{AB} + r_{BC}}{1 + 4r_{AB}r_{BC}}. \qquad (1.17)$$

Similar to the Haldane map function, when using the Kosambi map function, the genetic distance x should be measured in Morgan, not in centiMorgan.

Kosambi map function often fits data better than Haldane map function because it takes into account interference. Haldane map function has an attractive Markovian property, which can be used elegantly to analyze multiple loci. Figure 1.2 illustrates the difference between the two map functions. For the same map distance between two loci, the recombination fraction calculated based on the Haldane map function is always smaller than that calculated based on the Kosambi map function.

The Haldane and Kosambi map functions are the most popular ones used in genetic mapping. There are many other map functions available in the literature. Most of them are extensions of these two map functions. These additional map functions can be found in Liu (1998). It is important to understand that the map functions apply to diploid organisms. For polyploid organisms, e.g., autotetraploids, the map function can be different due to problems of multiple dosage of allelic inheritance, the null allele, allelic segregation distortion, and mixed bivalent and quadrivalent pairing in meiosis. For example, the maximum recombination frequency can be 0.75 in tetraploid species rather than 0.5 in diploid. Details of tetraploid map function and map construction can be found in the seminal studies conducted by Luo et al. (2004) and Luo et al. (2006).

Chapter 2
Recombination Fraction

Recombination fraction (also called recombination frequency) between two loci is defined as the ratio of the number of recombined gametes to the total number of gametes produced. Recombination fraction, denoted by r throughout the book, however, has a domain of $0 \leq r \leq 0.5$, with $r = 0$ indicating perfect linkage and $r = 0.5$ meaning complete independence of the two loci. In most situations, gametes are not directly observable. Therefore, special mating designs are required to infer the number of recombined gametes. When a designed mating experiment cannot be carried out, data collected from pedigrees can be used for estimating recombination fractions. However, inferring the number of recombined gametes in pedigrees is much more complicated than that in designed mating experiments. This book only deals with designed mating experiments.

2.1 Mating Designs

Two mating designs are commonly used in linkage study, the backcross (BC) design and the F_2 design. Both designs require two inbred lines, which differ in both the phenotypic values of traits (if marker-trait association study is to be performed) and allele frequencies of marker loci used for constructing the linkage map. We will use two marker loci as an example to show the mating designs and methods for estimating recombination fraction. The BC design is demonstrated in Fig. 2.1. Let A and B be the two loci under investigation. Let A_1 and A_2 be the two alleles at locus A and B_1 and B_2 be the two alleles at locus B. Let P_1 and P_2 be the two parents that initiate the line cross. Since both parents are inbred, we can describe the two-locus genotype for P_1 and P_2 by $\frac{A_1B_1}{A_1B_1}$ and $\frac{A_2B_2}{A_2B_2}$, respectively. The hybrid progeny of cross between P_1 and P_2 is denoted by F_1 whose genotype is $\frac{A_1B_1}{A_2B_2}$. The horizontal line in the F_1 genotype separates the two parental gametes, i.e., A_1B_1 is the gamete from P_1 and A_2B_2 is the gamete from P_2. The F_1 hybrid crosses back to one of the parents

S. Xu, *Principles of Statistical Genomics*, DOI 10.1007/978-0-387-70807-2_2,
© Springer Science+Business Media, LLC 2013

Fig. 2.1 The backcross (BC) mating design. The BC progeny generated by $F_1 \times P_1$ is called BC_1, whereas the BC population generated by $F_1 \times P_2$ is called BC'_1

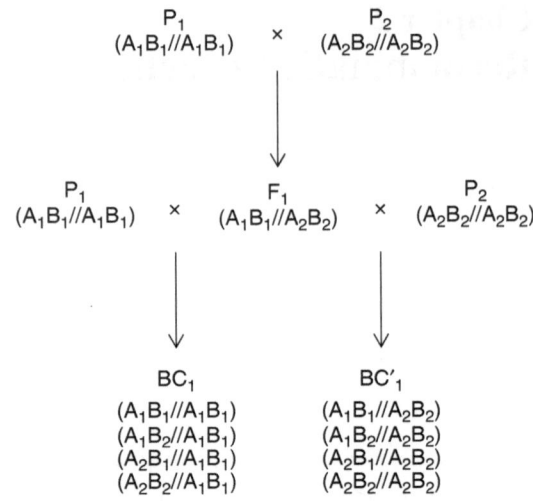

Table 2.1 Count data of two-locus genotypes collected from a BC_1 family

Genotype	Count	Frequency	Type
$\frac{A_1 B_1}{A_1 B_1}$	n_{11}	$\frac{1}{2}(1-r)$	Parental
$\frac{A_1 B_2}{A_1 B_1}$	n_{12}	$\frac{1}{2}r$	Recombinant
$\frac{A_2 B_1}{A_1 B_1}$	n_{21}	$\frac{1}{2}r$	Recombinant
$\frac{A_2 B_2}{A_1 B_1}$	n_{22}	$\frac{1}{2}(1-r)$	Parental

to generate multiple BC progeny, which will be used for linkage study. The BC population is a segregating population. Linkage analysis can only be conducted in such a segregating population. A segregating population is defined as a population that contains individuals with different genotypes. The two parental populations and the F_1 hybrid population are not segregating populations because individuals within each of the three populations are genetically identical. The BC progeny generated by $F_1 \times P_1$ is called BC_1, whereas the BC population generated by $F_1 \times P_2$ is called BC'_1.

We now use BC_1 progeny as an example to demonstrate the BC analysis. The gametes generated by the P_1 parent are all of the same type $A_1 B_1$. However, the F_1 hybrid can generate four different gametes and thus four distinguished genotypes. Let r be the recombination fraction between loci A and B. Let n_{ij} be the number of gametes of type $A_i B_j$ or the number of genotype of $\frac{A_i B_j}{A_1 B_1}$ kind for $i, j = 1, 2$. The four genotypes and their frequencies are given in Table 2.1. This table provides the data for the maximum likelihood estimation of recombination fraction. The maximum likelihood method will be described later.

The F_2 mating design requires mating of the hybrid with itself, called selfing and denoted by the symbol $\otimes$ (see Fig. 2.2 for the F_2 design). When selfing is impossible, e.g., in animals and self-incompatible plants, intercross between

Fig. 2.2 The F_2 mating design

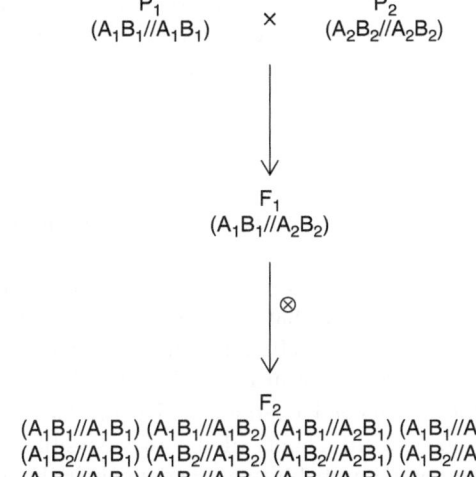

Table 2.2 The 16 possible genotypes and their observed counts in an F_2 family

	$A_1 B_1$	$A_1 B_2$	$A_2 B_1$	$A_2 B_2$
$A_1 B_1$	$\frac{A_1 B_1}{A_1 B_1}, n_{11}$	$\frac{A_1 B_1}{A_1 B_2}, n_{12}$	$\frac{A_1 B_1}{A_2 B_1}, n_{13}$	$\frac{A_1 B_1}{A_2 B_2}, n_{14}$
$A_1 B_2$	$\frac{A_1 B_2}{A_1 B_1}, n_{21}$	$\frac{A_1 B_2}{A_1 B_2}, n_{22}$	$\frac{A_1 B_2}{A_2 B_1}, n_{23}$	$\frac{A_1 B_2}{A_2 B_2}, n_{24}$
$A_2 B_1$	$\frac{A_2 B_1}{A_1 B_1}, n_{31}$	$\frac{A_2 B_1}{A_1 B_2}, n_{32}$	$\frac{A_2 B_1}{A_2 B_1}, n_{33}$	$\frac{A_2 B_1}{A_2 B_2}, n_{34}$
$A_2 B_2$	$\frac{A_2 B_2}{A_1 B_1}, n_{41}$	$\frac{A_2 B_2}{A_1 B_2}, n_{42}$	$\frac{A_2 B_2}{A_2 B_1}, n_{43}$	$\frac{A_2 B_2}{A_2 B_2}, n_{44}$

different F_1 individuals initiated from the same cross is required. The progeny of selfing F_1 or intercross between two F_1 hybrids is called an F_2 progeny. An F_2 family consists of multiple F_2 progeny. The F_2 family represents another segregating population for linkage analysis. Recall that an F_1 hybrid can generate four possible gametes for loci A and B jointly. Therefore, selfing of F_1 can generate 16 possible genotypes, as illustrated in Table 2.2. Let n_{ij} be the number of individuals combining the ith gamete from one parent and the jth gamete from the other parent, for $i, j = 1, \ldots, 4$. The frequencies of all the 16 possible genotypes are listed in Table 2.3. This table is the basis from which the maximum likelihood estimation of recombination fraction will be derived.

2.2   Maximum Likelihood Estimation of Recombination Fraction

In a BC design, the four types of gametes are distinguishable. Therefore, the recombination fraction can be directly calculated by taking the ratio of the number of recombinants to the total number of gametes. We use BC_1 as an example to demonstrate the method. The count data are given in Table 2.1. Let $n_p = n_{11} + n_{22}$ be the number of individuals carrying the parental gametes and $n_r = n_{12} + n_{21}$ be the number of recombinants. The estimated recombination fraction between loci A and B is simply

$$\hat{r} = \frac{n_{12} + n_{21}}{n_{11} + n_{12} + n_{21} + n_{22}} = \frac{n_r}{n_r + n_p}. \tag{2.1}$$

We use a hat above r to indicate estimation of r. The true value of recombination fraction is not known, but if the sample size is infinitely large, the estimated r will approach to the true value, meaning that the estimation is unbiased.

We now prove that $\hat{r}$ is the maximum likelihood estimate (MLE) of r. We introduce the ML method because it provides a significance test on the hypothesis that $r = 0.5$. To construct the likelihood function, we need a probability model, a sample of data and a parameter. The probability model is the binomial distribution, the data are the counts of the two possible genotypes, and the parameter is r. The binomial probability of the data given the parameter is

$$\Pr(n_r, n_p | r) = \frac{n!}{n_r! n_p!} \left(\frac{1}{2}\right)^{n_r + n_p} r^{n_r} (1 - r)^{n_p}, \tag{2.2}$$

where $n = n_r + n_p$ is the sample size. The value of r for $0 \leq r \leq 0.5$ that maximizes the probability is the MLE of r. Two issues need to be emphasized here for any maximum likelihood analysis, including this one. First, the probability involves a factor that does not depend on the parameter,

$$\text{const} = \frac{n!}{n_r! n_p!} \left(\frac{1}{2}\right)^n. \tag{2.3}$$

It is a constant with respect to the parameter r. This constant is irrelevant to the ML analysis and thus should be ignored. Secondly, the r value that maximizes a monotonic function of the probability also maximizes this probability. For computational convenience, we can maximize the logarithm of the probability. Therefore, it is the log likelihood function that is maximized in the ML analysis. The log likelihood function is defined as

$$L(r) = n_r \ln r + n_p \ln(1 - r). \tag{2.4}$$

To find the MLE of r, we need to find the derivative of $L(r)$ with respect to r,

$$\frac{d}{dr} L(r) = \frac{n_r}{r} - \frac{n_p}{1 - r}. \tag{2.5}$$

Letting $\frac{d}{dr}L(r) = 0$ and solving for r, we have

$$\hat{r} = \frac{n_r}{n_r + n_p}. \tag{2.6}$$

which is identical to that given in (2.1).

2.3   Standard Error and Significance Test

A parameter is a fixed but unknown quantity. The estimate of the parameter, however, is a variable because it varies from one sample to another. As the sample size increases, the estimate will approach to the true value of the parameter, provided that the estimate is unbiased. The deviation of the estimate from the true parameter can be measured by the standard error of the estimate. In this section, we will learn a method to calculate the standard error of $\hat{r}$. To calculate the standard error, we need the second derivative of the log likelihood function with respect to r and obtain a quantity called information, from which the variance of the estimated r can be approximated. Let us call the first derivative of $L(r)$ with respect to r the score function, denoted by $S(r)$,

$$S(r) = \frac{d}{dr}L(r) = \frac{n_r}{r} - \frac{n_p}{1-r}. \tag{2.7}$$

The second derivative of $L(r)$ with respect to r is called the Hessian matrix, denoted by $H(r)$,

$$H(r) = \frac{d}{dr}S(r) = \frac{d^2}{dr^2}L(r) = \frac{n_p}{(1-r)^2} - \frac{n_r}{r^2}. \tag{2.8}$$

Although $H(r)$ is a single variable, we still call it a matrix because in subsequent chapters we will deal with multiple dimension of parameters, in which case $H(r)$ is a matrix. From $H(r)$, we can find the information of r, which is

$$I(r) = -E[H(r)] = \frac{E(n_r)}{r^2} - \frac{E(n_p)}{(1-r)^2}. \tag{2.9}$$

The symbol E represents expectation of the data given the parameter value. Here, the data are referred to n_r and n_p, not n, which is the sample size (a constant). Suppose that we know the true parameter r, what is the expected number of recombinants if we sample n individuals? This expected number is $E(n_r) = rn$. The expected number of the parental types is $E(n_p) = (1-r)n$. Therefore, the information is

$$I(r) = -E[H(r)] = \frac{rn}{r^2} - \frac{(1-r)n}{(1-r)^2} = \frac{n}{r(1-r)}. \tag{2.10}$$

The variance of the estimated r takes the inverse of the information, with the true parameter replaced by $\hat{r}$,

$$\text{var}(\hat{r}) \approx I^{-1}(\hat{r}) = \frac{\hat{r}(1 - \hat{r})}{n}. \tag{2.11}$$

Therefore, the standard error of $\hat{r}$ is

$$\text{se}(\hat{r}) = \sqrt{\text{var}(\hat{r})} = \sqrt{\frac{\hat{r}(1 - \hat{r})}{n}}. \tag{2.12}$$

The standard error is inversely proportional to the square root of the sample size and thus approaches zero as n becomes infinitely large.

When we report the estimated recombination fraction, we also need to report the estimation error in a form like $\hat{r} \pm \text{se}(\hat{r})$. In addition to the sample size, the estimation error is also a function of the recombination fraction, with the maximum error occurring at $r = \frac{1}{2}$, i.e., when the two loci are unlinked. To achieve the same precision of estimation, it requires a larger sample to estimate a recombination fraction between two loosely linked loci than between two closely linked loci.

Because of the sampling error, even two unlinked loci may look like being linked as the estimated r may be superficially smaller than 0.5. How small an $\hat{r}$ is sufficiently small so that we can claim that the two loci are linked in the same chromosome? This requires a significance test.

The null hypothesis for such a test is denoted by $H_0 : r = \frac{1}{2}$. Verbally, H_0 is stated that the two loci are not linked. The alternative hypothesis is $H_A : r < 1/2$, i.e., the two loci are linked on the same chromosome. When the sample size is sufficiently large, we can always use the z-test to decide which hypothesis should be accepted. Here, we will use the usual likelihood ratio test statistic to declare the statistical significance of $\hat{r}$. Let $L(r)|_{r=\hat{r}} = L(\hat{r})$ be the log likelihood function evaluated at the MLE of r using (2.4). Let $L(r)|_{r=\frac{1}{2}} = L(1/2)$ be the log likelihood function evaluated under the null hypothesis. The likelihood ratio test statistic is defined as

$$\lambda = -2[L(1/2) - L(\hat{r})]. \tag{2.13}$$

where

$$L(\hat{r}) = n_r \ln \hat{r} + n_p \ln(1 - \hat{r}). \tag{2.14}$$

and

$$L(1/2) = -n \ln 2 = -0.6931n. \tag{2.15}$$

If the null hypothesis is true, λ will approximately follow a chi-square distribution with one degree of freedom. Therefore, if $\lambda > \chi^2_{1,1-\alpha}$, we will claim that the two loci are linked, where $\chi^2_{1,1-\alpha}$ is the $(1 - \alpha) \times 100$ percentile of the central χ^2_1 distribution and α is the type I error determined by the investigator. In human linkage studies, people often use LOD (log of odds) score instead. The relationship between LOD and λ is

$$\text{LOD} = \frac{\lambda}{2 \ln(10)} \approx 0.2171\lambda. \tag{2.16}$$

Conventionally, LOD > 3 is used as a criterion to declare a significant linkage. This converts to a likelihood ratio criterion of $\lambda > 3 \times \ln(100) = 13.81551$. The LOD criterion has an intuitive interpretation. An LOD of k means that the alternative model (linkage) is 10^k times more likely than the null model.

2.4   Fisher's Scoring Algorithm for Estimating r

The F_2 mating design is demonstrated in Fig. 2.2. The ML analysis described for the BC_1 mating design is straightforward. The MLE of r has an explicit form. In fact, there is no need to invoke the ML analysis for the BC design other than to demonstrate the basic principle of the ML analysis. To estimate r using an F_2 design, the likelihood function is constructed using the same probability model (multinomial distribution), but finding the MLE of r is complicated. Therefore, we will resort to some special maximization algorithms. The algorithm we will learn is the Fisher's scoring algorithm (Fisher 1946).

Let us look at the genotype table (Table 2.2) and the table of genotype counts and frequencies (Table 2.3) for the F_2 design. If we were able to observe all the 16 possible genotypes, the same ML analysis used in the BC design would apply here to the F_2 design. Unfortunately, some of the genotypes listed in Table 2.2 are not distinguishable from others. For example, genotypes $\frac{A_1 B_1}{A_1 B_2}$ and $\frac{A_1 B_2}{A_1 B_1}$ are not distinguishable. These two genotypes appear to be the same because they both carry an $A_1 B_1$ gamete and an $A_1 B_2$ gamete. However, the origins of the two gametes are different for the two genotypes. Furthermore, the four genotypes in the minor diagonal of Table 2.2 actually represent four different linkage phases of the same observed genotype (double heterozygote). If we consider the origins of the alleles, there are four possible genotypes for each locus. However, the two configurations of the heterozygote are not distinguishable. Therefore, there are only three observable genotypes for each locus, making a total of nine observable joint

Table 2.3 The counts (in parentheses) and frequencies of the 16 possible genotypes in an F_2 family

	$A_1 B_1$	$A_1 B_2$	$A_2 B_1$	$A_2 B_2$
$A_1 B_1$	$(n_{11})\,\frac{1}{4}(1-r)^2$	$(n_{12})\,\frac{1}{4}r(1-r)$	$(n_{13})\,\frac{1}{4}r(1-r)$	$(n_{14})\,\frac{1}{4}(1-r)^2$
$A_1 B_2$	$(n_{21})\,\frac{1}{4}r(1-r)$	$(n_{22})\,\frac{1}{4}r^2$	$(n_{23})\,\frac{1}{4}r^2$	$(n_{24})\,\frac{1}{4}r(1-r)$
$A_2 B_1$	$(n_{31})\,\frac{1}{4}r(1-r)$	$(n_{32})\,\frac{1}{4}r^2$	$(n_{33})\,\frac{1}{4}r^2$	$(n_{34})\,\frac{1}{4}r(1-r)$
$A_2 B_2$	$(n_{41})\,\frac{1}{4}(1-r)^2$	$(n_{42})\,\frac{1}{4}r(1-r)$	$(n_{43})\,\frac{1}{4}r(1-r)$	$(n_{44})\,\frac{1}{4}(1-r)^2$

Table 2.4 The nine observed genotypes and their counts in an F_2 population

	$B_1 B_1$	$B_1 B_2$	$B_2 B_2$
$A_1 A_1$	$A_1 A_1 B_1 B_1 \ (m_{11})$	$A_1 A_1 B_1 B_2 \ (m_{12})$	$A_1 A_1 B_2 B_2 \ (m_{13})$
$A_1 A_2$	$A_1 A_2 B_1 B_1 \ (m_{21})$	$A_1 A_2 B_1 B_2 \ (m_{22})$	$A_1 A_2 B_2 B_2 \ (m_{23})$
$A_2 A_2$	$A_2 A_2 B_1 B_1 \ (m_{31})$	$A_2 A_2 B_1 B_2 \ (m_{32})$	$A_2 A_2 B_2 B_2 \ (m_{33})$

two-locus genotypes, as shown in Table 2.4. Let m_{ij} be the counts of the joint genotype combining the ith genotype of locus A and the jth genotype of locus B, for $i, j = 1, \ldots, 3$. These counts are the data from which a likelihood function can be constructed.

Before we construct the likelihood function, we need to find the probability for each of the nine observed genotypes. These probabilities are listed in Table 2.5. The count data in the second column and the frequencies in the third column of Table 2.5 are what we need to construct the log likelihood function, which is

$$
\begin{aligned}
L(r) = \sum_{i=1}^{3} \sum_{j=1}^{3} m_{ij} \ln(q_{ij}) \\
= [2(m_{11} + m_{33}) + m_{12} + m_{21} + m_{23} + m_{32}] \ln(1 - r) \\
+ [2(m_{13} + m_{31}) + m_{12} + m_{21} + m_{23} + m_{32}] \ln(r) \\
+ m_{22} \ln[r^2 + (1 - r)^2].
\end{aligned}
\tag{2.17}
$$

The derivative of $L(r)$ with respect to r is

$$
\begin{aligned}
S(r) = \frac{d}{dr} L(r) \\
= -\frac{2(m_{11} + m_{33})}{1 - r} + \frac{(m_{12} + m_{21} + m_{23} + m_{32})(1 - 2r)}{r(1 - r)} \\
- \frac{2m_{22}(1 - 2r)}{1 - 2r + 2r^2} + \frac{2(m_{13} + m_{31})}{r}.
\end{aligned}
\tag{2.18}
$$

The MLE of r is obtained by setting $S(r) = 0$ and solving for r. Unfortunately, there is no explicit solution for r. Therefore, an iterative algorithm is resorted to solve for r. Before introducing the Fisher's scoring algorithm (Fisher 1946), we first try the Newton method, which also requires the second derivative of $L(r)$ with respect to r,

Table 2.5 Frequencies of the nine observed genotypes in an F_2 population

Genotype		Count		Probability	
$A_1 A_1 B_1 B_1$	$= \frac{A_1 B_1}{A_1 B_1}$	$m_{11} =$	n_{11}	$q_{11} =$	$\frac{1}{4}(1-r)^2$
$A_1 A_1 B_1 B_2$	$= \frac{A_1 B_1}{A_1 B_2}, \frac{A_1 B_2}{A_1 B_1}$	$m_{12} =$	$n_{12} + n_{21}$	$q_{12} =$	$\frac{1}{2}r(1-r)$
$A_1 A_1 B_2 B_2$	$= \frac{A_1 B_2}{A_1 B_2}$	$m_{13} =$	n_{22}	$q_{13} =$	$\frac{1}{4}r^2$
$A_1 A_2 B_1 B_1$	$= \frac{A_1 B_1}{A_2 B_1}, \frac{A_2 B_1}{A_1 B_1}$	$m_{21} =$	$n_{13} + n_{31}$	$q_{21} =$	$\frac{1}{2}r(1-r)$
$A_1 A_2 B_1 B_2$	$= \frac{A_1 B_1}{A_2 B_2}, \frac{A_1 B_2}{A_2 B_1}$	$m_{22} =$	$n_{14} + n_{23} +$	$q_{22} =$	$\frac{1}{2}[r^2 + (1-r)^2]$
	$\frac{A_2 B_1}{A_1 B_2}, \frac{A_2 B_2}{A_1 B_1}$		$n_{32} + n_{41}$		
$A_1 A_2 B_2 B_2$	$= \frac{A_1 B_2}{A_2 B_2}, \frac{A_2 B_2}{A_1 B_2}$	$m_{23} =$	$n_{24} + n_{42}$	$q_{23} =$	$\frac{1}{2}r(1-r)$
$A_2 A_2 B_1 B_1$	$= \frac{A_2 B_1}{A_2 B_1}$	$m_{31} =$	n_{33}	$q_{31} =$	$\frac{1}{4}r^2$
$A_2 A_2 B_1 B_2$	$= \frac{A_2 B_1}{A_2 B_2}, \frac{A_2 B_2}{A_2 B_1}$	$m_{32} =$	$n_{34} + n_{43}$	$q_{32} =$	$\frac{1}{2}r(1-r)$
$A_2 A_2 B_2 B_2$	$= \frac{A_2 B_2}{A_2 B_2}$	$m_{33} =$	n_{44}	$q_{33} =$	$\frac{1}{4}(1-r)^2$

$$H(r) = \frac{d}{dr}S(r) = \frac{d^2}{dr^2}L(r)$$

$$= -\frac{2(m_{11} + m_{33})}{(1-r)^2} - \frac{(m_{12} + m_{21} + m_{23} + m_{32})(1 - 2r + 2r^2)}{r^2(1-r)^2}$$

$$+ \frac{8m_{22}r(1-r)}{(1 - 2r + 2r^2)^2} - \frac{2(m_{13} + m_{31})}{r^2}. \tag{2.19}$$

The Newton method starts with an initial value of r, denoted by $r^{(t)}$ for $t = 0$, and update the value by

$$r^{(t+1)} = r^{(t)} - \frac{S(r^{(t)})}{H(r^{(t)})}. \tag{2.20}$$

The iteration process stops if

$$|r^{(t+1)} - r^{(t)}| \le \epsilon, \tag{2.21}$$

where ϵ is a small positive number, say 10^{-8}.

The derivation of the Newton method is very simple. It uses the Taylor series expansion to approximate the score function. Let $r^{(0)}$ be the initial value of r. The score function $S(r)$ can be approximated in the neighborhood of $r^{(0)}$ by

$$S(r) = S(r^{(0)}) + \frac{d}{dr}S(r^{(0)})(r - r^{(0)}) + \frac{1}{2!}\frac{d^2}{dr^2}S(r^{(0)})(r - r^{(0)})^2 + \cdots$$

$$\approx S(r^{(0)}) + \frac{d}{dr}S(r^{(0)})(r - r^{(0)}). \tag{2.22}$$

The approximation is due to ignorance of the higher order terms of the Taylor series. Recall that $H(r^{(0)}) = \frac{d}{dr} S(r^{(0)})$ and thus

$$S(r) \approx S(r^{(0)}) + H(r^{(0)})(r - r^{(0)}). \tag{2.23}$$

Letting $S(r) = 0$ and solving for r, we get

$$r = r^{(0)} - \frac{S(r^{(0)})}{H(r^{(0)})}. \tag{2.24}$$

We have moved from $r^{(0)}$ to r, one step closer to the true solution. Let $r = r^{(t+1)}$ and $r^{(0)} = r^{(t)}$. The Newton's equation of iteration (2.20) is obtained by substituting r and $r^{(0)}$ into (2.24).

The Newton method does not behave well when r is close to zero or 0.5 for the reason that $H^{-1}(r)$ can be easily overflowed. The Fisher's scoring method is a modified version of the Newton method for avoiding the overflow problem. As such, the method behaves well in all range of the parameter in the legal domain $0 \leq r \leq \frac{1}{2}$. In the Fisher's scoring method, the second derivative involved in the iteration is simply replaced by the so-called expectation of the second derivative. The iteration equation becomes

$$r^{(t+1)} = r^{(t)} - \frac{S(r^{(t)})}{\mathrm{E}[H(r^{(t)})]}, \tag{2.25}$$

where

$$\mathrm{E}[H(r^{(t)})] = -\frac{2n[1 - 3r^{(t)} + 3(r^{(t)})^2]}{r^{(t)}(1 - r^{(t)})[1 - 2r^{(t)} + 2(r^{(t)})^2]}. \tag{2.26}$$

Let $I(r^{(t)}) = -\mathrm{E}[H(r^{(t)})]$ be the Fisher's information. The iteration process can be rewritten as

$$r^{(t+1)} = r^{(t)} + I^{-1}(r^{(t)})S(r^{(t)}). \tag{2.27}$$

Assume that the iteration converges at the $t + 1$ iteration. The MLE of r is $\hat{r} = r^{(t+1)}$. The method provides an automatic way to calculate the variance of the estimate,

$$\mathrm{var}(\hat{r}) \approx I^{-1}(\hat{r}) = \frac{\hat{r}(1 - \hat{r})(1 - 2\hat{r} + 2\hat{r}^2)}{2n(1 - 3\hat{r} + 3\hat{r}^2)}, \tag{2.28}$$

where $n = \sum_{i=1}^{3} \sum_{j=1}^{3} m_{ij}$ is the sample size. Note that when $\hat{r} \to 0$, $\hat{r}^2$ becomes negligible and $1 - 2\hat{r} \approx 1 - 3\hat{r}$, leading to $1 - 2\hat{r} + 2\hat{r}^2 \approx 1 - 3\hat{r} + 3\hat{r}^2$. Therefore,

$$\mathrm{var}(\hat{r}) \approx \frac{\hat{r}(1 - \hat{r})}{2n}. \tag{2.29}$$

Comparing this variance with the one in the BC design shown in (2.11), we can see that the variance has been reduced by half. Therefore, using the F_2 design is more efficient than the BC design.

2.5 EM Algorithm for Estimating r

The EM algorithm was developed by Dempster et al. (1977) for handling missing data problems. The algorithm repeatedly executes an E-step and an M-step for iterations. The E-step stands for expectation and the M-step for maximization. The problem of estimating recombination fraction in F_2 can be formulated as a missing value problem and thus solved by the EM algorithm. The derivation of the EM algorithm is quite involved and will be introduced later when we deal with a simpler problem. We now only give the final equation of the EM iteration. Recall that the F_1 hybrid can produce four possible gametes, two of them are of parental type ($A_1 B_1$ and $A_2 B_2$) and the other two are recombinants ($A_1 B_2$ and $A_2 B_1$). Therefore, an F_2 progeny can be classified into one of three categories in terms of the number of recombinant gametes contained: 0, 1, or 2. From Table 2.3, we can see that each of the following observed genotypes carries one recombinant gamete: $A_1 A_1 B_1 B_2$, $A_1 A_2 B_1 B_1$, $A_1 A_2 B_2 B_2$, and $A_2 A_2 B_1 B_2$, and each of the following observed genotypes carries two recombinant gametes: $A_1 A_1 B_2 B_2$ and $A_2 A_2 B_1 B_1$. Let $n_1 = m_{12} + m_{21} + m_{23} + m_{32}$ be the number of individuals of category 1 and $n_2 = m_{13} + m_{31}$ be the number of individuals of category 2. The double heterozygote $A_1 A_2 B_1 B_2$ is an ambiguous genotype because it may carry 0 recombinant gamete, ($\frac{A_1 B_1}{A_2 B_2}$, $\frac{A_2 B_2}{A_1 B_1}$), or two recombinant gametes, ($\frac{A_1 B_2}{A_2 B_1}$, $\frac{A_2 B_1}{A_1 B_2}$). The number of double heterozygote individuals that carry two recombinant gametes is $n_{23} + n_{32}$. Unfortunately, this number is not observable. If it were, we would be able to take the ratio of the number of recombinant gametes to the total number of gametes in the F_2 progeny ($2n$) to get the estimated recombination fraction right away,

$$\hat{r} = \frac{1}{2n}[2(n_{23} + n_{32} + n_2) + n_1] \qquad (2.30)$$

The EM algorithm takes advantage of this simple expression by substituting the missing values ($n_{23} + n_{32}$) by its expectation. The expectation, however, requires knowledge of the parameter, which is what we want to estimate. Therefore, iterations are required. To calculate the expectation, we need the current value of r, denoted by $r^{(t)}$, and the number of double heterozygote individuals (m_{22}). Recall that the overall proportion of the double heterozygote is $\frac{1}{2}[r^2 + (1 - r)^2]$, where $\frac{1}{2}r^2$ represents the proportion of individuals carrying two recombinant gametes and $\frac{1}{2}(1 - r)^2$ represents the proportion of individuals carrying no recombinant gametes. The conditional expectation of $n_{23} + n_{32}$ is

$$\mathrm{E}(n_{23} + n_{32}) = \frac{(r^{(t)})^2}{(r^{(t)})^2 + (1 - r^{(t)})^2} m_{22} = w^{(t)} m_{22}. \qquad (2.31)$$

The iterative equation may be written as

$$r^{(t+1)} = \frac{1}{2n}\{2[\mathrm{E}(n_{23} + n_{32}) + n_2] + (n_1)\}. \qquad (2.32)$$

The final equation of the EM iteration becomes

$$r^{(t+1)} = \frac{1}{2n} \left[2(w^{(t)} m_{22} + m_{13} + m_{31}) + (m_{12} + m_{21} + m_{23} + m_{32}) \right]. \quad (2.33)$$

Calculating $E(n_{23} + n_{32})$ using (2.31) represents the E-step, and updating $r^{(t+1)}$ using (2.32) represents the M-step of the EM algorithm. The final result of the EM algorithm is so simple, yet it behaves extremely well with regard to the small number of iterations required for convergence and the insensitiveness to the initial value of r. A drawback of the EM algorithm is the difficulty in calculating the standard error of the estimate. Since the solution is identical to the Fisher's scoring method, the variance (square of the standard error) of the estimate given in (2.28) can be used as the variance of the EM estimate.

To test the hypothesis of no linkage, $r = \frac{1}{2}$, we will use the same likelihood ratio test statistic, as described in the BC design. The log likelihood value under the null model, however, needs to be evaluated in a slightly different way, that is, $L(\frac{1}{2}) = \sum_{i=1}^{3} \sum_{j=1}^{3} m_{ij} \ln(q_{ij})$, where q_{ij} is a function of $r = \frac{1}{2}$ (see Table 2.5). The log likelihood value under the alternative model is evaluated at $r = \hat{r}$, using $L(\hat{r}) = \sum_{i=1}^{3} \sum_{j=1}^{3} m_{ij} \ln(\hat{q}_{ij})$, where $\hat{q}_{ij}$ is a function of $r = \hat{r}$ (see Table 2.5).

Chapter 3
Genetic Map Construction

Gene loci are grouped into different chromosomes. Within the same chromosome, the loci are linearly arranged because the chromosome is a string-like structure. The distribution of loci among chromosomes and the order of loci within chromosomes are called the genetic map. The data used to construct the genetic map are the genotypes of these loci. From the genotypic data, we can estimate all pairwise recombination fractions. These estimated recombination fractions are used to construct the linkage map. Construction of the genetic map may be better called reconstruction of genetic map because the true genetic map is already present and we simply do not know about it. This is similar to the situation where phylogeny construction is more often called phylogeny reconstruction because we are not constructing the phylogeny of species; rather, we infer the existing phylogeny using observed data. Of course, the inferred map may not be the true one if the sample size is not sufficiently large. Map construction is the first step toward gene mapping (locating functional genes). There are two steps in genetic map construction. The first step is to classify markers into linkage groups according to the pairwise LOD scores or the likelihood ratio test statistics. A convenient rule is that all markers with pairwise LOD scores greater than 3 are classified into the same linkage group. A more efficient grouping rule may be chosen using a combination of LOD score and the recombination fraction. For example, loci A and B may be grouped together if $\text{LOD}_{AB} > 3$ and $r_{AB} < 0.45$. Grouping markers into the same linkage group is straightforward, and no additional technique is required other than comparing the LOD score of each pair of markers to a predetermined LOD criterion. If we choose a more stringent criterion, some markers may not be assigned into any linkage groups. These markers are called satellite markers. On the other hand, if we choose a less stringent criterion, markers on different chromosomes may be assigned into the same linkage group. The second step of genetic mapping is to find the optimal orders of the markers within the same linkage groups. In this chapter, we only discuss the second step of map construction.

S. Xu, *Principles of Statistical Genomics*, DOI 10.1007/978-0-387-70807-2_3,
© Springer Science+Business Media, LLC 2013

3.1   Criteria of Optimality

Given the estimated pairwise recombination fractions for m loci on the same linkage group, we want to find the optimal order of the loci. There are $m!/2$ possible ways to arrange the m loci. The factorial of m gives the total number of permutations of all m loci. However, the orientation of a linkage map is irrelevant. For instance, ABC and CBA are considered the same order for loci A, B, and C, as far as the relative positions of the loci are concerned.

We will first define a criterion of "optimality" and then select the particular order that minimizes or maximizes the criterion. The simplest and also the most commonly used criterion is the sum of adjacent recombination coefficients (sar). The criterion is defined as

$$\text{sar} = \sum_{i=1}^{m-1} \hat{r}_{i(i+1)} \tag{3.1}$$

where $\hat{r}_{i(i+1)}$ is the recombination fraction between loci i and $i + 1$ for $i = 1, \ldots, m - 1$, where i and $i + 1$ are two adjacent loci. For m loci, there are $m - 1$ adjacent recombination fractions. If there is no estimation error for each of the adjacent recombination fraction, the true sar should have the minimum value compared with any other orders. Consider the following example of three loci with the correct order of ABC. The sar value for this correct order is

$$\text{sar}_{ABC} = r_{AB} + r_{BC} \tag{3.2}$$

If we evaluate an alternative order, say ACB, we found that

$$\text{sar}_{ACB} = r_{AC} + r_{BC} \tag{3.3}$$

Remember that the true order is ABC so that $r_{AC} = r_{AB} + r_{BC} - 2r_{AB}r_{BC}$ assuming that there is no interference. Substituting r_{AC} into the above equation, we get

$$\text{sar}_{ACB} = r_{AB} + r_{BC} - 2r_{AB}r_{BC} + r_{BC}$$
$$= r_{AB} + r_{BC} + r_{BC}(1 - 2r_{AB}) \tag{3.4}$$

Because $r_{BC}(1 - 2r_{AB}) \geq 0$, we conclude that $\text{sar}_{ACB} \geq \text{sar}_{ABC}$. In reality, we always use estimated recombination fractions, which are subject to estimation errors, and thus the marker order with minimum sar may not be the true order.

Similar to sar, we may use sad (sum of adjacent distances) as the criterion, which is defined as

$$\text{sad} = \sum_{i=1}^{m-1} \hat{x}_{i(i+1)} \tag{3.5}$$

where $\hat{x}_{i(i+1)}$ is the estimated additive distance between loci i and $i + 1$ and is converted from the estimated recombination fraction using either the Haldane or Kosambi map function. Similar to the sar criterion, the order of loci that minimizes sad is the optimal order.

The sum of adjacent likelihoods (*sal*) is another criterion for map construction. Note that the likelihood refers to the log likelihood. In contrast to *sar*, the optimal order should be the one which maximizes *sal*. Define $L(\hat{r}_{i(i+1)})$ as the log likelihood value for the recombination fraction between loci i and $i + 1$. The *sal* is defined as

$$\text{sal} = \sum_{i=1}^{m-1} L(\hat{r}_{i(i+1)}) \qquad (3.6)$$

Both *sad* and *sal* are additive, which is a property required by the branch and bound algorithm for searching the optimal order of markers (see next section).

3.2   Search Algorithms

3.2.1   Exhaustive Search

Exhaustive search is an algorithm in which all possible orders are evaluated. As a result, it guarantees to find the optimal order. Recall that for m loci, the total number of orders to be evaluated is $n = m!/2$. The number of orders (n) grows quickly as m increases, as shown in the following table.

m	n
2	1
3	3
4	12
5	60
6	360
7	2,520
8	20,160
9	181,440
10	1,814,400

The algorithm will use up the computing resource quickly as m increases. Therefore, this algorithm is rarely used when $m > 10$. When writing the computer code to evaluate the orders, we want to make sure that all possible orders are evaluated. This can be done using the following approach. Assume that there are five loci, denoted by A, B, C, D, and E, that need to be ordered. First, we arbitrarily choose two loci, say A and B, to initiate the map. We then add locus C to the existing map. There are three possible places where we can put C in the existing map: CAB, ACB, and ABC. For each of the three orders of the three-locus map, we add locus D. For example, to add locus D to the existing order ACB, we need to evaluate the following four possible orders: DACB, ADCB, ACDB, and ACBD. We then add locus E (the last locus) to each of the four orders of the four-locus map.

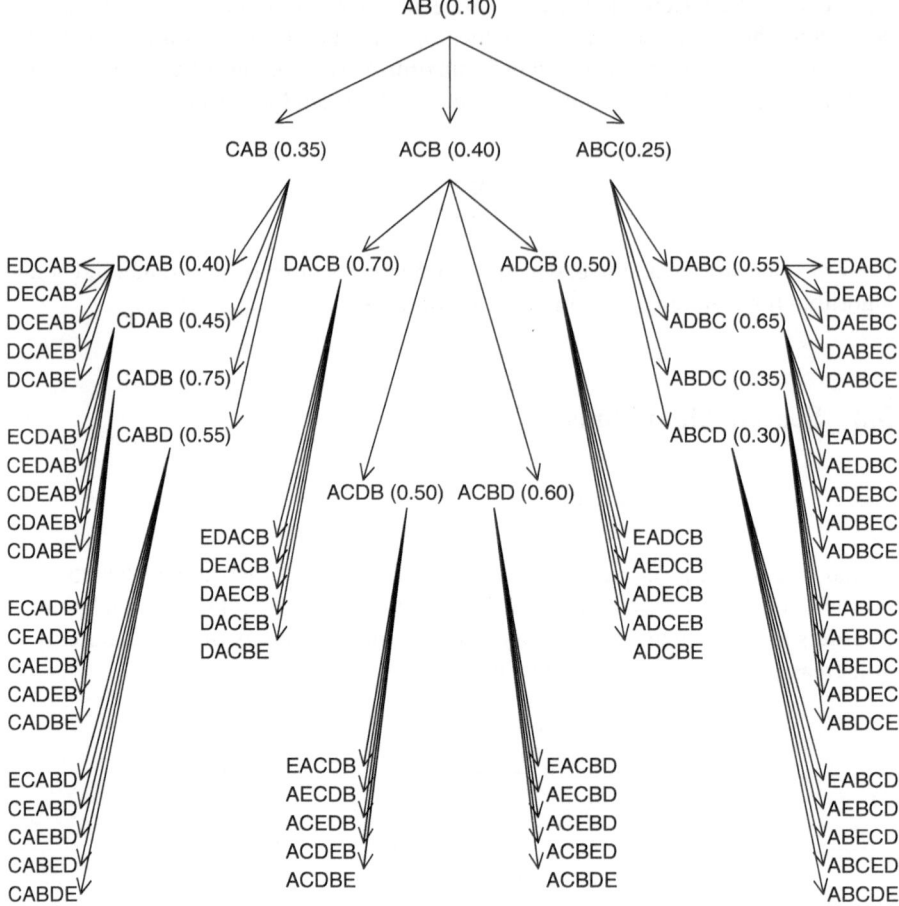

Fig. 3.1 All possible orders (60) of a map with five loci

For example, locus E can be added to the existing order DACB in five different places, leading to EDACB, DEACB, DAECB, DACEB, and DACBE. We can see that all the $3 \times 4 \times 5 = 5!/2 = 60$ possible orders have been evaluated so far (see Fig. 3.1). This ends the exhaustive search.

3.2.2 Heuristic Search

When the number of loci is too large to permit the exhaustive search, the optimal order can be sought via a heuristic approach that sacrifices the guarantee of optimality in favor of the reduced computing time. With a heuristic search, one starts with an arbitrary order and evaluates this particular order. The order is then rearranged in a random fashion and reevaluated. If the new order is "better" than the initial order, the new order is accepted. If the new order is "worse" than the initial order, the initial order is kept. This completes one cycle of the search. The second cycle of the search starts with a rearrangement of the order of the previous cycle. The rearranged order is then accepted or rejected depending on whether or not the value of the order has been improved. The process continues until no further improvement is achieved for a certain number of consecutive cycles, e.g., 50 cycles. This method is also called the greedy method because it is too greedy to "climb up" the hill. It is likely to end up with a local optimum instead of a global one. Therefore, there is no guarantee that the method finds the global optimal order.

The so-called random rearrangement may be conducted in several different ways. One is called complete rearrangement or global rearrangement. This is done by selecting a completely different order by random permutation. Information of the previous order has no effect on the selection of the new order. This approach is conceptually simple but may not be efficient. For example, if a previous order is already close to the optimal one, a complete random rearrangement may be far worse than this order, leading to many cycles of random rearrangements before an improved order appears. The other way of rearranging the order is called partial or local rearrangement. This is done by randomly rearranging a subset of the loci. Let $m_s (2 \leq m_s < m)$ be the size of the subset. Although m_s may be chosen in a arbitrary fashion, $m_s = 3$ may be a convenient choice. First, we randomly choose a triplet from the m loci. We then rearrange the three loci within their existing positions and leave the order of the remaining loci intact. There are $3! = 6$ possible ways to rearrange the three loci, and all of them are evaluated. The best one of the six is chosen as a candidate new order for reevaluation. If this new order is better than the order in the previous cycle, the order is updated; otherwise, the previous order is carried over to the next cycle.

The result of heuristic search depends on the initial order selected to start the search. We will use the five-locus example to demonstrate a simple way to choose the initial map order. Let A, B, C, D, and E be the five loci. We start with two most closely linked loci, say loci A and D. We then add a third locus to the existing two-locus map. The third locus is chosen such that it has the minimum average recombination fractions from loci A and D. Assume that locus C satisfies this criterion. We then add locus C to the existing map AD. There are three places where locus C can be added: CAD, ACD, and ADC. Choose the best of the three orders as the optimal map, say ACD. We then choose a next locus to add to the existing map ACD, using the same criterion, i.e., minimum average recombination fractions from loci A, C, and D. Assume that locus E satisfies this criterion. There are four places that locus E can be inserted into the existing map ACD, which are EACD, AECD, ACED, and ACDE. Assume that EACD is the best of the four orders. Finally, we add locus B (the last locus) to the four-locus map. We evaluate all the five different orders: BEACD, EBACD, EABCD, EACBD, and EACDB. Assume that BEACD is the best of the five orders. This order (BEACD) can be used as the initial order to start the heuristic search.

3.2.3   Simulated Annealing

Simulated annealing is a method which examines a much larger subset of the possible orders. The method was developed to prevent the solution from being trapped into a local optimum. Like the heuristic search, we start with an arbitrary order and evaluate the order using *sar*, *sad*, or $-sal$ as the criterion. Note that *sal* is replaced by $-sal$ because the method always searches for the minimum value of the criterion. The score of the initial order is denoted by E_0. The order is then subject to local rearrangement in a random fashion. This involves Monte Carlo simulation for the order. Note that the rearrangement should be local rather than global. Assume that the score for the new order is E_1. If the new order is "better" than the initial order, i.e., $E_1 < E_0$, the new order is accepted. If the new order is "worse" than the initial order, i.e., $E_1 > E_0$, it is accepted with a probability,

$$\alpha = \exp\left(-\frac{E_1 - E_0}{k_b T}\right) \tag{3.7}$$

where k_b is a physical constant called Boltsman constant and T corresponds to the temperature. Once the new order is accepted, we replace E_0 by E_1 and continue the search for another order. This sampling strategy was proposed by Metropolis et al. (1953) and thus also is referred to as the Metropolis algorithm. By trial and error, it is found that $k_b = 0.95$ usually works well. However, different values should be chosen if $k_b = 0.95$ does not. The value of the temperature T can be chosen arbitrarily, say $T = 2$ or any other values. For m loci, $100m$ new orders should be examined for each value of T, and then T should be changed to $k_b T$ (the temperature has been lowered) at this point. With a pseudocode notation, the change of temperature is expressed as $T = k_b T$ when the temperature is decided to change. The algorithm stops after 100 rearrangements have failed to provide a better order. When we write the computer code to simulate the event of accepting or rejecting a new order, we do not care about whether $E_1 < E_0$ or $E_1 > E_0$. We simply let the new order to be accepted with probability α, which is defined as

$$\alpha = \min\left[1, \exp\left(-\frac{E_1 - E_0}{k_b T}\right)\right] \tag{3.8}$$

If $E_1 < E_0$, i.e., the new order is better than the old order, $\alpha = 1$, meaning that the new order is always accepted. If $E_1 > E_0$, then $\alpha < 1$; the probability of accepting the new order is not 100 %. When the temperature T gets lower, it makes the acceptance of a worse order harder. This can be shown by looking at the profile of the acceptance probability as a function of the deviation of the new order E_1 from the current order E_0 (see Fig. 3.2). The initial length of the map is $E_0 = 3$. The new length E_1 ranges from 2 to 6. The Boltsman constant is $k_b = 0.95$. The three lines represent three different T values ($T = 2, 0.5, 0.1$). When $E_1 \leq E_0$, the probability of acceptance is 1. After E_1 passes $E_0 = 3$ ($E_1 > E_0$), i.e., the new order is

Fig. 3.2 Change of
acceptance probability as E_1
deviates from $E_0 = 3.0$

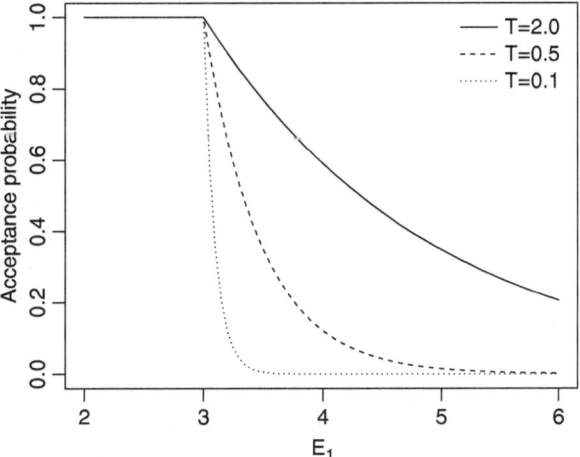

worse than the current order, the acceptance probability starts to decrease but very
slowly for $T = 2$ (high temperature). As the temperature cools down ($T = 0.5$),
the acceptance probability decreases more sharply, meaning that it is hard to accept
a worse order. When $T = 0.1$, very low temperature, the acceptance probability
decreases very sharply, making the acceptance of a worse order extremely difficult.
In the end, only a better order gets accepted, and no worse order will be accepted.
This will end the search.

The intention of allowing a worse order to be accepted is to prevent the algorithm
from settling down at a local optimal order and ignoring a global optimum elsewhere
in the space of possible orders. Simulated annealing was set up in a language from
the observation that when liquids are cooled very slowly, they will crystallize in a
state of minimum energy (Metropolis et al. 1953).

3.2.4 Branch and Bound

The branch and bound method is often used in search for evolutionary trees (also
called phylogenies). The method was first developed by Land and Doig (1960).
It is adopted here to search for the optimal order of loci. This algorithm is not the
exhaustive search, but it guarantees to find the global optimum order. There will be
occasions when it would require examination of all orders, but generally, it requires
examination of only a small subset of all possible orders. The criterion of evaluation
must be "additive." This property will make sure that the map length for a particular
order with k loci cannot be shortened by adding another locus to the existing map
of k loci. Both *sad* and *−sal* follow the additive rule and thus can be used as the
length of a map for the branch and bound search. However, *sar* cannot be used
here because it is not additive. Let us assume that there are four loci, ABCD, to be

ordered. Suppose that the first two loci to be considered are AB. The next locus C can be inserted in one of three positions, corresponding to orders CAB, ACB, and ABC, respectively. The fourth locus D can be inserted in four different positions for each of the three-locus orders. There will be $4!/2 = 12$ possible orders after locus D is inserted. In general, we start with two loci (one possible order) and insert the third locus to the two-locus map (three possible places to insert the third locus). When the ith locus ($i = 3, \ldots, m$) is inserted into the map of $i - 1$ loci, there will be i branch points (places) to insert the ith locus. Overall, there are $3 \times 4 \times \cdots \times m = (1 \times 2 \times 3 \times 4 \times \cdots \times m)/(1 \times 2) = m!/2$ possible orders. This process is similar to a tree growing process, as illustrated in Fig. 3.1 for the example of five loci, except that this tree is drawn upside down with the root at the top. Each tip of the tree represents a particular order of the map for all the m loci, called a child. Each branch point represents an order with $m-1$ or a lower number of loci, called a parent. The initial order of two loci is the root of the tree, called the ancestor. Each member of the tree (including the ancestor, the parents, and the children) is associated with an *sad* value, i.e., the length of the member.

If all the possible orders were evaluated, the method would be identical to the exhaustive search. The branch and bound method, however, starts with an arbitrarily chosen order of the m locus map (a child) and assigns the length of this order to E_0, called the upper bound. It is more efficient to select the shortest map order found from a heuristic search as the initial upper bound. Once an upper bound is assigned, it is immediately known that the optimal order cannot have a value greater than E_0. Let T_0 be the map order for the selected child (the length has been chosen as the upper bound). The branch and bound algorithm starts evaluating all the siblings of T_0. The upper bound will be replaced by the shortest length of the siblings in the family if the current T_0 is not the shortest one. Once all the siblings of the current family are evaluated, we backtrack to the parent and evaluate all the siblings of the parent (the uncles of T_0). Remember that all members in the parental generation have $m - 1$ loci. Any uncles whose scores are longer that E_0 will be disqualified for further evaluation because they will not produce children with scores shorter than E_0 due to the property that inserting additional loci cannot possibly decrease the score. Therefore, we can dispense with the evaluation of all children that descend from those disqualified uncles in the search and immediately backtrack and proceed down a different path. Only uncles whose scores are shorter than E_0 will be subject to further evaluation. The upper bound will be updated if a shorter member is found in the uncle's families. Once all the uncles and their families are evaluated, we backtrack to the great grandparent and the siblings of the grandparent and evaluate the families of all the siblings of the grandparent. The process continues until all qualified families have been evaluated. The upper bound E_0 is constantly updated to ensure that it holds the length of the shortest map order among the orders evaluated so far. Constantly updating the upper bound is important, as it may enable other search paths to be terminated more quickly.

The following example is used to demonstrate the branch and bound algorithm. Let A, B, C, and D be four loci with unknown order. The recombination fractions are stored in the upper triangular positions of the matrix given in Table 3.1. The additive

Table 3.1 Recombination fractions and additive distances for four marker loci

	A	B	C	D
A		$r_{AB}(0.0906)$	$r_{AC}(0.1967)$	$r_{AD}(0.2256)$
B	$x_{AB}(0.10)$		$r_{BC}(0.1296)$	$r_{BD}(0.1648)$
C	$x_{AC}(0.25)$	$x_{BC}(0.15)$		$r_{CD}(0.0476)$
D	$x_{AD}(0.30)$	$x_{BD}(0.20)$	$x_{CD}(0.05)$	

Table 3.2 The 12 possible orders of a four locus map and their *sad* scores

Order	Map	*sad* score
1	DCAB	0.40
2	CDAB	0.45
3	CADB	0.75
4	CABD	0.55
5	DACB	0.70
6	ADCB	0.50
7	ACDB	0.50
8	ACBD	0.60
9	DABC	0.55
10	ADBC	0.65
11	ABDC	0.35
12	ABCD	0.30

distances converted from the recombination fractions using the Haldane map function are stored in the lower triangular positions of the matrix (Table 3.1). The 12 possible orders of the loci are given in Table 3.2 along with the *sad* score for each order. For example, the *sad* score for order CABD is

$$\text{sad}_{DBAC} = x_{BD} + x_{AB} + x_{AC} = 0.20 + 0.10 + 0.25 = 0.55. \quad (3.9)$$

From Table 3.2, we can see that the optimal order is order 12, i.e., ABCD, because its *sad* score (0.30) is minimum among all other orders. This would be the result of exhaustive search because we had evaluated all the 12 possible orders. Let us pretend that we had not looked at Table 3.2 and we want to proceed with the branch and bound method to search for the optimal order.

The entire tree of four loci is given in Fig. 3.3. Each child has four loci and a length given in parentheses. Note that order ACBD has a length 0.60. The five children of ACBD do not belong to this tree of four loci. They are presented here to indicate that the tree of five loci can be expanded from the tree of four loci in this way. A randomly selected order, say DCAB, is used as T_0 whose *sad* score is used as the upper bound, $E_0 = \text{sad}_{DCAB} = 0.40$. All the siblings of DCAB are evaluated for the *sad* scores. It turns out that $\text{sad}_{DCAB} = 0.40$ is the shortest order in the family, and thus E_0 cannot be improved. The search is backtracked to CAB (the parent of DCAB), and the two siblings of the parent are evaluated, with scores of $\text{sad}_{ACB} = 0.40$ and $\text{sad}_{ABC} = 0.25$. Because $\text{sad}_{ACB} = \text{sad}_{DCAB} = E_0 = 0.40$,

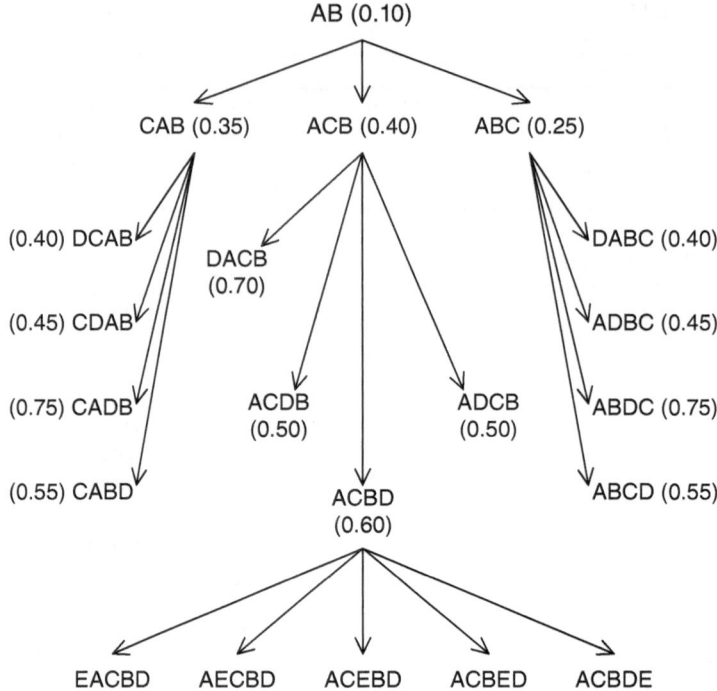

Fig. 3.3 All possible orders (12) of a map with four loci. The five-locus children of order ACBD are also given to show that the tree can be expanded this way for more loci

it is concluded that the optimal order cannot occur in the "lineage" descending from ACB because adding one more locus cannot possibly make the *sad* score shorter. Therefore, only children of ABC are qualified for further evaluation. The shortest order occurs in this family, and it is ABCD with $sad_{ABCD} = 0.30$. We only evaluate two families out of three $(2/3)$ to find the optimal order by using the branch and bound algorithm.

Let us use the *sad* of a different order as the upper bound to start the search and show that the branch and bound algorithm may end up with evaluating all possible orders. Assume that we choose the score of ACDB as the upper bound, i.e., $E_0 = sad_{ACDB} = 0.50$. We first evaluated all the siblings of ACDB and found that the upper bound cannot be improved. We then backtracked to the parent of ACDB, which is ACB. The parent has two siblings, CAB and ABC; both are shorter than E_0, and thus both should be further evaluated. However, which of the lineages is evaluated first can make a difference regarding the efficiency of the search. Assume that the lineage under CAB is evaluated first. This leads to an improved upper bound $E_0 = sad_{DCAB} = 0.40$. This upper bound is identical to the length of the parent of the family that we started the search. This family would not have been evaluated if we had chosen $E_0 = 0.40$ as the upper bound in the beginning. Unfortunately, it was too late that we had already evaluated this family. Since ABC is shorter than $E_0 = 0.40$, the lineage under ABC is subject to further evaluation. The shortest order occurs in this lineage, which is ABCD with sad$_{ABCD} = 0.30$. All the 12 possible orders have been evaluated before the optimal order is found. However, if we had evaluated the lineage under ABC first, we would not have to evaluate the lineage under CAB, leading to a 1/3 cut of the computing load.

Finally, if the upper bound is assigned a value from a member in the ABC lineage, say $E_0 = $ sad$_{ABDC} = 0.35$, the upper bound is immediately updated as $E_0 = $ sad$_{ABCD} = 0.30$. This upper bound immediately disqualifies the other two lineages, leading to a 2/3 reduction of the number of orders for evaluation.

The branch and bound algorithm guarantees to find the shortest order, but the efficiency depends on the upper bound chosen and the sequence in which the paths are visited.

3.3   Bootstrap Confidence of a Map

In phylogeny analysis, an empirical confidence can be put on each internal branch of a particular phylogeny via bootstrap samplings (Felsenstein 1985). This idea can be adopted here for map construction. A map with m markers have $m - 1$ internal segments, similar to the internal branches of a phylogenetic tree. We can put a confidence on each segment. Let us assume that C–D–B–A–E is the optimal order we found. We want to put a bootstrap confidence on segment D–B. First, we draw a large number of bootstrap samples, say N. Each bootstrap sample contains n randomly sampled progeny from the original sample (map population) with replacement. This means that in a bootstrap sample, some progeny may be drawn several times while others may not be drawn at all. For each bootstrap sample, we estimate all the pairwise recombination fractions and construct a map (find the optimal order of the markers for that particular bootstrap sample). In the end, we will have N different maps, one from each bootstrap sample. We then count the number of maps that have segment D–B, i.e., D and B are joined together. The proportion of the maps that reserve this segment is the confidence of this segment. Let $N = 100$ be the number of bootstrap samples and $N_{D-B} = 95$ be the number of samples reserving segment D–B; the confidence for segment D–B is $N_{D-B}/N = 0.95$. Each segment of the map can be put a confidence using the same approach.

The way to place a bootstrap confidence for a segment of a map described here appears to be different from the bootstrap confidence of an internal branch of a phylogenetic tree. We simply adopted the idea of phylogenetic bootstrap analysis, not the way of confidence assignment. To fully adopt the bootstrap confidence assignment, we need to find the number of bootstrap samples that partition the loci into {C,D} and {B,A,E} subsets and also reserve the D–B segment. That number divided by $N = 100$ would give the confidence for the D–B segment.

Chapter 4
Multipoint Analysis of Mendelian Loci

Each Mendelian locus occupies a specific point on a chromosome. A linkage analysis requires two or more Mendelian loci and thus involves two or more points. When a linkage analysis involves two Mendelian loci, as we have seen in Chap. 2 for estimating the recombination fraction between two loci, the analysis is called two-point analysis. When more than two Mendelian loci are analyzed simultaneously, the method is called multipoint analysis (Jiang and Zeng 1997). Multipoint analysis can extract more information from the data if markers are not fully informative, e.g., missing genotypes, dominance alleles, and so on.

When there is no interference between the crossovers of two consecutive chromosome segments, the joint distribution of genotypes of marker loci is Markovian. We can imagine that the entire chromosome behaves like a Markov chain, in which the genotype of one locus depends only on the genotype of the "previous" locus. A Markov chain has a direction, but a chromosome has no meaningful direction. Its direction is defined in an arbitrary fashion. Therefore, we can use either a forward Markov chain or a backward Markov chain to define a chromosome, and the result will be identical, regardless of which direction has been taken.

A Markov chain is used to derive the joint distribution of all marker genotypes. The joint distribution is eventually used to construct a likelihood function for estimating multiple recombination fractions. Given the recombination fractions, one can derive the conditional distribution of the genotype of a locus bracketed by two marker loci given the genotypes of the markers. The conditional distribution is fundamentally important in genetic mapping for complex traits, a topic to be discussed in a later chapter.

4.1 Joint Distribution of Multiple-Locus Genotype

When three loci are considered jointly, the method is called three-point analysis. Theory developed for three-point analysis applies to arbitrary number of loci.

4.1.1   BC Design

Let ABC be three ordered loci on the same chromosome with pairwise recombination fractions denoted by r_{AB}, r_{BC}, and r_{AC}. We can imagine that these loci form a Markov chain as either A $\longrightarrow$ B $\longrightarrow$ C or A $\longleftarrow$ B $\longleftarrow$ C. The direction is arbitrary. Each locus represents a discrete variable with two or more distinct values (states). For an individual from a BC population, each locus takes one of two possible genotypes and thus two states. Let $A_1 A_1$ and $A_1 A_2$ be the two possible genotypes for locus A, $B_1 B_1$ and $B_1 B_2$ be the two possible genotypes for locus B, and $C_1 C_1$ and $C_1 C_2$ be the two possible genotypes for locus C. For convenience, each state is assigned a numerical value. For example, $A = 1$ or $A = 2$ indicates that an individual takes genotype $A_1 A_1$ or $A_1 A_2$. Let us take A $\longrightarrow$ B $\longrightarrow$ C as the Markov chain; the joint distribution of the three-locus genotype is

$$\Pr(A, B, C) = \Pr(A) \Pr(B|A) \Pr(C|B), \tag{4.1}$$

where $\Pr(A = 1) = \Pr(A = 2) = \frac{1}{2}$ assuming that there is no segregation distortion. The conditional probabilities, $\Pr(B|A)$ and $\Pr(C|B)$, are called the transition probabilities between loci A and B and between loci B and C, respectively. The transition probabilities depend on the genotypes of the two loci and the recombination fractions between the two loci. These transition probabilities can be found from the following 2×2 transition matrix:

$$T_{AB} = \begin{bmatrix} \Pr(B = 1|A = 1) & \Pr(B = 2|A = 1) \\ \Pr(B = 1|A = 2) & \Pr(B = 2|A = 2) \end{bmatrix}. \tag{4.2}$$

Because $\Pr(B = 1|A = 1) = \Pr(B = 2|A = 2) = 1 - r_{AB}$ represents the probability of no recombination between the two loci and $\Pr(B = 2|A = 1) = \Pr(B = 1|A = 2) = r_{AB}$ represents the probability of recombination between the two loci, the exact form of the transition matrix between loci A and B is

$$T_{AB} = \begin{bmatrix} T_{AB}(1, 1) & T_{AB}(1, 2) \\ T_{AB}(2, 1) & T_{AB}(2, 2) \end{bmatrix} = \begin{bmatrix} 1 - r_{AB} & r_{AB} \\ r_{AB} & 1 - r_{AB} \end{bmatrix}, \tag{4.3}$$

where $T_{AB}(k, l) \; \forall k, l = 1, 2$ denotes the kth row and the lth column of matrix T_{AB}. It is now obvious that $T_{AB}(k, l) = \Pr(B = l|A = k)$. Note that we have used a special notation "$\forall k, l = 1, 2$" to indicate that k and l each takes a value from 1 to 2. Verbally, "$\forall k, l = 1, 2$" means "for all $k = 1, 2$ and $l = 1, 2$". When using this kind of notation, we should particularly pay attention to the positions of k and l in $T_{AB}(k, l) = \Pr(B = l|A = k)$. It is a conditional probability that $B = l$ given $A = k$. Replacing the conditional probabilities by the elements of the transition matrix, we rewrite the joint probability of the three-locus genotype as

$$\Pr(A, B, C) = \frac{1}{2} T_{AB}(A, B) T_{BC}(B, C). \tag{4.4}$$

For example, the probability that $A = 1$, $B = 2$, and $C = 2$ is

$$\Pr(A = 1, B = 2, C = 2) = \frac{1}{2}T_{AB}(1,2)T_{BC}(2,2) = \frac{1}{2}r_{AB}(1 - r_{BC}).$$

This joint probability can be written in matrix notation. Let us use a 2×2 diagonal matrix D_A to denote the genotype of locus A. This matrix is defined as

$$D_A = \begin{bmatrix} 1 & 0 \\ 0 & 0 \end{bmatrix} \text{ for } A = 1 \text{ and } D_A = \begin{bmatrix} 0 & 0 \\ 0 & 1 \end{bmatrix} \text{ for } A = 2.$$

Diagonal matrices D_B and D_C are defined similarly for loci B and C, respectively. The original data are in the form of genotype indicator variables, A, B, and C, but the new form of the data is represented by the diagonal matrices. Let us define a 2×1 unity vector by $J = [1 \quad 1]'$. The joint distribution given in (4.4) is rewritten in matrix notation as

$$\Pr(A, B, C) = \frac{1}{2}J'D_A T_{AB} D_B T_{BC} D_C J. \tag{4.5}$$

One can verify that

$$\Pr(A = 1, B = 2, C = 2)$$

$$= \frac{1}{2}\begin{bmatrix} 1 & 1 \end{bmatrix}\begin{bmatrix} 1 & 0 \\ 0 & 0 \end{bmatrix}\begin{bmatrix} 1 - r_{AB} & r_{AB} \\ r_{AB} & 1 - r_{AB} \end{bmatrix}\begin{bmatrix} 0 & 0 \\ 0 & 1 \end{bmatrix}\begin{bmatrix} 1 - r_{BC} & r_{BC} \\ r_{BC} & 1 - r_{BC} \end{bmatrix}\begin{bmatrix} 0 & 0 \\ 0 & 1 \end{bmatrix}\begin{bmatrix} 1 \\ 1 \end{bmatrix}$$

$$= \frac{1}{2}r_{AB}(1 - r_{BC}).$$

4.1.2 F_2 Design

Taking into consideration the order of the two alleles carried by an F_2 individual, we have four possible genotypes: A_1A_1, A_1A_2, A_2A_1, and A_2A_2. The first and the last genotypes are homozygotes, while the second and third genotypes are heterozygotes. The two forms of heterozygote represent two different origins of the alleles. They are indistinguishable from each other. Therefore, we adopt a special notation, (A_1A_2), to denote the unordered heterozygote. The alleles and genotypes for the other loci are expressed using similar notation. Let $A = k, \forall k = 1, \ldots, 4$ be an indicator variable to indicate the four genotypes of locus A. Variables B and C are similarly defined for loci B and C, respectively. The joint probability of the three-locus genotype is $\Pr(A, B, C) = \Pr(A)\Pr(B|A)\Pr(C|B)$ where $\Pr(A=k)=\frac{1}{4}, \forall k = 1, \ldots, 4$. $\Pr(B|A)$ and $\Pr(C|B)$ are the transition probabilities from locus A to locus B and from locus B to locus C, respectively. The transition probabilities from locus A to locus B can be found from the following 4×4 transition matrix:

$$T_{AB} = \begin{bmatrix} (1-r_{AB})^2 & (1-r_{AB})r_{AB} & r_{AB}(1-r_{AB}) & r_{AB}^2 \\ (1-r_{AB})r_{AB} & (1-r_{AB})^2 & r_{AB}^2 & r_{AB}(1-r_{AB}) \\ r_{AB}(1-r_{AB}) & r_{AB}^2 & (1-r_{AB})^2 & 1-r_{AB})r_{AB} \\ r_{AB}^2 & r_{AB}(1-r_{AB}) & 1-r_{AB})r_{AB} & (1-r_{AB})^2 \end{bmatrix}. \quad (4.6)$$

The transition matrix from locus B to locus C is denoted by T_{BC}, which is equivalent to matrix (4.6) except that the subscript $_{AB}$ is replaced by subscript $_{BC}$.

Note that this transition matrix is obtained by the Kronecker square (denoted by a superscript [2]) of a 2×2 transition matrix,

$$H_{AB} = \begin{bmatrix} 1-r_{AB} & r_{AB} \\ r_{AB} & 1-r_{AB} \end{bmatrix}, \quad (4.7)$$

that is,

$$T_{AB} = \begin{bmatrix} 1-r_{AB} & r_{AB} \\ r_{AB} & 1-r_{AB} \end{bmatrix}^{[2]} = \begin{bmatrix} 1-r_{AB} & r_{AB} \\ r_{AB} & 1-r_{AB} \end{bmatrix} \otimes \begin{bmatrix} 1-r_{AB} & r_{AB} \\ r_{AB} & 1-r_{AB} \end{bmatrix}.$$

The 4×4 transition matrix (4.6) may be called the zygotic transition matrix, and the 2×2 transition matrix (4.7) may be called the gametic transition matrix. That the zygotic transition matrix is the Kronecker square of the gametic transition matrix is very intuitive because a zygote is the product of two gametes. Let $T_{AB}(k,l)$ be the kth row and the lth column of the 4×4 transition matrix T_{AB}, $\forall k,l = 1,\ldots,4$. The joint probability of the three-locus genotype is expressed as

$$\Pr(A, B, C) = \frac{1}{4}T_{AB}(A, B)T_{BC}(B, C). \quad (4.8)$$

For example, the joint three-locus genotype $A_1A_1B_1B_2C_2C_1$ is numerically coded as $A = 1$, $B = 2$, and $C = 3$, whose probability is

$$\Pr(A = 1, B = 2, C = 3) = \frac{1}{4}T_{AB}(1,2)T_{BC}(2,3) = \frac{1}{4}(1 - r_{AB})r_{AB}r_{BC}^2.$$

In practice, people will never observe a three-locus genotype like $A_1A_1B_1B_2C_2C_1$ because the two forms of the heterozygote are not distinguishable. The joint three-locus genotype $A_1A_1(B_1B_2)(C_1C_2)$ is actually what we can observe. The numerical code for the first locus is $A = 1$, but the codes for loci B and C are ambiguous. For example, locus B can be coded as either $B = 2$ or $B = 3$ with an equal

Fig. 4.1 Four-way (FW) cross mating design

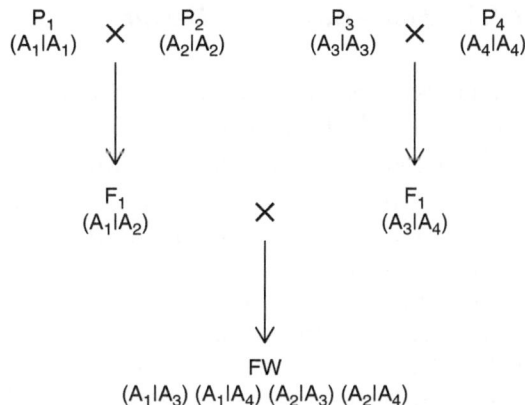

probability. This ambiguous situation is denoted by $B = (2, 3)$. Similar notation applies to locus C as $C = (2, 3)$. The joint distribution for $A_1 A_1 (B_1 B_2)(C_1 C_2)$ is

$$\Pr[A = 1, B = (2,3), C = (2,3)] = \frac{1}{4} \sum_{k=2}^{3} \left[T_{AB}(1, k) \sum_{l=2}^{3} T_{BC}(k, l) \right]$$

$$= \frac{1}{2} r_{AB}(1 - r_{AB})[r_{BC}^2 + (1 - r_{BC})^2].$$

Again, the joint distribution of the three-locus genotype (4.8) can be expressed in matrix notation. We now use a 4×4 diagonal matrix to denote the genotype of a locus. For locus A, this diagonal matrix is defined as

$$D_A = \begin{bmatrix} 1 & 0 & 0 & 0 \\ 0 & 0 & 0 & 0 \\ 0 & 0 & 0 & 0 \\ 0 & 0 & 0 & 0 \end{bmatrix}, \quad D_A = \begin{bmatrix} 0 & 0 & 0 & 0 \\ 0 & 1 & 0 & 0 \\ 0 & 0 & 1 & 0 \\ 0 & 0 & 0 & 0 \end{bmatrix} \text{ and } D_A = \begin{bmatrix} 0 & 0 & 0 & 0 \\ 0 & 0 & 0 & 0 \\ 0 & 0 & 0 & 0 \\ 0 & 0 & 0 & 1 \end{bmatrix},$$

for $A = 1, A = (2, 3)$ and $A = 4$, respectively. Verbally, matrix D_A is a diagonal matrix with unity values for the diagonal elements corresponding to the positions pointed by the value of A. Having defined these diagonal matrices for all loci, we can rewrite the joint distribution of the three-locus genotype as

$$\Pr(A, B, C) = \frac{1}{4} J' D_A T_{AB} D_B T_{BC} D_C J, \tag{4.9}$$

where J is now a 4×1 vector of unity, rather than a 2×1 vector as in the BC design (Fig. 2.1).

4.1.3 Four-Way Cross Design

A four-way cross design involves two different crosses and four different inbred parents. Let $F_1^{(12)}$ be the hybrid progeny derived from the cross of P_1 and P_2 and $F_1^{(34)}$ be the progeny derived from the cross of P_3 and P_4. The cross between $F_1^{(12)}$ and $F_1^{(34)}$ is called the four-way cross. Such a design is called the four-way cross design (FW) as illustrated in Fig. 4.1. Let $A_k A_k B_k B_k C_k C_k$ be the three-locus genotype for the k parent, $\forall k = 1, \ldots, 4$. The three-locus genotypes for $F_1^{(12)}$ and $F_1^{(34)}$ are $A_1 A_2 B_1 B_2 C_1 C_2$ and $A_3 A_4 B_3 B_4 C_3 C_4$, respectively. Consider a single locus, say locus A. An FW progeny can take one of the four genotype: $A_1 A_3$, $A_1 A_4$, $A_2 A_3$, and $A_3 A_4$. Let $A = 1, \ldots, 4$ denote the numerical code for each of the four genotypes. The joint three-locus genotype is still expressed by (4.9) with the same transition matrices as defined earlier in the F_2 design. The diagonal matrices, D_A, D_B, and D_C, are defined similarly to those in the F_2 design except that the second and third genotypes are distinguishable. The numerical code of $A = k$ is translated into a D_A matrix whose elements are all zero except that the kth row and the kth column are unity. For example, the joint probability that $A = 3$, $B = 1$, and $C = 4$ is

$$\Pr(A = 3, B = 1, C = 4) = \frac{1}{4} J' D_A T_{AB} D_B T_{BC} D_C J$$

$$= \frac{1}{4} T_{AB}(3, 1) T_{BC}(1, 4)$$

$$= \frac{1}{4} r_{AB}(1 - r_{AB}) r_{BC}^2.$$

4.2 Incomplete Genotype Information

4.2.1 Partially Informative Genotype

The FW cross design described earlier represents a situation where all the four genotypes in the progeny are distinguishable. In reality, it is often that not all genotypes are distinguishable. This may happen when two or more of the grandparents carry the same allele at the locus of interest. The consequence is that the F_1 hybrid initiated by the first level of the cross may be homozygous or the two F_1 parents may have the same genotype. Assume that $F_1^{(34)}$ has a genotype of $A_3 A_3$, which is homozygous. This may be caused by a cross between two parents, both of which are fixed at A_3 allele. Regardless of the reason that causes the homozygosity of the F_1 hybrid, let us focus on the genotypes of the two F_1 parents and consider the four possible genotypes of the FW progeny. Assume that $F_1^{(12)}$ and $F_1^{(34)}$ have genotypes of $A_1 A_2$ and $A_3 A_3$, respectively.

The four possible genotypes of the progeny are A_1A_3, A_1A_3, A_2A_3, and A_2A_3. The first and the second genotypes are not distinguishable, although the A_3 allele carried by the two genotypes has different origins. This situation applies to the third and fourth genotypes. Considering the allelic origins, we have four ordered genotypes, but we only observe two distinguishable genotypes. This phenomenon is called incomplete information for the genotype. Such a genotype is called partially informative genotype. If we observe a genotype A_1A_3, the numerical code for the genotype is $A = (1, 2)$. In matrix notation, it is represented by

$$D_A = \begin{bmatrix} 1 & 0 & 0 & 0 \\ 0 & 1 & 0 & 0 \\ 0 & 0 & 0 & 0 \\ 0 & 0 & 0 & 0 \end{bmatrix}.$$

If an observed genotype is A_2A_3, the numerical code becomes $A = (3, 4)$, represented by

$$D_A = \begin{bmatrix} 0 & 0 & 0 & 0 \\ 0 & 0 & 0 & 0 \\ 0 & 0 & 1 & 0 \\ 0 & 0 & 0 & 1 \end{bmatrix}.$$

If both parents are homozygous and fixed to the same allele, say A_1, then all the four genotypes of the progeny have the same observed form, A_1A_1. The numerical code for the genotype is $A = (1, 2, 3, 4)$, a situation called no information. Such a locus is called uninformative locus and usually excluded from the analysis. The diagonal matrix representing the genotype is simply a 4×4 identity matrix.

The following is an example showing how to calculate the three-locus joint genotype using the FW cross approach with partial information. Let $A_1A_3B_2B_3C_1C_1$ and $A_4A_4B_2B_3C_1C_2$ be the three-locus genotypes for two parents. The linkage phases of markers in the parents are assumed to be known so that the order of the two alleles within a locus is meaningful. In fact, the phase-known genotypes of the parents are better denoted by $\frac{A_1B_2C_1}{A_3B_3C_1}$ and $\frac{A_4B_2C_1}{A_4B_3C_2}$, respectively, for the two parents. Assume that a progeny has a genotype of $A_3A_4B_2B_2C_1C_1$. We want to calculate the probability of observing such a progeny given the genotypes of the parents. First, we examine each single-locus genotype to see which one of the four possible genotypes this individual belongs to. For locus A, the parental genotypes are A_1A_2 and A_4A_4. The four possible genotypes of a progeny are A_1A_4, A_1A_4, A_3A_4, and A_3A_4, respectively. The single-locus genotype of the progeny is A_3A_4, matching the third and fourth genotypes, and thus $A = (3, 4)$. For locus B, the parental genotypes are B_2B_3 and B_2B_3. The four possible genotypes of a progeny are B_2B_2, B_2B_3, B_3B_2, and B_3B_3, respectively. The single-locus genotype B_2B_2 for the progeny matches the first genotype and thus $B = 1$. For locus C, the parental genotypes are C_1C_1 and C_1C_2. The four possible genotypes of a progeny are C_1C_1, C_1C_2, C_1C_1, and C_1C_2, respectively. The single-locus genotype of the progeny C_1C_1 matches the first and the third genotypes and thus $C = (1, 3)$. In summary, the numerical codes for the three loci are $A = (3, 4)$, $B = 1$, and $C = (1, 3)$, respectively. We now convert the three single-locus genotypes into their corresponding diagonal matrices,

$$
D_A = \begin{bmatrix} 0 & 0 & 0 & 0 \\ 0 & 0 & 0 & 0 \\ 0 & 0 & 1 & 0 \\ 0 & 0 & 0 & 1 \end{bmatrix}, \; D_B = \begin{bmatrix} 1 & 0 & 0 & 0 \\ 0 & 0 & 0 & 0 \\ 0 & 0 & 0 & 0 \\ 0 & 0 & 0 & 0 \end{bmatrix} \text{ and } D_C = \begin{bmatrix} 1 & 0 & 0 & 0 \\ 0 & 0 & 0 & 0 \\ 0 & 0 & 1 & 0 \\ 0 & 0 & 0 & 0 \end{bmatrix}.
$$

Substituting these matrices into (4.9), we have

$$
\begin{aligned}
\Pr[A = (3, 4), B = 1, C = (1, 3)] &= \frac{1}{4} J' D_A T_{AB} D_B T_{BC} D_C J \\
&= \frac{1}{4} [T_{AB}(3, 1) + T_{AB}(4, 1)][T_{BC}(1, 1) + T_{BC}(1, 3)] \\
&= \frac{1}{4} r_{AB}(1 - r_{BC})
\end{aligned}
$$

4.2.2   BC and F$_2$ Are Special Cases of FW

The four-way cross design is a general design where the BC and F_2 designs are special cases of the general design with partial information. For example, the two parents of the BC$_1$ design have genotypes of $A_1 A_2$ and $A_1 A_1$, respectively. If we treat a BC progeny as a special FW progeny, the four possible genotypes are $A_1 A_1$, $A_1 A_1$, $A_2 A_1$, and $A_2 A_1$, only two distinguishable observed types. If a progeny has a genotype $A_1 A_1$, the numerical code of the genotype in terms of an FW cross is $A = (1, 2)$. If a progeny has a genotype of $A_2 A_1$, its numerical codes become $A = (3, 4)$. The two parents of a BC$_1'$ design have genotypes of $A_1 A_2$ and $A_2 A_2$, respectively. In terms of an FW cross, the four possible genotypes are $A_1 A_2$, $A_1 A_2$, $A_2 A_2$, and $A_2 A_2$. Again, there are only two distinguishable genotypes. The two parents of an F_2 design have genotypes of $A_1 A_2$ and $A_1 A_2$, respectively. If we treat an F_2 progeny as a special FW progeny, the four possible genotypes are $A_1 A_1$, $A_1 A_2$, $A_2 A_1$, and $A_2 A_2$, only three distinguishable genotypes. The numerical codes for the two types of homozygote are $A = 1$ and $A = 4$, respectively, whereas the numerical code for the heterozygote is $A = (2, 3)$. In summary, when the general FW design is applied to a BC design, only two of the four possible genotypes are distinguishable, and the numerical codes are $A = (1, 2)$ for one observed genotype and $A = (3, 4)$ for the other observed genotype. When the general FW design is applied to the F_2 design, the two forms of heterozygote are not distinguishable. When coding the genotype, we use $A = (2, 3)$ to represent the heterozygote and $A = 1$ and $A = 4$ to represent the two types of homozygote, respectively. The transition matrices remain the same as those used in an FW cross design.

We have learned the BC design in Sec. 4.1.1 using the 2×2 transition matrix. When using the FW design for the BC problem, we have combined the first and second genotypes to form the first observable genotype and combined the third and fourth genotypes to form the second observable genotype for the BC design. It can be shown that the joint probability calculated by the Markov chain with two states (using the 2×2 transition matrix) and that calculated by the Markov chain with four states (the 4×4 transition matrix) are identical.

The F_2 design we learned earlier can be handled by combining the second and third genotypes into the observed heterozygote. The 4×4 transition matrix is converted into a 3×3 transition matrix,

$$T_{AB} = \begin{bmatrix} (1 - r_{AB})^2 & 2(1 - r_{AB})r_{AB} & r_{AB}^2 \\ (1 - r_{AB})r_{AB} & r_{AB}^2 + (1 - r_{AB})^2 & (1 - r_{AB})r_{AB} \\ r_{AB}^2 & 2(1 - r_{AB})r_{AB} & (1 - r_{AB})^2 \end{bmatrix}.$$

The joint probability of multiple-locus genotype for an F_2 individual can be calculated using a Markov chain with the 3×3 transition matrix. The numerical code for a genotype must be redefined in the following way. The three defined genotypes, $A_1 A_1$, $A_1 A_2$, and $A_2 A_2$, are numerically coded by $A = 1$, $A = 2$, and $A = 3$, respectively.

In matrix notation, the three genotypes are denoted by

$$D_A = \begin{bmatrix} 1 & 0 & 0 \\ 0 & 0 & 0 \\ 0 & 0 & 0 \end{bmatrix}, \ D_A = \begin{bmatrix} 0 & 0 & 0 \\ 0 & 1 & 0 \\ 0 & 0 & 0 \end{bmatrix} \text{ and } D_A = \begin{bmatrix} 0 & 0 & 0 \\ 0 & 0 & 0 \\ 0 & 0 & 1 \end{bmatrix},$$

respectively.

The general FW design using a Markov chain with four states is computationally more intensive when applied to BC and F_2 designs compared to the specialized BC (with 2×2 transition matrix) and F_2 (with 3×3 transition matrix) algorithm. However, the difference in computing times is probably unnoticeable given the current computing power. In addition, the 3×3 transition matrix is not symmetrical, a factor that may easily cause a programming error. Therefore, the general FW design is recommended for all line crossing experiments.

4.2.3 Dominance and Missing Markers

A dominance marker is a type of marker whose heterozygous genotype cannot be distinguished from one of the two homozygous genotypes. Therefore, dominance markers cannot be used in a BC design. However, partial information can be extracted from dominance markers in an F_2 design. Consider locus A with four possible genotypes in an F_2 population under a biallelic system, alleles A_1 vs A_2. The four ordered genotypes are $A_1 A_1$, $A_1 A_2$, $A_2 A_1$, and $A_2 A_2$. Dominance can be found in two directions. If A_1 is dominant over A_2, we cannot distinguish the three genotypes, A_1A_1, A_1A_2, and A_2A_1. If A_2 is dominant over A_1, however, we cannot distinguish the three genotypes, A_1A_2, A_2A_1, and A_2A_2. Therefore, we can only observe two possible genotypes for a particular locus. The two possible genotypes are represented by A_1A_* and A_2A_2 if A_1 dominates over A_2, or A_2A_* and A_1A_1 if A_2 dominates over A_1. Allele A_* is a wild card and can be either A_1 or A_2. When A_1 dominates over A_2, we use $A = (1, 2, 3)$ to code genotype A_1A_* and $A = 4$ to code genotype A_2A_2. If A_2 dominates over A_1, we use $A = 1$ to code genotype A_1A_1 and $A = (2, 3, 4)$ to code genotype A_2A_*. The numerical code for each locus is then converted into an appropriate diagonal matrix, D_A, D_B, or D_C, for calculating the joint probability of a joint three-locus genotype.

If the genotype for a locus, say locus A, is missing, the numerical code for the locus is $A = (1, 2, 3, 4)$, and the corresponding diagonal matrix D_A is simply a 4×4 identity matrix. Missing marker genotypes are treated the same way as genotypes of uninformative loci.

4.3   Conditional Probability of a Missing Marker Genotype

An important application of the three-point analysis to genetic mapping is to calculate the probability of genotype of a locus conditional on genotypes of flanking markers. Note that flanking markers are the two nearby markers of a locus, one in each side. Consider three loci, ABC, where A and C are two markers with known genotypes and B is a locus whose genotype is not observable. The conditional probability of genotype of locus B is

$$\Pr(B|A, C) = \frac{\Pr(A, B, C)}{\Pr(A, C)}. \tag{4.10}$$

The joint probability of the three-locus genotype in the numerator can be rewritten as

$$\Pr(A, B, C) = \Pr(B)\Pr(A, C|B) = \Pr(B)\Pr(A|B)\Pr(C|B).$$

We are able to write $\Pr(A, C|B) = \Pr(A|B)\Pr(C|B)$ because conditional on the genotype of B, the genotypes of A and C are independent due to the Markovian property of Mendelian loci. The joint probability of the two-locus genotype in the denominator of (4.10) is expressed as

$$\Pr(A, C) = \sum_{B=1}^{4} \Pr(A, B, C) = \sum_{B=1}^{4} \Pr(B)\Pr(A|B)\Pr(C|B).$$

Eventually, the conditional probability is expressed as

$$\Pr(B|A, C) = \frac{\Pr(B)\Pr(A|B)\Pr(C|B)}{\sum_{B=1}^{4}\Pr(B)\Pr(A|B)\Pr(C|B)} \tag{4.11}$$

We realize that $\Pr(A|B)$ and $\Pr(C|B)$ are the transition probabilities and $\Pr(B = k) = \frac{1}{4}, \forall k = 1, \ldots, 4$, is the marginal probability. The conditional probability expressed this way (4.11) is an expression of Bayes' theorem.

We now use matrix notation to express the conditional probability. Assume that we want to calculate $\Pr(B = k|A, C), \forall k = 1, \ldots, 4$, where the genotypes of loci A and C are known and represented by matrices D_A and D_C. Since marker B is treated as a missing marker, its genotype is represented by $D_B = I_{4 \times 4}$, an identity matrix. The matrix version of the numerator of (4.11) is

$$\Pr(B = k)\Pr(A|B = k)|\Pr(C|B = k) = \frac{1}{4}J'D_A T_{AB} D_{(k)} T_{BC} D_C J, \quad (4.12)$$

where $D_{(k)}$ is a diagonal matrix with all elements equal to zero except the element at the kth row and the kth column, which is unity. The matrix expression of the denominator of (4.11) is

$$\sum_{B=1}^{4} \Pr(B = k)\Pr(A|B = k)|\Pr(C|B = k) = \frac{1}{4}J'D_A T_{AB} D_B T_{BC} D_C J. \quad (4.13)$$

Therefore, the matrix expression of the conditional probability is

$$\Pr(B = k|A, C) = \frac{J'D_A T_{AB} D_{(k)} T_{BC} D_C J}{J'D_A T_{AB} D_B T_{BC} D_C J}. \quad (4.14)$$

We now use an F_2 progeny as an example to show how to calculate the conditional probabilities of a locus given genotypes of the flanking markers. Let $A_1 A_1$ and $(C_1 C_2)$ be the genotypes of loci A and C, respectively. Recall that $(C_1 C_2)$ means that locus C is heterozygous, which has two forms, $C_1 C_2$ and $C_2 C_1$. We want to calculate the conditional probability that locus B is $B_1 B_1$. The numerical codes for the genotypes of A and C are $A = 1$ and $C = (2, 3)$, respectively, which are translated into matrices of $D_A = D_{(1)}$ and $D_C = D_{(2)} + D_{(3)}$, respectively. Let $D_B = I_{4 \times 4}$ because locus B is a missing marker. The numerator and the denominator of the conditional probability are

$$J'D_A T_{AB} D_{(1)} T_{BC} D_C J = T_{AB}(1, 1)(T_{BC}(1, 2) + T_{BC}(1, 3))$$

$$= 2(1 - r_{AB})^2 r_{BC}(1 - r_{BC})$$

and

$$J'D_A T_{AB} D_B T_{BC} D_C J = \sum_{k=1}^{4} T_{AB}(1, k)(T_{BC}(k, 2) + T_{BC}(k, 3))$$

$$= 2r_{AC}(1 - r_{AC})$$

respectively. Therefore, the conditional probability is

$$\Pr[B = 1|A = 1, C = (2,3)] = \frac{2(1 - r_{AB})^2 r_{BC}(1 - r_{BC})}{2r_{AC}(1 - r_{AC})}$$

$$= \frac{(1 - r_{AB})^2 r_{BC}(1 - r_{BC})}{r_{AC}(1 - r_{AC})}$$

where

$$r_{AC} = r_{AB}(1 - r_{BC}) + r_{BC}(1 - r_{AB})$$

For the same genotypes of marker A and C, what is the conditional probability that marker B is heterozygous? This probability is represented by

$$\Pr[B = (2,3)|A = 1, C = (2,3)] = \frac{J' D_A T_{AB}(D_{(2)} + D_{(3)}) T_{BC} D_C J}{J' D_A T_{AB} D_B T_{BC} D_C J}$$

4.4 Joint Estimation of Recombination Fractions

The three-locus genotype distribution can be used to estimate r_{AB} and r_{BC} jointly. Again, let ABC be the three ordered loci under consideration. Assume that we have collected n progeny from a line cross family. The family can be a BC, an F_2, or an FW, but all represented by the generalized FW family so that the 4×4 transition matrix between consecutive markers applies to all designs. Let A^i be the numerical code for the genotype of individual i at locus A, $\forall i = 1, \ldots, n$, where A^i can take a subset of $\{1, 2, 3, 4\}$, depending on the actual genotype of individual i. The three-locus genotype is denoted by $A^i B^i C^i$. The corresponding diagonal matrices for the individual locus genotypes are denoted by D_A^i, D_B^i, and D_C^i, respectively. The joint three-locus genotype for individual i is

$$\Pr(A^i B^i C^i) \propto J' D_A^i T_{AB} D_B^i T_{BC} D_C^i J. \tag{4.15}$$

The equal sign is replaced by the sign of "proportional to" because the expression in the right-hand side of the equation differs from that in the left-hand side by a constant factor $(\frac{1}{4})$. The log likelihood function of the recombination fractions established from all the n individuals is

$$L(r_{AB}, r_{BC}) = \sum_{i=1}^{n} \ln(J' D_A^i T_{AB} D_B^i T_{BC} D_C^i J). \tag{4.16}$$

Explicit solutions for the ML estimates of the recombination fractions are possible if there are no missing genotypes of the markers. In this case, the above log likelihood function can be rewritten as

$$L(r_{AB}, r_{BC}) = \sum_{i=1}^{n} \ln T_{AB}(A_i, B_i) + \sum_{i=1}^{n} \ln T_{BC}(B_i, C_i). \qquad (4.17)$$

The first term is simply a function of r_{AB}, and the second term is a function of r_{BC}, which are denoted by $L(r_{AB})$ and $L(r_{BC})$, respectively. Therefore, the log likelihood function for the three-point analysis is simply the sum of the two pairwise log likelihood functions,

$$L(r_{AB}, r_{BC}) = L(r_{AB}) + L(r_{BC}). \qquad (4.18)$$

As a consequence, the three-point analysis provides identical results for the estimated recombination fractions as the pairwise analysis. Therefore, when markers are all fully informative, there is no reason to invoke the three-point analysis. The three-point analysis, however, can extract additional information from the data if partially informative markers are present or there are missing marker genotypes. One reason for the increased efficiency of the three-point analysis is the incorporation of the marker order. For the pairwise analysis of three markers, one would have to estimate r_{AC} also from the same data. However, the three-point analysis treats the estimated r_{AC} as a function of the other two recombination fractions, i.e., $\hat{r}_{AC} = \hat{r}_{AB} + \hat{r}_{BC} - 2\hat{r}_{AB}\hat{r}_{BC}$. Therefore, information about the order of the three markers has been incorporated implicitly in the three-point analysis.

In general, there is no explicit solution for the joint estimate of the two recombination fractions, unless all markers are fully informative and there are no missing marker genotypes. A general numerical algorithm, e.g., the simplex method of Nelder and Mead (1965), can be adopted here to search for the MLE of the parameters. For problems with two clearly bounded parameters, such as this one with ($0 < r_{AB}, r_{BC} < 0.5$), we may even use the simple grid search algorithm, which guarantees that the global optimal solutions for the parameters are obtained.

4.5   Multipoint Analysis for m Markers

We have just learned the three-point analysis ($m = 3$) as a special case of the general multipoint analysis. We now extend the methods to situations where $m > 3$. Let us use $j = 1, \ldots, m$ to index the locus. We now have $m - 1$ consecutive recombination fractions and thus $m - 1$ transition matrices. The recombination fraction between loci j and $j + 1$ is denoted by $r_{j(j+1)}$, and the corresponding transition matrix is denoted by $T_{j(j+1)}$. Let D_j be the diagonal matrix for the genotype of locus j. We now use $G_j = k, \forall k = 1, \ldots, 4$, to denote the numerical code for the genotype of the jth locus. Recall that there are four possible genotypes for the generalized four-way cross design. Again, D_j is a matrix version of the numerical code for the genotype of locus j with $D_j = D_{(k)}$ for $G_j = k$. For an ambiguous genotype like $G_j = (2, 3)$ or $G_j = (1, 2, 3, 4)$, the corresponding diagonal matrix is denoted by $D_j = D_{(2)} + D_{(3)}$ or $D_j = D_{(1)} + D_{(2)} + D_{(3)} + D_{(4)} = I_{4 \times 4}$, respectively.

We now discuss the joint distribution for the m locus genotype, the conditional distribution of a missing marker genotype given the observed genotypes of $m - 1$ markers, and the log likelihood function for jointly estimating $m - 1$ recombination fractions using m markers. The joint distribution of the m locus genotype is denoted by

$$\Pr(G_1, G_2, \ldots, G_m) = \frac{1}{4} J' D_1 T_{12} D_2 \ldots T_{(j-1)j} D_j T_{j(j+1)} \ldots D_{m-1} T_{(m-1)m} D_m J.$$
(4.19)

Assume that the genotype of the jth marker is missing. The conditioning probability of $G_j = k$ given the genotypes of all the $m - 1$ markers is

$$\Pr(G_j = k | G_1, \ldots, G_m)$$

$$= \frac{J' D_1 T_{12} D_2 \ldots T_{(j-1)j} D_{(k)} T_{j(j+1)} \ldots D_{m-1} T_{(m-1)m} D_m J}{J' D_1 T_{12} D_2 \ldots T_{(j-1)j} D_j T_{j(j+1)} \ldots D_{m-1} T_{(m-1)m} D_m J}.$$
(4.20)

Recall that $D_j = I_{4 \times 4}$ because j is the missing marker. The probability that $G_j = (2, 3)$ is simply obtained by substituting $D_{(k)}$ in the numerator of the above equation by $D_{(2)} + D_{(3)}$. Let D_j^i be the matrix representation of G_j for individual i for $i = 1, \ldots, n$. The log likelihood function for estimating $\theta = \{r_{12}, r_{23}, \ldots, r_{(m-1)m}\}$ is

$$L(\theta) = \sum_{i=1}^{n} \ln J' D_1^i T_{12} D_2^i \ldots T_{(j-1)j} D_j^i T_{j(j+1)} \ldots D_{m-1}^i T_{(m-1)m} D_m^i J. \quad (4.21)$$

One property of the multipoint analysis is that

$$\Pr(G_j = k | G_1, \ldots, G_m) = \Pr(G_j = k | G_{j-1}, G_{j+1}), \quad (4.22)$$

if markers $j - 1$ and $j + 1$ are fully informative. Verbally, this property is stated as "the genotype of a marker only depends on the genotypes of the flanking markers." This can be proved by the following argument. If loci $j - 1$ and $j + 1$ are fully informative, the numerator of (4.20) can be rewritten as

$$H_l \times (J' D_{j-1} T_{(j-1)j} D_{(k)} T_{j(j+1)} D_{j+1} J) \times H_r \quad (4.23)$$

and the denominator of (4.20) can be rewritten as

$$H_l \times (J' D_{j-1} T_{(j-1)j} D_j T_{j(j+1)} D_{j+1} J) \times H_r, \quad (4.24)$$

where

$$H_l = J' D_1 T_{12} D_2 \ldots D_{j-1} J$$

and

$$H_r = J'D_{j+1}\ldots D_{m-1}T_{(m-1)m}D_m J.$$

Note that H_l and H_r are scalars and they appear in both the numerator and the denominator. Therefore, they are canceled out in the conditional probability, leaving

$$\Pr(G_j = k|G_1,\ldots,G_m) = \frac{H_l \times (J'D_{j-1}T_{(j-1)j}D_{(k)}T_{j(j+1)}D_{j+1}J) \times H_r}{H_l \times (J'D_{j-1}T_{(j-1)j}D_j T_{j(j+1)}D_{j+1}J) \times H_r}$$

$$= \frac{J'D_{j-1}T_{(j-1)j}D_{(k)}T_{j(j+1)}D_{j+1}J}{J'D_{j-1}T_{(j-1)j}D_j T_{j(j+1)}D_{j+1}J}$$

$$= \Pr(G_j = k|G_{j-1},G_{j+1}), \tag{4.25}$$

which is the conditional probability we have learned in the three-point analysis.

4.6   Map Construction with Unknown Recombination Fractions

The multipoint analysis described so far has only been used when the order of the markers is known, in which only $m - 1$ recombination fractions are estimated. Recombination fractions between nonconsecutive markers are irrelevant and thus are not estimated. The recombination fraction between any two nonconsecutive markers can be obtained using Haldane map function if such information is required. Taking the map ABCD for example, the multipoint analysis only provides estimates for r_{AB}, r_{BC}, and r_{CD}. One can obtain the remaining three recombination fractions by $r_{AC} = r_{AB} + r_{BC} - 2r_{AB}r_{BC}$, $r_{AD} = r_{AC} + r_{CD} - 2r_{AC}r_{CD}$, and $r_{BD} = r_{BC} + r_{CD} - 2r_{BC}r_{CD}$. Alternatively, one may convert the recombination fractions into additive distances and join the additive distances to make an additive map, from which all pairwise recombination fractions can be calculated using the Haldane map function.

For those understudied species, marker orders may be unknown. The multipoint method provides a mechanism to order markers and estimate recombination fractions simultaneously. The marker order and the estimated recombination fractions under that order should be the joint ML estimates if such an order and the estimated recombination fractions under that order generate the maximum likelihood value compared to all other orders.

Part II
Analysis of Quantitative Traits

Chapter 5
Basic Concepts of Quantitative Genetics

Quantitative genetics is a special branch of genetics, which is concerned with the inheritance of the differences between individuals that are measured in degree rather than in kind. These individual differences are referred to as quantitative differences or quantitative traits. Formally, a quantitative trait is defined as a trait whose value varies continuously across individuals (Falconer and Mackay 1996; Lynch and Walsh 1998). The phenotype of a quantitative trait measured from an individual is not determined by genes alone; it is also determined by environmental variants. The proportion of the phenotypic variance explained by the segregation of a single gene is usually small. However, the contribution of all these small-effect genes collectively is significant to the variation of the phenotype. Genes controlling the variation of a quantitative trait are called quantitative trait loci (QTL). Note that the term QTL defined in this book is used for both the singular and plural forms, e.g., one QTL for weight and two QTL for height. In the quantitative genetics literature, QTL represents the singular form and QTLs is used as the plural form. No matter how small a QTL is, it segregates just like a regular Mendelian locus. For small-effect QTL, we simply cannot observe the segregation and must resort to statistical methods to infer the segregation. Most statistical methods applied in quantitative genetics require specific genetic models, which will be the focus of this chapter.

5.1 Gene Frequency and Genotype Frequency

Throughout the entire book, we consider only diploid organisms. A diploid organism carries two copies of the homologous genome, one from the paternal parent and the other from the maternal parent. Each copy of the genome is called a haploid. Each locus of the genome, therefore, contains two alleles, one from each parent. Although each individual carries at most two different alleles, the entire population may have many different alleles, called multiallelic population. For simplicity, we only consider a population that contains two different alleles, called a biallelic

S. Xu, *Principles of Statistical Genomics*, DOI 10.1007/978-0-387-70807-2_5,
© Springer Science+Business Media, LLC 2013

population. Let A_1 and A_2 be the two alleles of locus A in the population of interest. In the biallelic population, there are only three possible genotypes, denoted by A_1A_1, A_1A_2 and A_2A_2, respectively. Depending on the structure and the mating system of the population, the population may have different proportions of the three genotypes. Let $P_{11} = \Pr(A_1A_1)$, $P_{12} = \Pr(A_1A_2)$, and $P_{22} = \Pr(A_2A_2)$ be the frequencies of the three genotypes. The genotypes A_1A_2 and A_1A_1 contain one and two copies of allele A_1, respectively. Therefore, the frequency of allele A_1 in the population is $p_1 = \Pr(A_1) = P_{11} + \frac{1}{2}P_{12}$. The allelic frequency for A_2 is then $p_2 = \Pr(A_2) = P_{22} + \frac{1}{2}P_{12}$. These relationships hold regardless the population history and structure. However, to express genotypic frequencies as functions of the allele frequencies, some assumptions are required. In a large random-mating population, there is a unique relationship between genotype frequencies and gene frequencies, which is represented by $P_{11} = p_1^2$, $P_{12} = 2p_1p_2$, and $P_{22} = p_2^2$. This can be interpreted as independence of the two alleles joining together to form the genotype. The frequency of the heterozygote is $2p_1p_2$ because it contains two configurations of the same genotype, that is, A_1A_2 and A_2A_1, representing two different origins of the gametes. This particular relationship is represented by the binomial expansion,

$$(p_1 + p_2)^2 = p_1^2 + 2p_1p_2 + p_2^2 \tag{5.1}$$

corresponding to the event of

$$(A_1 + A_2)^2 = A_1A_1 + 2(A_1A_2) + A_2A_2 \tag{5.2}$$

If such a population undergoes no selection, no mutation, and no migration, the gene frequencies and genotypic frequencies will remain constant from generation to generation. Such a population is said to be in Hardy–Weinberg equilibrium (Hardy 1908; Weinberg 1908; Li 1955). If a large population is not in Hardy–Weinberg equilibrium, one generation of random mating will suffice to lead the population to Hardy–Weinberg equilibrium.

The Hardy–Weinberg equilibrium for a population with $k(k > 2)$ alleles is represented by $P_{ij} = 2p_ip_j$ for $i \neq j$ and $P_{ii} = p_i^2$ for $i = j$, where $P_{ij} = \Pr(A_iA_j)$ and $p_i = \Pr(A_i)$ for $i, j = 1, \ldots, k$.

Gene frequencies and genotypic frequencies are properties of a population. The genes studied are usually related to fitness and thus determine the adaption of the population to environmental changes and the evolution of the population. These are contents of population genetics. In quantitative genetics, we are interested in genes that determine the expression of quantitative traits. Therefore, we must first assign some value to a genotype and a value to an allele. These values are called genetic effects.

5.2   Genetic Effects and Genetic Variance

Each individual of a population has a phenotypic value for a particular quantitative trait. Assume that we can observe the genotypes of all individuals in the population. The genetic effect for genotype A_1A_1 is defined as the average phenotypic value of all individuals bearing genotype A_1A_1. This genotypic value is denoted by G_{11}. Similar notation applies to genotypes A_1A_2 and A_2A_2. The reason that G_{11} takes the average phenotypic value is explained as follows. Let Y_{11} be the phenotypic value for an individual with genotype A_1A_1, which can be expressed as

$$Y_{11} = G_{11} + E_{11}. \tag{5.3}$$

where E_{11} is a random environmental deviation. The environmental deviation varies from one individual to another, even though all the individuals have the same genotypic value. When we take the average value across all individuals of type A_1A_1, the equation becomes

$$\bar{Y}_{11} = G_{11} + \bar{E}_{11}. \tag{5.4}$$

For a sufficient number of individuals collected from this genotype, we have $\bar{E}_{11} \approx 0$ because positive and negative deviations tend to cancel out each other. This leads to $\bar{Y}_{11} = G_{11}$.

We now define three parameters as functions of the three genotypic values,

$$\mu = \frac{1}{2}(G_{11} + G_{22})$$

$$a = G_{11} - \frac{1}{2}(G_{11} + G_{22})$$

$$d = G_{12} - \frac{1}{2}(G_{11} + G_{22}), \tag{5.5}$$

where μ is called the midpoint value, a the additive effect, and d the dominance effect. The three genotypic values are then expressed as

$$G_{11} = \mu + a$$
$$G_{12} = \mu + d$$
$$G_{22} = \mu - a \tag{5.6}$$

We then express each genotypic value as a deviation from the midpoint value

$$\Phi_{11} = G_{11} - \mu = a$$
$$\Phi_{12} = G_{12} - \mu = d$$
$$\Phi_{22} = G_{22} - \mu = -a \tag{5.7}$$

Under Hardy–Weinberg equilibrium, the population mean of the genotypic values (expressed as deviations from the midpoint value) is

$$
\begin{aligned}
\mu_G = E(\Phi) &= P_{11}\Phi_{11} + P_{12}\Phi_{12} + P_{22}\Phi_{22} \\
&= p_1^2 a + 2 p_1 p_2 d + p_2^2(-a) \\
&= (p_1 - p_2)a + 2 p_1 p_2 d
\end{aligned}
\tag{5.8}
$$

and the variance of the genotypic values is

$$
\sigma_G^2 = \mathrm{var}(\Phi) = E(\Phi^2) - E^2(\Phi),
\tag{5.9}
$$

where

$$
\begin{aligned}
E(\Phi^2) &= P_{11}\Phi_{11}{}^2 + P_{12}\Phi_{12}{}^2 + P_{22}\Phi_{22}{}^2 \\
&= p_1^2 a^2 + 2 p_1 p_2 d^2 + p_2^2(-a)^2 \\
&= (p_1^2 + p_2^2)a^2 + 2 p_1 p_2 d^2
\end{aligned}
\tag{5.10}
$$

After some algebraic manipulations, we have

$$
\sigma_G^2 = 2 p_1 p_2 [a + (p_2 - p_1)d]^2 + (2 p_1 p_2 d)^2.
\tag{5.11}
$$

5.3   Average Effect of Allelic Substitution

A single locus of an individual consists of two alleles, one from each of the two parents. When the individual reproduces, the two alleles will go to different gametes. The gametes will reunite in the next generation to form the genotypes of the next generation. Therefore, a genotype cannot be inherited from generation to generation. It is the allele (haplotype) that is passed from one generation to another. Therefore, we need to define the effect of an allele. Let us look at the following 2×2 table for the definition of the allelic effect (Table 5.1).

Table 5.1 Definitions for allelic effects and dominance deviations

	A_1 (p_1)	A_2 (p_2)	
A_1 (p_1)	$A_1 A_1$ (p_1^2)	$A_1 A_2$ $(p_1 p_2)$	α_1
	$\Phi_{11} - \mu_G = \alpha_1 + \alpha_1 + \delta_{11}$	$\Phi_{12} - \mu_G = \alpha_1 + \alpha_2 + \delta_{12}$	
A_2 (p_2)	$A_2 A_1$ $(p_2 p_1)$	$A_2 A_2$ (p_2^2)	
	$\Phi_{12} - \mu_G = \alpha_2 + \alpha_1 + \delta_{12}$	$\Phi_{22} - \mu_G = \alpha_2 + \alpha_2 + \delta_{22}$	α_2
	α_1	α_2	

Table 5.2 Breeding values and dominance deviations of the three genotypes in a Hardy–Weinberg population

Genotype	Frequency	Genotypic value	Breeding value	Dominance deviation
$A_1 A_1$	p_1^2	a	$2 p_2 \alpha$	$-2 p_2^2 d$
$A_1 A_2$	$2 p_1 p_2$	d	$(p_2 - p_1) \alpha$	$2 p_1 p_2 d$
$A_2 A_2$	p_2^2	$-a$	$-2 p_1 \alpha$	$-2 p_1^2 d$

The effect of allele A_1 is defined as

$$\alpha_1 = \frac{(\Phi_{11} - \mu_G) p_1^2 + (\Phi_{12} - \mu_G) p_2 p_1}{p_1} = p_2[a + d(p_2 - p_1)] \qquad (5.12)$$

and the effect of allele A_2 is defined as

$$\alpha_2 = \frac{(\Phi_{12} - \mu_G) p_1 p_2 + (\Phi_{22} - \mu_G) p_2^2}{p_2} = -p_1[a + d(p_2 - p_1)] \qquad (5.13)$$

The difference between the two allelic effects is called the average effect of allelic substitution, denoted by α,

$$\alpha = \alpha_1 - \alpha_2 = a + (p_2 - p_1)d \qquad (5.14)$$

The sum of the two allelic effects included in a genotype is called the "breeding value," which is the expected genotypic value of the progeny of an individual bearing this genotype. Therefore, the breeding value for genotype $A_1 A_1$ is $A_{11} = 2\alpha_1 = 2 p_2 \alpha$. The breeding values for the other two genotypes are $A_{12} = \alpha_1 + \alpha_2 = (p_2 - p_1)\alpha$ and $A_{22} = 2\alpha_2 = -p_1 \alpha$, respectively. The deviations of the actual genotypic values from the breeding values are called dominance deviations. The three dominance deviations are

$$\delta_{11} = \Phi_{11} - \mu_G - A_{11} = -2 p_2^2 d$$

$$\delta_{12} = \Phi_{12} - \mu_G - A_{12} = 2 p_1 p_2 d$$

$$\delta_{22} = \Phi_{22} - \mu_G - A_{22} = -2 p_1^2 d \qquad (5.15)$$

The genotypic values (G), the breeding values (A), and the dominance deviations (D) for the three genotypes are listed in Table 5.2.

5.4 Genetic Variance Components

One can verify that the expectations of both the breeding values and the dominance deviations are zero, i.e.,

$$E(A) = P_{11} A_{11} + P_{12} A_{12} + P_{22} A_{22}$$

$$= p_1^2 (2 p_2 \alpha) + 2 p_1 p_2 [(p_2 - p_1)\alpha] + p_2^2 (-2 p_1 \alpha) = 0 \qquad (5.16)$$

and

$$E(D) = P_{11}\delta_{11} + P_{12}\delta_{12} + P_{22}\delta_{22}$$
$$= p_1^2(-2p_2^2d) + 2p_1p_2(2p_1p_2d) + p_2^2(-2p_1^2d) = 0. \tag{5.17}$$

This leads to

$$\sigma_A^2 = E(A^2) = p_1^2(2p_2\alpha)^2 + 2p_1p_2[(p_2 - p_1)\alpha]^2 + p_2^2(-2p_1\alpha)^2 = 2p_1p_2\alpha^2 \tag{5.18}$$

and

$$\sigma_D^2 = E(D^2) = p_1^2(-2p_2^2d)^2 + 2p_1p_2(2p_1p_2d)^2 + p_2^2(-2p_1\alpha)^2 = (2p_1p_2d)^2 \tag{5.19}$$

Looking at the genetic variance given in (5.11), we found that the first part is σ_A^2 and the second part is σ_D^2. Therefore,

$$\sigma_G^2 = \sigma_A^2 + \sigma_D^2, \tag{5.20}$$

i.e., the total genetic variance has been partitioned into an additive variance component and a dominance variance component.

5.5 Heritability

The phenotypic value Y can be expressed by the following linear model:

$$Y = G + E = A + D + E, \tag{5.21}$$

where E is an environmental error with mean zero and variance σ_E^2. The phenotypic variance $\sigma_P^2 = \text{var}(Y)$ is

$$\sigma_P^2 = \sigma_G^2 + \sigma_E^2 = \sigma_A^2 + \sigma_D^2 + \sigma_E^2. \tag{5.22}$$

The phenotypic variance contributed by the total genetic variance is called broad-sense heritability, denoted by

$$H^2 = \frac{\sigma_A^2 + \sigma_D^2}{\sigma_A^2 + \sigma_D^2 + \sigma_E^2}, \tag{5.23}$$

while the proportion contributed by the additive variance is called narrow-sense heritability, denoted by

$$h^2 = \frac{\sigma_A^2}{\sigma_A^2 + \sigma_D^2 + \sigma_E^2}. \tag{5.24}$$

The narrow-sense heritability reflects the proportion of variance that is heritable. Therefore, it is a very important parameter to consider in developing a selection program for genetic improvement.

5.6 An F$_2$ Family Is in Hardy–Weinberg Equilibrium

An F$_2$ family initiated from the cross of lines $A_1 A_1$ and $A_2 A_2$ is in Hardy–Weinberg equilibrium, and thus, the theory developed in this chapter applies to an F$_2$ family. The allele frequencies are $p_1 = p_2 = \frac{1}{2}$, and the genotypic frequencies are $P_{11} = p_1^2 = \frac{1}{4}$, $P_{12} = 2p_1 p_2 = \frac{1}{2}$, and $P_{22} = p_2^2 = \frac{1}{4}$. The average effect of allelic substitution is $\alpha = a + (p_2 - p_1)d = a$. Therefore,

$$\sigma_G^2 = 2p_1 p_2 \alpha^2 + (2p_1 p_2 d)^2 = \frac{1}{2}a^2 + \frac{1}{4}d^2. \tag{5.25}$$

The same result can be obtained from a different perspective. The genotypic value of an F$_2$ individual can be expressed as

$$G = \mu + Za + Wd, \tag{5.26}$$

where

$$Z = \begin{cases} +1 \text{ for } A_1 A_1 \text{ with probability } \frac{1}{4} \\ 0 \text{ for } A_1 A_2 \text{ with probability } \frac{1}{2} \\ -1 \text{ for } A_2 A_2 \text{ with probability } \frac{1}{4} \end{cases} \tag{5.27}$$

and

$$W = \begin{cases} 0 \text{ for } A_1 A_1 \text{ with probability } \frac{1}{4} \\ 1 \text{ for } A_1 A_2 \text{ with probability } \frac{1}{2} \\ 0 \text{ for } A_2 A_2 \text{ with probability } \frac{1}{4} \end{cases} \tag{5.28}$$

The genotypic variance is partitioned into

$$\sigma_G^2 = \sigma_Z^2 a^2 + \sigma_W^2 d^2 = \frac{1}{2}a^2 + \frac{1}{4}d^2. \tag{5.29}$$

where

$$\begin{aligned} \sigma_Z^2 &= \mathrm{E}(Z^2) - \mathrm{E}^2(Z) \\ &= \left[\frac{1}{4}(+1)^2 + \frac{1}{2}(0)^2 + \frac{1}{4}(-1)^2\right] - \left[\frac{1}{4}(+1) + \frac{1}{2}(0) + \frac{1}{4}(-1)\right]^2 \\ &= \frac{1}{2} - 0 = \frac{1}{2} \end{aligned} \tag{5.30}$$

and

$$\sigma_W^2 = E(W^2) - E^2(W)$$

$$= \left[\frac{1}{4}(0)^2 + \frac{1}{2}(1)^2 + \frac{1}{4}(0)^2\right] - \left[\frac{1}{4}(0) + \frac{1}{2}(1) + \frac{1}{4}(0)\right]^2$$

$$= \frac{1}{2} - \frac{1}{4} = \frac{1}{4} \qquad\qquad (5.31)$$

Note that the covariance between Z and W is zero. Otherwise, a term $2\text{cov}(Z, W)ad$ should be added to the genotype variance.

Chapter 6
Major Gene Detection

When a quantitative trait is controlled by the segregation of a major gene and the genotypes of the major gene can be observed, the effect of the major gene can be estimated and tested. In reality, the genotypes of a major gene cannot be observed. We normally evaluate a candidate gene whose genotypes can be measured using a particular molecular technique. We may have some reason to believe that the gene has a function on the variation of a quantitative trait. We can even evaluate a DNA marker whose genotypes are known but with unknown function on the trait. If this DNA marker is closely linked to a major gene, the effect of the gene can also be estimated and tested through the marker. In either case, major gene analysis means estimation and test of the effect of an observed Mendelian locus on a quantitative trait.

Major gene detection is more often conducted in designed line crossing experiments. In humans, forest trees, and some large animals where designed crossing experiments are infeasible, pedigree data can be used for major gene detection (Elston and Steward 1971). In this chapter, we only discuss major gene detection in line crossing experiments.

6.1 Estimation of Major Gene Effect

6.1.1 BC Design

We will first discuss the backcross (BC) design and then extend the model to the F_2 design. Let P_1 and P_2 be the two inbred parents with A_1A_1 and A_2A_2 being the genotypes of the major gene for the two inbred lines, respectively. The F_1 hybrid has a genotype of A_1A_2 for the major gene. There are two types of BC design, depending on which parent the F_1 hybrid is backcrossed to. Let us assume that P_1 is the backcrossed parent. The mating type is represented by $A_1A_2 \times A_1A_1$, and the BC progeny has two possible genotypes, A_1A_1 and A_1A_2, each with a 50%

S. Xu, *Principles of Statistical Genomics*, DOI 10.1007/978-0-387-70807-2_6,
© Springer Science+Business Media, LLC 2013

of the chance in the absence of segregation distortion. Recall that the genotypic values of A_1A_1 and A_1A_2 are denoted by a and d, respectively. In the BC family, there are two segregating genotypes, which are not enough for us to estimate two genotypic values. Therefore, one cannot estimate a and d separately using a BC design. In the absence of dominance, i.e., $d = 0$, we can estimate the additive effect a. We use a general linear model to describe the relationship between the genetic effect of the major gene and the phenotypic value. Let y_j be the phenotypic value of a quantitative trait for individual j, $\forall j = 1, \ldots, n$, where n is the sample size. The linear model is expressed as

$$y_j = \mu + X_j a + \epsilon_j, \tag{6.1}$$

where μ is a constant and ϵ_j is a residual error with an assumed $N(0, \sigma^2)$ distribution. The independent variable, X_j, is an indicator variable, defined as

$$X_j = \begin{cases} 1 & \text{for } A_1A_1 \\ 0 & \text{for } A_1A_2 \end{cases} \tag{6.2}$$

There are three parameters in the model, μ, a, and σ^2. We have assumed that $d = 0$ so that the trait is only controlled by the additive effect. If $d \neq 0$, this effect will not disappear. Using the above linear model (6.1), this effect will be absorbed by μ and a, that is, the population mean and the additive effect in the presence of d in fact are confounded effects with $\mu = \mu + d$ and $a = a - d$. In other words, in the presence of dominance effect, the BC design can only estimate and test $a - d$, the difference between the additive effect and the dominance effect.

More often, people use a different scale to define the X variable, such as

$$X_j = \begin{cases} +1 & \text{for } A_1A_1 \\ -1 & \text{for } A_1A_2 \end{cases} \tag{6.3}$$

In this case, the major gene effect is different from that defined under the original scale (6.2). Although the estimated major gene effects are different under different scales of X, the test statistics will be the same, and thus, the difference in the scale of X does not affect the result of major gene detection.

The three parameters in the BC design, μ, a, and σ^2, can be estimated using the least-squares (LS) method. Once we code the genotypes into numerical values of variable X, the genetic model (6.1) becomes a standard regression model. We now change the notation to follow the standard regression analysis. Let $b_0 = \mu$ be the intercept and $b_1 = a$ be the regression coefficient. The regression model is

$$y_j = b_0 + X_j b_1 + \epsilon_j. \tag{6.4}$$

The least-squares estimates of the parameters are

$$\hat{b}_1 = \frac{\sum_{j=1}^{n}(X_j - \bar{X})(y_j - \bar{y})}{\sum_{j=1}^{n}(X_j - \bar{X})^2}, \tag{6.5}$$

for the regression coefficient (also called the genetic effect because $b_1 = a$),

$$\hat{b}_0 = \bar{y} - \bar{X}\hat{b}_1,$$ (6.6)

for the intercept (also called the population mean because $b_0 = \mu$), and

$$\hat{\sigma}^2 = \frac{\sum_{j=1}^{n}(y_j - \hat{b}_0 - X_j\hat{b}_1)^2}{n-2}.$$ (6.7)

for the residual error variance, where $\bar{X}$ and $\bar{y}$ are the sample means of variables X and y, respectively.

6.1.2 F$_2$ Design

The F_2 progeny are generated by selfing the F_1 hybrid of a cross or intercrossing different F_1 individuals that are derived from the same cross. The F_1 hybrid has a genotype of A_1A_2 at the major gene. The mating type $A_1A_2 \times A_1A_2$ will generate three different genotypes in the F_2 family, A_1A_1, A_1A_2, and A_2A_2, with the expected Mendelian ratio $1 : 2 : 1$ for the three genotypes if absence of segregation distortion is assumed. Recall that the three genotypes can be assigned the following genotypic values, a, d, and $-a$, respectively. There are three genotypes, sufficient to estimate both a and d. The genetic model can be expressed as

$$y_j = \mu + Z_j a + W_j d + \epsilon_j.$$ (6.8)

The two independent variables are genotype indicators, defined as

$$Z_j = \begin{cases} +1 & \text{for } A_1A_1 \\ 0 & \text{for } A_1A_2 \\ -1 & \text{for } A_2A_2 \end{cases}$$ (6.9)

and

$$W_j = \begin{cases} 0 & \text{for } A_1A_1 \\ 1 & \text{for } A_1A_2 \\ 0 & \text{for } A_2A_2 \end{cases}$$ (6.10)

Redefining the genetic model under the standard regression framework, we have

$$y_j = b_0 + X_{j1}b_1 + X_{j2}b_2 + \epsilon_j$$ (6.11)

where $b_0 = \mu$ is a constant, $b_1 = a$ is the additive effect, and $b_2 = d$ is the dominance effect. Variable $X_{j1} = Z_j$ is the additive genetic effect indicator, and $X_{j2} = W_j$ is the dominance effect indicator. Define $b = \{b_0, b_1, b_2\}$ as a 3×1 vector for the regression coefficients (including the intercept) and $X_j = \begin{bmatrix} X_{j0} & X_{j1} & X_{j2} \end{bmatrix}$ be an 1×3 vector where $X_{j0} = 1$ for $j = 1, \ldots, n$. The above linear model may be further simplified as

$$y_j = X_j b + \epsilon_j. \tag{6.12}$$

We can express the LS estimates of b in a single simultaneous equation set, i.e.,

$$\hat{b} = \left[\sum_{j=1}^{n} X_j^T X_j \right]^{-1} \left[\sum_{j=1}^{n} X_j^T y_j \right]. \tag{6.13}$$

Details of the above LS estimate of b are shown below:

$$\begin{bmatrix} \hat{b}_0 \\ \hat{b}_1 \\ \hat{b}_2 \end{bmatrix} = \begin{bmatrix} n & \sum_{j=1}^{n} X_{j1} & \sum_{j=1}^{n} X_{j2} \\ \sum_{j=1}^{n} X_{j1} & \sum_{j=1}^{n} X_{j1}^2 & \sum_{j=1}^{n} X_{j1} X_{j2} \\ \sum_{j=1}^{n} X_{j2} & \sum_{j=1}^{n} X_{j1} X_{j2} & \sum_{j=1}^{n} X_{j2}^2 \end{bmatrix}^{-1} \begin{bmatrix} \sum_{j=1}^{n} y_j \\ \sum_{j=1}^{n} X_{j1} y_j \\ \sum_{j=1}^{n} X_{j2} y_j \end{bmatrix}. \tag{6.14}$$

The estimated residual variance is

$$\hat{\sigma}^2 = \frac{1}{n-3} \sum_{j=1}^{n} (y_j - X_j \hat{b})^T (y_j - X_j \hat{b})$$

$$= \frac{1}{n-3} \sum_{j=1}^{n} y_j^T (y_j - X_j \hat{b}). \tag{6.15}$$

The second expression of the above equation can be verified using the following equivalence:

$$\hat{b}^T \sum_{j=1}^{n} X_j^T y_j = \hat{b}^T \left(\sum_{j=1}^{n} X_j^T X_j \right) \hat{b} \tag{6.16}$$

which is due to the following least-squares equation used to derive the LS estimation for parameter b:

$$\sum_{j=1}^{n} X_j^T y_j = \left(\sum_{j=1}^{n} X_j^T X_j \right) \hat{b} \tag{6.17}$$

6.2   Hypothesis Tests

6.2.1   BC Design

Once the parameters are estimated, they are subject to statistical tests. Hypothesis test for $a = 0$ is equivalent to $b_1 = 0$ in the BC design. The test can be accomplished using either the t-test or the F-test. However, the t-test may be confusing due to the difference between one-tailed and two-tailed tests, while the F-test does not have such a concern. In major gene detection, we rarely use one-tailed t-test. For the two-tailed t-test, the result is identical to the F-test. Therefore, we only concentrate on the F-test statistic for major gene detection.

To perform a hypothesis test, we first need to know the precision of the estimated parameter, i.e., the precision of $\hat{a}$ in the BC design. The precision of an estimated parameter may also be called the information. It is easy to understand the information than the precision because the reciprocal of the information of an estimated parameter is often used as the variance of the estimated parameter. Therefore, we need to find the variance of $\hat{a}$ first before we can conduct a hypothesis test.

The variance of $\hat{b}_1$, and thus the variance of $\hat{a}$, has the following expression:

$$\text{var}(\hat{b}_1) = \frac{\hat{\sigma}^2}{\sum_{j=1}^{n}(X_j - \bar{X})^2} \tag{6.18}$$

where $\hat{\sigma}^2$ is the estimated residual error variance. It is important to understand that when $\hat{a}$ is reported, it is often accompanied by the standard error of $\hat{a}$, denoted by $se(\hat{a})$ or $s_{\hat{a}}$, in the format $\hat{a} \pm s_{\hat{a}}$. The standard error, however, is simply the square root of the variance of the estimate. The F-test statistic is

$$F = \frac{\hat{b}_1^2}{\text{var}(\hat{b}_1)} \tag{6.19}$$

The critical value for the F to compare is $F_{1,n-2,1-\alpha}$, where $\alpha = 0.05$ or 0.01 is the type I error rate set by the investigator. The critical value is the $(1-\alpha) \times 100$ percentile of the $F_{1,n-2}$ distribution. The subscripts 1 and $n-2$ of the F-distribution are the numerator and denominator degrees of freedom, respectively.

6.2.2 F₂ Design

In the F$_2$ design, there are two genetic effects to be tested. A complete analysis of F$_2$ design requires three different hypothesis tests. The first hypothesis is that there is no genetic effect for the major gene on the quantitative trait. This is denoted by $H_0 : a = d = 0$ or $H_0 : b_1 = b_2 = 0$. Once H_0 is rejected, we can further test $H_a : a = 0, d \neq 0$, which is also denoted by $H_a : b_1 = 0, b_2 \neq 0$. The hypothesis for no dominance effect is denoted by $H_d : d = 0, a \neq 0$ or equivalently denoted by $H_d : b_2 = 0, b_1 \neq 0$. Like the hypothesis test in the BC design, we need the estimation errors of the regression coefficients. The estimation errors are drawn from the so-called variance–covariance matrix of the estimated regression coefficients.

Let us define var($\hat{b}$) as a 3×3 variance–covariance matrix,

$$\text{var}(\hat{b}) = \text{var}\begin{bmatrix} \hat{b}_0 \\ \hat{b}_1 \\ \hat{b}_2 \end{bmatrix} = \begin{bmatrix} \text{var}(\hat{b}_0) & \text{cov}(\hat{b}_0, \hat{b}_1) & \text{cov}(\hat{b}_0, \hat{b}_2) \\ \text{cov}(\hat{b}_0, \hat{b}_1) & \text{var}(\hat{b}_1) & \text{cov}(\hat{b}_1, \hat{b}_2) \\ \text{cov}(\hat{b}_0, \hat{b}_2) & \text{cov}(\hat{b}_1, \hat{b}_2) & \text{var}(\hat{b}_2) \end{bmatrix} \qquad (6.20)$$

This 3×3 matrix can be obtained using

$$\text{var}(\hat{b}) = V = \left[\sum_{j=1}^{n} X_j^T X_j \right]^{-1} \hat{\sigma}^2. \qquad (6.21)$$

To test the first hypothesis H_0, we use the following F-test statistic:

$$F = \frac{1}{2} \begin{bmatrix} \hat{b}_1 & \hat{b}_2 \end{bmatrix} \begin{bmatrix} \text{var}(\hat{b}_1) & \text{cov}(\hat{b}_1, \hat{b}_2) \\ \text{cov}(\hat{b}_1, \hat{b}_2) & \text{var}(\hat{b}_2) \end{bmatrix}^{-1} \begin{bmatrix} \hat{b}_1 \\ \hat{b}_2 \end{bmatrix} \qquad (6.22)$$

Under the null hypothesis, i.e., H_0 is true, the F-test statistic follows an F-distribution with 2 degrees of freedom for the numerator and $n - 3$ degrees of freedom for the denominator. Therefore, the critical value used to declare statistical significance is $F_{2,n-3,1-\alpha}$, which is the $(1 - \alpha) \times 100$ percentile of the $F_{2,n-3}$ distribution. The hypotheses H_a and H_d are tested using the following F-test statistics, respectively:

$$F_a = \frac{\hat{b}_1^2}{\text{var}(\hat{b}_1)} \qquad (6.23)$$

$$F_d = \frac{\hat{b}_2^2}{\text{var}(\hat{b}_2)} \qquad (6.24)$$

Under the null hypotheses, each one of F_a and F_d will follow an $F_{1,n-3}$ distribution. Therefore, the critical value for the test statistics to compare is $F_{1,n-3,1-\alpha}$.

In general, an F-test statistic is constructed as a quadratic form of the estimated parameters. We can show that a subset of vector b can be expressed as linear combinations of vector b. For example, $\{b_1, b_2\}$ is a subset of vector b. This subset can be expressed as

$$\begin{bmatrix} b_1 \\ b_2 \end{bmatrix} = \begin{bmatrix} 0 & 1 & 0 \\ 0 & 0 & 1 \end{bmatrix} \begin{bmatrix} b_0 \\ b_1 \\ b_2 \end{bmatrix} = Lb, \qquad (6.25)$$

where L is a 2×3 subset selection matrix. Similarly, b_1 and b_2 can be expressed as $L_a b$ and $L_d b$, respectively, where $L_a = [0\ 1\ 0]$ and $L_d = [0\ 0\ 1]$. The variance–covariance matrix for $L\hat{b}$ is var($L\hat{b}$) = LVL^T. Similarly, var($\hat{b}_1$) = var($L_a\hat{b}$) = $L_a VL_a^T$ and var($\hat{b}_2$) = var($L_d\hat{b}$) = $L_d VL_d^T$. Let $r = \text{rank}(L) = 2$, $r_a = \text{rank}(L_a) = 1$, and $r_d = \text{rank}(L_d) = 1$. The F-test statistic can be expressed by

$$F = \frac{1}{r} b^T L^T (LVL^T)^{-1} Lb \tag{6.26}$$

for the overall test,

$$F_a = \frac{1}{r_a} b^T L_a^T (L_a VL_a^T)^{-1} L_a b \tag{6.27}$$

for testing the additive effect and

$$F_d = \frac{1}{r_d} b^T L_d^T (L_d VL_d^T)^{-1} L_d b \tag{6.28}$$

for testing the dominance effect.

When n is sufficiently large, people often use the Wald-test statistic (Wald 1943), which is

$$W = b^T L^T (LVL^T)^{-1} Lb = r\, F. \tag{6.29}$$

Under the null hypothesis $H_0 : Lb = 0$, W approximately follows the χ_r^2 distribution, and hence, $\chi_{r,1-\alpha}^2$ is used as the critical value for declaring statistical significance. When testing the additive effect or the dominance effect (but not both), similar Wald test can be used. The degree of freedom for either test is $r_a = r_d = 1$. Therefore, the Wald-test and F-test statistics are identical because of the unity degree of freedom. Furthermore, when the degree of freedom is unity, the F-test statistic is equivalent to the squared t-test statistic.

6.3   Scale of the Genotype Indicator Variable

Genotype indicator variables in the F_2 design are denoted by Z for the additive effect and W for the dominance effect. These two variables are defined in an arbitrary manner. The locations (means) and scales (variances) of these variables affect the estimated values of the regression coefficients (including the intercept) but do not affect the significance tests. Recall that the genetic model for the phenotypic value y for an F_2 population is

$$y = \mu + Za + Wd + \epsilon \tag{6.30}$$

Within this F_2 population, y, Z, W, and ϵ are variables, whereas μ, a, and d are constants (but unknown). This is because the observed values of y, Z, W, and ϵ vary across individuals, but the values of μ, a, and d are the same for all individuals. The expectation and variance of an individual y in an F_2 population are

$$E(y) = \mu + E(Z)a + E(W)d \tag{6.31}$$

and

$$\mathrm{var}(y) = \mathrm{var}(Z)a^2 + \mathrm{var}(W)d^2 + \sigma^2. \tag{6.32}$$

respectively. If the central locations of variables Z and W are defined differently from what have been defined before, $E(Z)$ and $E(W)$ will change, leading to a different estimate of μ, but the estimated a and d remain the same. If the scales of Z and W are defined differently from those defined before, the estimated μ will remain the same but the estimated a and d will change. However, var(y) and σ^2 are independent of the locations and the scales of Z and W, so are var$(Za) =$ var$(Z)a^2$ and var$(Wd) =$ var$(W)d^2$. If we increase the scale for Z, the estimated a must be decreased to maintain a constant var(Za). This also applies to variable W. In the hypothesis test for a BC design, we learned that the variance of the estimated genetic effect is

$$\text{var}(\hat{b}_1) = \frac{\sigma^2}{\sum_{j=1}^{n}(X_j - \bar{X})^2} = \frac{\sigma^2}{(n-1)\text{var}(X)}, \tag{6.33}$$

where

$$\text{var}(X) = \frac{1}{n-1}\sum_{j=1}^{n}(X_j - \bar{X})^2.$$

This equation suggests that a large variance (scale) for the independent variable will decrease the variance of the estimated regression coefficient and thus increase the precision of the estimation. However, one cannot arbitrarily choose a large scale for the independent variable and hope to increase the precision of the estimation. The scales of the independent variables are chosen for convenience of parameter estimation and interpretation, not for increase of estimation precision.

We now demonstrate that scales of the independent variables will not affect significance tests of the regression coefficients. Let $X = \{1, Z, W\}$ be the $n \times 3$ design matrix where each of the three components is an $n \times 1$ vector. Let us rescale variables Z and W by $c_z Z$ and $c_w W$ where c_z and c_w are some positive numbers (constants). Let $X^* = \{1, c_z X, c_w W\}$ be the rescaled design matrix, which can be rewritten as $X^* = XC$ where $C = \text{diag}\{1, c_z, c_w\}$. The original model can be reformulated in the new scale,

$$y = Xb + \epsilon = XCC^{-1}b + \epsilon = (XC)(C^{-1}b) + \epsilon = X^*b^* + \epsilon, \tag{6.34}$$

where $b^* = C^{-1}b$ is the regression coefficients defined in the new scale. Let $\hat{b}^*$ be the LS solution for b^* and

$$\text{var}(\hat{b}^*) = V^* = [(X^*)^T X^*]^{-1}\hat{\sigma}^2 = C^{-1}(X^T X)^{-1}C^{-1}\hat{\sigma}^2 = C^{-1}VC^{-1}. \tag{6.35}$$

The null hypothesis for the rescaled regression coefficients is $H_0 : Lb^* = 0$. The F-test statistic is

$$F = \frac{1}{r}(b^*)^T L^T (LV^*L^T)^{-1}Lb^*$$

$$= \frac{1}{r}b^T C^{-1}L^T (LC^{-1}VC^{-1}L^T)^{-1}LC^{-1}b$$

$$= \frac{1}{r}b^T L^T C^{-1}(C^{-1}LVL^T C^{-1})^{-1}C^{-1}Lb$$

$$= \frac{1}{r}b^T L^T C^{-1}C(LVL^T)^{-1}CC^{-1}Lb$$

$$= \frac{1}{r}b^T L^T (LVL^T)^{-1}Lb, \tag{6.36}$$

which is identical to the F-statistic for $H_0 : Lb = 0$ defined in the original scale. Therefore, the scales of the independent variables do not affect the significance test.

The scales for Z and W defined in this chapter are adopted from Falconer and Mackay (1996). Let $\sigma_z^2 = \text{var}(Z)$, $\sigma_w^2 = \text{var}(W)$, and $\sigma_{zw} = \text{cov}(Z, W)$. We can show that

$$\sigma_z^2 = E(Z^2) - E^2(Z) = \left[\frac{1}{4}(1)^2 + \frac{1}{2}(0)^2 + \frac{1}{4}(-1)^2\right] - \left[\frac{1}{4}(1) + \frac{1}{2}(0) + \frac{1}{4}(-1)\right]^2$$

$$= \left[\frac{1}{2}\right] - [0]^2 = \frac{1}{2},$$

$$\sigma_w^2 = E(W^2) - E^2(W) = \left[\frac{1}{4}(0)^2 + \frac{1}{2}(1)^2 + \frac{1}{4}(0)^2\right] - \left[\frac{1}{4}(0) + \frac{1}{2}(1) + \frac{1}{4}(0)\right]^2$$

$$= \left[\frac{1}{2}\right] - \left[\frac{1}{2}\right]^2 = \frac{1}{4}$$

and

$$\sigma_{zw} = E(ZW) - E(Z)E(Z) = \left[\frac{1}{4}(1)(0) + \frac{1}{2}(0)(1) + \frac{1}{4}(-1)(0)\right] - [0]\left[\frac{1}{2}\right]$$

$$= 0 - 0 = 0.$$

The total genetic variance is

$$\sigma_G^2 = \sigma_z^2 a^2 + \sigma_w^2 d^2 + 2\sigma_{zw}ad = \frac{1}{2}a^2 + \frac{1}{4}d^2. \tag{6.37}$$

We can see that the scales defined this way are nice in terms of biological interpretation but not attractive in terms of statistical convenience because the independent variables are not centered, although they are orthogonal. The following scales for Z and W appear to be odd, but they are statistically more attractive.

$$Z = \begin{cases} +\sqrt{2} & \text{for } A_1 A_1 \\ 0 & \text{for } A_1 A_2 \\ -\sqrt{2} & \text{for } A_2 A_2 \end{cases} \tag{6.38}$$

and

$$W = \begin{cases} -1 & \text{for } A_1 A_1 \\ +1 & \text{for } A_1 A_2 \\ -1 & \text{for } A_2 A_2 \end{cases} \tag{6.39}$$

We can prove that $E(Z) = E(W) = 0$ and $\sigma_z^2 = \sigma_w^2 = 1$. Variables Z and W with such properties are said to be in a standardized scale. Furthermore, $\sigma_{zw} = 0$, i.e., Z and W are orthogonal. As a result,

$$\sigma_G^2 = a^2 + d^2, \tag{6.40}$$

which is the mathematical attractiveness of the new scale. One should be cautious that the nice properties for the new scale only apply to an F_2 population and the major gene must follow the Mendelian segregation ratio.

The magnitudes and the biological interpretations of the additive effect (a) and the dominance effect (d) depend on the scales we choose for variables Z and W. This may have frustrated many experimental quantitative geneticists. An alternative way for major gene detection that may eliminate the above frustration is to estimate the genotypic value for each of the three genotypes. In the first step, we can utilize the general linear model to estimate the genotypic values. This step is independent of the scales of the independent variables. In the second step, linear contrasts are established to estimate and test the additive and dominance effects. Let $\beta = \{\beta_1, \beta_2, \beta_3\}$ be a vector for the means of the three genotypes, $A_1 A_1$, $A_1 A_2$, and $A_2 A_2$. Let X_{j1}, X_{j2}, and X_{j3} be the genotype indicators for the three genotypes. Each one of them takes either 0 or 1, depending on which genotype individual j takes. If j has a genotype $A_1 A_1$, $X_{j1} = 1$, $X_{j2} = 0$, and $X_{j3} = 0$. If individual j has a genotype $A_1 A_2$, $X_{j1} = 0$, $X_{j2} = 1$, and $X_{j3} = 0$. If individual j takes genotype $A_1 A_2$, $X_{j1} = 0$, $X_{j2} = 0$, and $X_{j3} = 1$. The linear model for y_j is

$$y_j = X_{j1} \beta_1 + X_{j2} \beta_2 + X_{j3} \beta_3 + \epsilon_j. \tag{6.41}$$

In matrix notation,

$$y = X\beta + \epsilon. \tag{6.42}$$

The LS estimate of β is

$$\hat{\beta} = (X^T X)^{-1} X^T y \tag{6.43}$$

and the estimated residual variance is

$$\hat{\sigma}^2 = \frac{1}{n-3}(y - X\hat{\beta})^T (y - X\hat{\beta}) \tag{6.44}$$

The variance–covariance matrix for $\hat{\beta}$ is

$$\text{var}(\hat{\beta}) = V = (X^T X)^{-1} \hat{\sigma}^2. \tag{6.45}$$

Once we have the LS estimate of β, the additive and dominance effects can be converted using any convenient scale. For example, to find μ (the midpoint value), a, and d in the original scale (Falconer and Mackay 1996), we take the following transformations:

$$\hat{\mu} = \frac{1}{2}(\hat{\beta}_1 + \hat{\beta}_3)$$

$$\hat{a} = \hat{\beta}_1 - \frac{1}{2}(\hat{\beta}_1 + \hat{\beta}_3)$$

$$\hat{d} = \hat{\beta}_2 - \frac{1}{2}(\hat{\beta}_1 + \hat{\beta}_3) \tag{6.46}$$

In matrix notation, we have

$$\begin{bmatrix} \hat{\mu} \\ \hat{a} \\ \hat{d} \end{bmatrix} = \begin{bmatrix} \frac{1}{2} & 0 & \frac{1}{2} \\ \frac{1}{2} & 0 & -\frac{1}{2} \\ -\frac{1}{2} & 1 & -\frac{1}{2} \end{bmatrix} \begin{bmatrix} \hat{\beta}_1 \\ \hat{\beta}_2 \\ \hat{\beta}_3 \end{bmatrix} \tag{6.47}$$

Let $b = \{\mu, a, d\}$ be the vector of regression coefficients defined in the original scale. In compact matrix notation, it is expressed as $b = A\beta$, where A is the 3×3 transformation matrix given above. The variance–covariance matrix of $\hat{b}$ is $\text{var}(\hat{b}) = \text{var}(A\hat{\beta}) = A\text{var}(\hat{\beta})A^T = AVA^T$. Hypothesis tests for b can be performed using the same technique we have learned before.

When formulating the genotypic values as the original parameters, researchers have their own freedom to choose the A matrix. For example, in the orthogonal and standardized scale defined earlier, the A matrix has a form shown in the following equation:

$$\begin{bmatrix} \hat{\mu} \\ \hat{a} \\ \hat{d} \end{bmatrix} = \begin{bmatrix} \frac{1}{4} & \frac{1}{2} & \frac{1}{4} \\ \frac{\sqrt{2}}{4} & 0 & -\frac{\sqrt{2}}{4} \\ -\frac{1}{4} & \frac{1}{2} & -\frac{1}{4} \end{bmatrix} \begin{bmatrix} \hat{\beta}_1 \\ \hat{\beta}_2 \\ \hat{\beta}_3 \end{bmatrix}. \tag{6.48}$$

6.4 Statistical Power

Before conducting an experiment for major gene detection or QTL mapping, investigators may want to know the sample size required to detect a gene that explains a certain percent of the phenotypic variance with a specific statistical power. Sometimes, the sample size may be fixed due to limitation of resources, but the investigators may be interested in the statistical power to detect a major gene that explains a certain percent of the phenotypic variance of a quantitative trait. Therefore, the statistical power and sample size are closely related. To study the statistical power, we also need to know the type I and type II errors, which are very important statistical concepts. They are also very closely related to the statistical

Table 6.1 Relationship of the type I error, the type II error, and the statistical power

	$\hat{H}$	
H	H_0	H_A
H_0	$\psi = \Pr(\hat{H} = H_0 \| H = H_0)$	$\alpha = \Pr(\hat{H} = H_A \| H = H_0)$
H_A	$\beta = \Pr(\hat{H} = H_0 \| H = H_A)$	$\omega = \Pr(\hat{H} = H_A \| H = H_A)$

power. In this chapter, we will learn how to calculate the statistical powers under the F-test framework. When the sample size is not too small, say $n > 30$, the F-test statistic is almost identical to the Wald test (Wald 1943). Since the Wald test is much more convenient to deal with in terms of power calculation, we will discuss the Wald test.

6.4.1   Type I Error and Statistical Power

In major gene detection, we always deal with two hypotheses under each test, the null hypothesis (H_0) and the alternative hypothesis (H_A). Let H be the true hypothesis, which can take either H_0 or H_A but not both. Let $\hat{H}$ be the estimated hypothesis (conclusion drawn from the data), which also takes either H_0 or H_A but not both. We can make two mistakes for each test. If $H = H_0$ but we conclude that $\hat{H} = H_A$, we then make a type I error, which means that the null hypothesis is true but we accept the alternative hypothesis. A type II error occurs if $H = H_A$ but we conclude that $\hat{H} = H_0$. The probability that we make a type I error is denoted by $\alpha = \Pr(\hat{H} = H_A | H = H_0)$. The probability that we make a type II error is denoted by $\beta = \Pr(\hat{H} = H_0 | H = H_A)$. Corresponding to the two errors, we can make two correct decisions. One correct decision is that we accept H_A while in fact H_A is true. The probability to make this correct decision is called the statistical power, denoted by $\omega = \Pr(\hat{H} = H_A | H = H_A) = 1 - \beta$. The other correct decision is that we accept H_0 while in fact H_0 is true. The probability that we make this correct decision is denoted by $\psi = \Pr(\hat{H} = H_0 | H = H_0) = 1 - \alpha$. There is no name for this probability because it is not something of interest. These four probabilities are summarized in Table 6.1.

6.4.2   Wald-Test Statistic

In the BC design, we have learned the F-test statistic for testing the hypothesis $H_0 : b_1 = 0$, i.e., $H_0 : a = 0$. Because we are testing a single genetic effect, the F-test is equivalent to the Wald test. For simplicity, let b be the estimated regression coefficient, i.e., the estimated additive effect, and a be the true additive effect. The Wald-test statistic is

$$W = \frac{b^2}{\sigma_b^2} \approx n\sigma_X^2 \frac{b^2}{\sigma^2}. \tag{6.49}$$

This relationship holds because

$$\sigma_b^2 = \frac{\sigma^2}{\sum_{j=1}^{n}(X_j - \bar{X})^2} \approx \frac{\sigma^2}{n\sigma_X^2}. \tag{6.50}$$

The variance of X is a constant, depending on the design of experiment. If H_0 : $a=0$ is true, the Wald-test statistic will follow a central chi-square distribution with one degree of freedom. Therefore, the critical value used to declare significance for the Wald test is $\chi_{1,1-\alpha}^2$, the $(1 - \alpha) \times 100$ percentile of the χ_1^2 distribution. If the alternative hypothesis, $H_A : a \neq 0$, is true, then the Wald test will follow a noncentral chi-square distribution with a noncentrality parameter

$$\delta = n\sigma_X^2 \frac{a^2}{\sigma^2}. \tag{6.51}$$

Note that the noncentrality parameter differs from the Wald-test statistic by replacing b by a, where b is the estimated regression coefficient and a is the true additive genetic effect. Before we study the power calculation, let us first introduce the central and noncentral chi-square distributions. The central chi-square is often called chi-square distribution for simplicity. Let χ^2 denote a chi-square variable. If χ^2 follows a central chi-square distribution with d degrees of freedom, the cumulative distribution function is denoted by $F(\chi^2|d, 0)$. If χ^2 follows a noncentral chi-square distribution with d degrees of freedom and noncentrality parameter δ, the cumulative distribution function is denoted by $F(\chi^2|d, \delta)$. The inverse function of the χ^2 distribution is called the quantile of the chi-square distribution, which can be described as follows. Let $p = F(\chi^2|d, 0)$ be the cumulative probability of the central chi-square distribution, then $\chi_{1,p}^2 = F^{-1}(p|d, 0)$ is the quantile or $p \times 100$ percentile of the central chi-square distribution. If $p = F(\chi^2|d, \delta)$ is the cumulative probability of the noncentral chi-square distribution, then $\chi_{1,p,\delta}^2 = F^{-1}(p|d, \delta)$ is the quantile of the noncentral chi-square distribution. For the noncentral chi-square distribution, there is another type of inverse function that is $\delta_p = F_{-1}(\chi^2|d, p)$, i.e., given the chi-square value and the probability, we can find the noncentrality parameter. We use F^{-1} and F_{-1} to express the two different inverse functions of the noncentral chi-square distribution. These inverse functions may be calculated as built-in functions in some software packages, e.g., SAS (SAS Institute 2008b).

Once we understand the noncentral chi-square distribution and its inverse functions, we can perform power calculation and find the minimum sample size required to detect a major gene. Let α be the type I error. The critical value used for the Wald-test statistic to compare is

$$\chi_{1,1-\alpha}^2 = F^{-1}(1 - \alpha|1, 0). \tag{6.52}$$

Given the noncentrality parameter (6.51), the type II error (β) is given by

$$\beta = F(\chi^2_{1,1-\alpha}|1,\delta).$$
(6.53)

The statistical power is simply

$$\omega = 1 - \beta = 1 - F(\chi^2_{1,1-\alpha}|1,\delta).$$
(6.54)

Given the power ω (or the type II error $\beta = 1 - \omega$), we can calculate the noncentrality parameter using

$$\delta_\beta = F_{-1}(\chi^2_{1,1-\alpha}|1,\beta).$$
(6.55)

The noncentrality parameter allows us to calculate the minimum sample size required to detect a major gene using the following relationship:

$$\delta_\beta = n\sigma^2_X \frac{a^2}{\sigma^2}.$$
(6.56)

Rearranging the equation, we get

$$n = \frac{\delta_\beta}{\sigma^2_X}\left(\frac{\sigma^2}{a^2}\right).$$
(6.57)

which is the minimum sample size required to detect a gene of size a with a power of $\omega = 1 - \beta$.

6.4.3   Size of a Major Gene

The size of a major gene is determined by the genetic effect, which is a under the additive model. However, it is only meaningful when expressed relative to the standard deviation of the residual error. In other words, the major gene effect influences the power and sample size calculation only through $\frac{a}{\sigma}$ (see (6.56) and (6.57)). Therefore, we often use the proportion of phenotypic variance contributed by the gene as a measurement of the size of the major gene. This proportion is often called the "heritability" of the major gene. Although "heritability of a gene" is an inappropriate phrase, we still use it for the very reason that people have adopted it ever since marker-trait association studies started. Under the additive model, the heritability of a major gene is

$$H^2 = \frac{V_G}{V_G + V_E} = \frac{\sigma^2_X a^2}{\sigma^2_X a^2 + \sigma^2} = \frac{\sigma^2_X \left(\frac{a^2}{\sigma^2}\right)}{\sigma^2_X \left(\frac{a^2}{\sigma^2}\right) + 1}$$
(6.58)

Rearranging the above equation leads to the following relationship:

$$\frac{a^2}{\sigma^2} = \frac{H^2}{\sigma_X^2(1 - H^2)}. \tag{6.59}$$

Substituting this into (6.57) yields

$$n = \frac{\delta_\beta(1 - H^2)}{H^2}. \tag{6.60}$$

We now discuss σ_X^2 because it depends on the designs of line crossing experiments. Let

$$y_j = \mu + X_j a + \epsilon_j \tag{6.61}$$

be the linear model for the major gene with additive effect only. For the F_2 mating design, the genotype indicator variable X may be coded as

$$X_j = \begin{cases} +1 & \text{for } A_1A_1 \\ 0 & \text{for } A_1A_2 \\ -1 & \text{for } A_2A_2 \end{cases} \tag{6.62}$$

For the BC mating design, the corresponding code is

$$X_j = \begin{cases} 1 & \text{for } A_1A_1 \\ 0 & \text{for } A_1A_2 \end{cases} \tag{6.63}$$

Under Mendelian segregation, i.e., 1:2:1 for F_2 and 1:1 for BC, the variance of X for the F_2 design is $\sigma_X^2 = \frac{1}{2}$, while for the BC design, this variance is $\sigma_X^2 = \frac{1}{4}$. If a is fixed (the same for both designs), the F_2 design only requires half the sample size as needed for the BC design to achieve the same power because the sample size is inversely proportional to σ_X^2 (see (6.57)). On the other hand, if H^2 is fixed (the same for both designs), the sample sizes required for the two designs are the same (see (6.60)). This phenomenon appears to contradict with the notion that the F_2 design is more efficient than the BC design. This contradiction can be explained as follows. The a^2 required for the BC design is twice as large as that required for the F_2 design to achieve the same H^2 due to the fact that $\sigma_X^2 a^2$ is the genetic variance contributed by the major gene. The inefficiency of the BC design has already been taken into account when H^2 is fixed.

6.4.4   Relationship Between W-test and Z-test

People may be more familiar with the Z-test than the W-test (full name of the W-test is the Wald test). The two-tailed Z-test is identical to the W-test. The Z-test statistic is

$$Z = \frac{|b|}{\sigma_b}. \tag{6.64}$$

Therefore, the relationship between the two test statistics is

$$W = Z^2 = \frac{b^2}{\sigma_b^2}.$$

(6.65)

The Z-test is also called the normal test. Let $p = \Phi(Z)$ be the standardized cumulative normal distribution function and $Z_p = \Phi^{-1}(p)$ be the inverse function (quantile) of the normal distribution. Let α be the type I error and β be the type II error. The Z-test statistic is significant if $Z > Z_{1-\alpha/2}$. The relationship between the type I and type II errors is

$$\beta = \Phi \left(Z_{1-\alpha/2} - \sqrt{n}\sigma_X \frac{a}{\sigma} \right).$$

(6.66)

Therefore, the statistical power is

$$\omega = 1 - \beta = 1 - \Phi \left(Z_{1-\alpha/2} - \sqrt{n}\sigma_X \frac{a}{\sigma} \right).$$

(6.67)

The sample size required to detect a major gene is

$$n = \frac{(Z_{1-\alpha/2} + Z_{1-\beta})^2}{\sigma_X^2} \left(\frac{\sigma^2}{a^2} \right)$$

(6.68)

or

$$n = \frac{(Z_{1-\alpha/2} + Z_{1-\beta})^2 (1 - H^2)}{H^2}.$$

(6.69)

Comparing this equation with (6.60), we can see that

$$\delta_\beta = (Z_{1-\alpha/2} + Z_{1-\beta})^2.$$

(6.70)

This provides a different way to calculate the noncentrality parameter of a noncentral chi-square distribution, i.e.,

$$\delta_\beta = F_{-1}(\chi_{1,1-\alpha}^2 | 1, \beta) = (Z_{1-\alpha/2} + Z_{1-\beta})^2.$$

(6.71)

This particular relationship only holds when the degree of freedom is one. In addition, $\chi_{1,1-\alpha}^2 = Z_{1-\alpha/2}^2$. Therefore, one can use either the Z-test or the W-test to perform power and sample size calculations, and the results are identical.

6.4.5   Extension to Dominance Effect

To calculate the statistical power for a model with both additive (a) and dominance (d) effects, one can only use the W-test statistic (the Z-test does not apply here). The W-test statistic is given in (6.29). The critical value under a type I error α is

$$\chi_{2,1-\alpha}^2 = F^{-1}(1 - \alpha | 2, 0).$$

(6.72)

The noncentrality parameter is

$$\delta = n\frac{\sigma_G^2}{\sigma^2},$$

(6.73)

where

$$\sigma_G^2 = \sigma_X^2 a^2 + \sigma_W^2 d^2$$

(6.74)

assuming that X and W are defined in such a way that they are not correlated. Given the noncentrality parameter, the power can be calculated using

$$\omega = 1 - \beta = 1 - F(\chi_{2,1-\alpha}^2 | 2, \delta).$$

(6.75)

Given the power ω, and thus the type II error $\beta = 1 - \omega$, the noncentrality parameter is

$$\delta_\beta = F_{-1}(\chi_{2,1-\alpha}^2 | 2, \beta).$$

(6.76)

Therefore, the sample size required to detect a major gene is

$$n = \delta_\beta \frac{\sigma^2}{\sigma_G^2}.$$

(6.77)

In summary, one simply modifies the following two steps for the dominance effect extension. The first step is to replace the noncentral chi-square distribution with one degree of freedom for the additive model by the noncentral chi-square distribution with two degrees of freedom for the model with both effects. The second step is to replace $\sigma_X^2 a^2$ for the additive model by $\sigma_G^2 = \sigma_X^2 a^2 + \sigma_W^2 d^2$ for the additive and dominance model.

Examples

Example 1: Calculate the statistical power to detect a major gene with $a = 5.0$ in an F_2 population of size $n = 200$ under a type I error rate of $\alpha = 0.05$, assuming that $\sigma^2 = 400$.

For an F_2 population, the variance of the genotype indicator variable is $\sigma_X^2 = \frac{1}{2}$. Given the parameters in the example, we found that

$$\chi_{1,1-0.05}^2 = F^{-1}(0.95 | 1, 0) = 3.84.$$

The noncentrality parameter is

$$\delta = n\sigma_X^2 \frac{a^2}{\sigma^2} = 200 \times \frac{1}{2} \times \frac{5^2}{20^2} = 6.25.$$

The statistical power is

$$\omega = 1 - F(\chi^2_{1,1-\alpha}|1,\delta) = 1 - F(3.84|1,6.25) = 1 - 0.2946 = 0.7054.$$

Example 2: Find the minimum sample size required to detect a gene that explains 3 % of the phenotypic variance with a 90 % power under a type I error rate of 5 %.

The parameter values for this question are $\alpha = 0.05$, $\beta = 1 - \omega = 1 - 0.9 = 0.10$, and $H^2 = 0.3$. From these values, we get

$$\chi^2_{1,1-0.05} = F^{-1}(0.95|1,0) = 3.84.$$

The noncentrality parameter is

$$\delta_\beta = F_{-1}(\chi^2_{1,1-\alpha}|1,\beta) = F_{-1}(3.84|1,0.1) = 10.505.$$

Therefore, the sample size required is

$$n = \frac{\delta_\beta(1 - H^2)}{H^2} = \frac{10.505 \times (1 - 0.03)}{0.03} = 340.$$

Chapter 7
Segregation Analysis

Quantitative traits, by definition, are controlled by the segregation of multiple genes. However, the continuous distribution of a quantitative trait does not require the segregation of too many genes. Segregation of just a few genes or even a single gene may be sufficient to generate a continuously distributed phenotype, provided that the environmental variant contributes substantial amount of the trait variation. It is often postulated that a quantitative trait may be controlled by one or a few "major genes" plus multiple modifier genes (genes with very small effects). Such a model is called oligogenic model, which is in contrast to the so called polygenic model where multiple genes with small and equal effects are assumed.

In this chapter, we will discuss a method to test the hypothesis that a quantitative trait is controlled by a single major gene even without observing the genotypes of the major gene. The method is called segregation analysis of quantitative traits. Although segregation analysis belongs to major gene detection, we discuss this topic separately from the previous topic to emphasize a slight difference between segregation analysis and the major gene detection discussed earlier. Here, we define major gene detection as an association study between a single-locus genotype with a quantitative trait where genotypes of the major gene are observed for all individuals. Segregation analysis, however, refers to a single-locus association study where genotypes of the major gene are not observed at all. Another reason for separating major gene detection from segregation analysis is that the statistical method and hypothesis test for segregation analysis can be quite different from those of the major gene detection.

7.1 Gaussian Mixture Distribution

We will use an F_2 population as an example to discuss the segregation analysis. Consider the three genotypes in the following order: A_1A_1, A_1A_2, and A_2A_2. Let $k = 1, 2, 3$ indicate the three ordered genotypes. The means of individuals bearing the three ordered genotypes are denoted by μ_1, μ_2, and μ_3, respectively.

S. Xu, *Principles of Statistical Genomics*, DOI 10.1007/978-0-387-70807-2_7,
© Springer Science+Business Media, LLC 2013

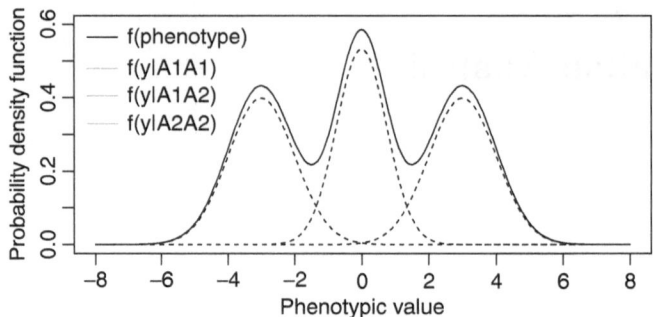

Fig. 7.1 Gaussian mixture with three components. The *solid line* represents the mixture distribution, while the three *dashed lines* represent the three components

Let y_j be the phenotypic value of individual j for $j = 1, \ldots, n$, where n is the sample size. Given that individual j has the kth genotype, the linear model for y_j is

$$y_j = \mu_k + \epsilon_j, \tag{7.1}$$

where $\epsilon_j \sim N(0, \sigma^2)$ and σ^2 is the residual error variance. The probability density of y_j conditional on the kth genotype is

$$f_k(y_j) = \frac{1}{\sqrt{2\pi}\sigma} \exp\left[-\frac{1}{2\sigma^2}(y_j - \mu_k)^2\right]. \tag{7.2}$$

In reality, the genotype of an individual is not observable, and thus, a mixture distribution is needed to describe the probability density of y_j. Let $\pi_k, \forall k = 1, 2, 3$, be the proportion of genotype k (also called the mixing proportion). Without any prior knowledge, π_k may be described by the Mendelian segregation ratio, i.e., $\pi_1 = \pi_3 = \frac{1}{2}\pi_2 = \frac{1}{4}$. Therefore, under the assumption of Mendelian segregation, the π_k's are constants, not parameters. The distribution of y_j is a mixture of three normal distributions, each is weighted by the Mendelian mixing proportion. The mixture distribution is demonstrated by Fig. 7.1.

The probability density of y_j is

$$f(y_j) = \sum_{k=1}^{3} \pi_k f_k(y_j). \tag{7.3}$$

The overall observed log likelihood function for parameters $\theta = \{\mu_1, \mu_2, \mu_3, \sigma^2\}$ is

$$L(\theta) = \sum_{j=1}^{n} \ln f(y_j) = \sum_{j=1}^{n} \ln\left[\sum_{k=1}^{3} \pi_k f_k(y_i)\right]. \tag{7.4}$$

Any numerical algorithms may be used to estimate the parameters. However, the EM algorithm (Dempster et al. 1977) appears to be the most convenient method for such a mixture model problem and thus will be introduced in this chapter.

7.2 EM Algorithm

The expectation-maximization (EM) algorithm was developed by Dempster et al. (1977) as a special numerical algorithm for finding the maximum likelihood estimates (MLE) of parameters. In contrast to the Newton–Raphson algorithm, the EM algorithm is not a general algorithm for MLE; rather, it can only be applied to some special problems. If the following two conditions hold, then we should consider using the EM algorithm. The first condition is that the maximum likelihood problem can be formulated as a missing value problem. The second condition is that if the missing values were not missing, the MLE would have a closed form solution or, at least, a mathematically attractive form of the solution. We now evaluate the mixture model problem to see whether the two conditions apply.

7.2.1 Closed Form Solution

We introduce a label η_j to indicate the genotype of individual j. The definition of η_j is

$$\eta_j = \begin{cases} 1 \text{ for } A_1 A_1 \\ 2 \text{ for } A_1 A_2 \\ 3 \text{ for } A_2 A_2 \end{cases} \tag{7.5}$$

Since the genotype of an individual is not observable, the label η_j is missing. Therefore, we can formulate the problem as a missing value problem. The missing values are the genotypes of the major gene and denoted by variable η_j for $j = 1, \ldots, n$. Therefore, the first condition for using the EM algorithm is met. If η_j is not missing, do we have a closed form solution for the parameters? Let us now define three more variables as functions of η_j. These three variables are called $\delta(\eta_j, 1)$, $\delta(\eta_j, 2)$, and $\delta(\eta_j, 3)$, and their values are defined as

$$\delta(\eta_j, k) = \begin{cases} 1 \text{ if } \eta_j = k \\ 0 \text{ if } \eta_j \neq k \end{cases} \tag{7.6}$$

for $k = 1, 2, 3$. We now use $\delta(\eta_j, k)$ to represent the missing values. If $\delta(\eta_j, k)$ were not missing, the linear model would be described by

$$y_j = \delta(\eta_j, 1)\mu_1 + \delta(\eta_j, 2)\mu_2 + \delta(\eta_j, 3)\mu_3 + \epsilon_j. \tag{7.7}$$

Let us define $\delta_j = [\delta(\eta_j, 1)\ \delta(\eta_j, 2)\ \delta(\eta_j, 3)]$ as a 1×3 vector and $\beta = [\mu_1\ \mu_2\ \mu_3]^T$ as a 3×1 vector. The linear model can be rewritten as

$$y_j = \delta_j \beta + \epsilon_j. \tag{7.8}$$

When $\epsilon_j \sim N(0, \sigma^2)$ is assumed, the maximum likelihood estimates of parameters are

$$\hat{\beta} = \left[\sum_{j=1}^{n} \delta_j^T \delta_j \right]^{-1} \left[\sum_{j=1}^{n} \delta_j^T y_j \right] \tag{7.9}$$

for the means and

$$\hat{\sigma}^2 = \frac{1}{n} \sum_{j=1}^{n} (y_j - \delta_j \beta)^2 \tag{7.10}$$

for the residual variance. We see that if the missing variables were not missing, the MLE of the parameters do have an attractive closed form solution. Since both requirements of the EM algorithm are met, we can adopt the EM algorithm to search for the MLE of parameters.

7.2.2 EM Steps

Before we derive the EM algorithm, let us show the expectation and maximization steps of the EM algorithm. The E-step involves calculating the expectations of all items containing the missing variables δ_j. The M-step is simply to estimate β and σ^2 using the closed form solutions given above with the items containing the missing variables replaced by the expectations obtained in the E-step, as shown below:

$$\beta = \left[\sum_{j=1}^{n} E(\delta_j^T \delta_j) \right]^{-1} \left[\sum_{j=1}^{n} E(\delta_j^T) y_j \right] \tag{7.11}$$

and

$$\sigma^2 = \frac{1}{n} \sum_{j=1}^{n} E[(y_j - \delta_j \beta)^2]. \tag{7.12}$$

We can see that the EM algorithm is better described by introducing the M-step first and then describing the E-step (in a reverse direction). The detail of the E-step is now given below:

$$E(\delta_j^T \delta_j) = \begin{bmatrix} E[\delta(\eta_j, 1)] & 0 & 0 \\ 0 & E[\delta(\eta_j, 2)] & 0 \\ 0 & 0 & E[\delta(\eta_j, 3)] \end{bmatrix}, \tag{7.13}$$

$$E(\delta_j^T)y_j = \begin{bmatrix} E[\delta(\eta_j, 1)]y_j \\ E[\delta(\eta_j, 2)]y_j \\ E[\delta(\eta_j, 3)]y_j \end{bmatrix} \qquad (7.14)$$

and

$$E[(y_j - \delta_j \beta)^2] = \sum_{k=1}^{3} E[\delta(\eta_j, k)](y_j - \mu_k)^2. \qquad (7.15)$$

Here, we only need to calculate $E[\delta(\eta_j, k)]$, which is the conditional expectation of $\delta(\eta_j, k)$ given the parameter values and the phenotypic value. The full expression of the conditional expectation should be $E[\delta(\eta_j, k)|y_j, \beta, \sigma^2]$, but we use $E[\delta(\eta_j, k)]$ as a short notation.

$$E[\delta(\eta_j, k)] = \frac{\pi_k f_k(y_j|\theta)}{\sum_{k'=1}^{3} \pi_{k'} f_{k'}(y_j|\theta)}. \qquad (7.16)$$

where $\pi_1 = \pi_3 = \frac{1}{2}\pi_2 = \frac{1}{4}$ is the Mendelian segregation ratio and $f_k(y_j|\theta) = N(y_j|\mu_k, \sigma^2)$ is the normal density. In summary, the EM algorithm is described by

- Initialization: set $t = 0$ and let $\theta = \theta^{(t)}$.
- E-step: calculate $E[\delta(\eta_j, k)|y_j, \theta^{(t)}]$.
- M-step: update $\beta^{(t+1)}$ and $\sigma^{2(t+1)}$.
- Iteration: set $t = t + 1$ and iterate between the E-step and the M-step.

The convergence criterion is

$$||\theta^{(t+1)} - \theta^{(t)}|| = \sqrt{(\theta^{(t+1)} - \theta^{(t)})'(\theta^{(t+1)} - \theta^{(t)})/\dim(\theta)} \leq \epsilon, \qquad (7.17)$$

where $\dim(\theta) = 4$ is the dimension of the parameter vector and ϵ is an arbitrarily small positive number, say 10^{-8}.

Once the three genotypic values are estimated, the additive and dominance effects are estimated using linear contrasts of the genotypic values, e.g.,

$$\begin{cases} \hat{a} = \hat{\beta}_1 - \frac{1}{2}(\hat{\beta}_1 + \hat{\beta}_3) \\ \hat{d} = \hat{\beta}_2 - \frac{1}{2}(\hat{\beta}_1 + \hat{\beta}_3) \end{cases}. \qquad (7.18)$$

7.2.3   Derivation of the EM Algorithm

The observed log likelihood function is given in (7.4). The MLE of θ is the (vector) value that maximizes this log likelihood function. The EM algorithm, however, does not directly maximize this likelihood function; instead, it maximizes the expectation of the complete-data log likelihood function with the expectation taken with respect to the missing variable $\delta(\eta_j, k)$. The complete-data log likelihood function is

$$L_c(\theta) = -\sum_{j=1}^{n}\left[\frac{1}{2\sigma^2}\sum_{k=1}^{3}\delta(\eta_j, k)(y_j - \mu_k)^2 + \sum_{k=1}^{3}\delta(\eta_j, k)\ln(\pi_k)\right]$$
$$-\frac{n}{2}\ln(\sigma^2) \tag{7.19}$$

The expectation of the complete-data log likelihood is $E_{\theta^{(t)}}[L_c(\theta)|y, \theta^{(t)}]$, which is denoted in short by $L(\theta|\theta^{(t)})$ and is defined as

$$L(\theta|\theta^{(t)}) = -\frac{n}{2}\ln(\sigma^2) - \frac{1}{2\sigma^2}\sum_{j=1}^{n}\sum_{k=1}^{3}E[\delta(\eta_j, k)](y_j - \mu_k)^2$$

$$+\sum_{j=1}^{n}\sum_{k=1}^{3}E[\delta(\eta_j, k)]\ln(\pi_k) \tag{7.20}$$

With the EM algorithm, the target likelihood function for maximization is neither the complete-data log likelihood function (7.19) nor the observed log likelihood function (7.4); rather, it is the expected complete-data log likelihood function (7.20). An alternative expression of the above equation is

$$L(\theta|\theta^{(t)}) = -\frac{n}{2}\ln(\sigma^2) - \frac{1}{2\sigma^2}\sum_{j=1}^{n}E[(y_j - \delta_j\beta)^2]$$

$$+\sum_{j=1}^{n}\sum_{k=1}^{3}E[\delta(\eta_j, k)]\ln(\pi_k). \tag{7.21}$$

The partial derivatives of $L(\theta|\theta^{(t)})$ with respect to β and σ^2 are

$$\frac{\partial}{\partial\beta}L(\theta|\theta^{(t)}) = \frac{1}{\sigma^2}E(\delta_j^T)y_j - \frac{1}{\sigma^2}\sum_{j=1}^{n}E(\delta_j^T\delta_j)\beta \tag{7.22}$$

and

$$\frac{\partial}{\partial\sigma^2}L(\theta|\theta^{(t)}) = -\frac{n}{2\sigma^2} + \frac{1}{2\sigma^4}\sum_{j=1}^{n}E[(y_j - \delta_j^T\beta)^2], \tag{7.23}$$

respectively. Setting $\frac{\partial}{\partial\beta}L(\theta|\theta^{(t)}) = \frac{\partial}{\partial\sigma^2}L(\theta|\theta^{(t)}) = 0$, we get

$$\beta = \left[\sum_{j=1}^{n}E(\delta_j^T\delta_j)\right]^{-1}\left[\sum_{j=1}^{n}E(\delta_j^T)y_j\right] \tag{7.24}$$

and

$$\sigma^2 = \frac{1}{n} \sum_{j=1}^{n} E[(y_j - \delta_j \beta)^2].$$ (7.25)

This concludes the derivation of the EM algorithm.

7.2.4 Proof of the EM Algorithm

The target likelihood function for maximization in the EM algorithm is the expectation of the complete-data log likelihood function. However, the actual MLE of θ is obtained by maximization of the observed log likelihood function. To prove that the EM solution of the parameters is indeed the MLE, we only need to show that the partial derivative of the expected complete-data likelihood is identical to the partial derivative of the observed log likelihood, i.e., $\frac{\partial}{\partial \theta} L(\theta|\theta^{(t)}) = \frac{\partial}{\partial \theta} L(\theta)$. If the two partial derivatives are the same, then the solutions must be the same because they both solve the same equation system, i.e., $\frac{\partial}{\partial \theta} L(\theta) = 0$.

Recall that the partial derivative of the expected complete-data log likelihood function with respect to β is

$$\frac{\partial}{\partial \beta} L(\theta|\theta^{(t)}) = \frac{1}{\sigma^2} E(\delta_j^T) y_j - \frac{1}{\sigma^2} E(\delta_j^T \delta_j) \beta,$$ (7.26)

which is a 3×1 vector as shown below:

$$\frac{\partial}{\partial \beta} L(\theta|\theta^{(t)}) = \left[\frac{\partial}{\partial \mu_1} L(\theta|\theta^{(t)}) \quad \frac{\partial}{\partial \mu_2} L(\theta|\theta^{(t)}) \quad \frac{\partial}{\partial \mu_3} L(\theta|\theta^{(t)}) \right]^T.$$

The kth component of this vector is

$$\frac{\partial}{\partial \mu_k} L(\theta|\theta^{(t)}) = \frac{1}{\sigma^2} E[\delta(\eta_j, k)] y_j - \frac{1}{\sigma^2} E[\delta^2(\eta_j, k)] \mu_k$$

$$= \frac{1}{\sigma^2} E[\delta(\eta_j, k)] y_j - \frac{1}{\sigma^2} E[\delta(\eta_j, k)] \mu_k$$

$$= \frac{1}{\sigma^2} E[\delta(\eta_j, k)](y_j - \mu_k)$$ (7.27)

The equation holds because $E[\delta(\eta_j, k)] = E[\delta^2(\eta_j, k)]$, a property for the Bernoulli distribution. We now evaluate the partial derivative of the expected complete-data log likelihood with respect to σ^2,

$$\frac{\partial}{\partial \sigma^2} L(\theta|\theta^{(t)}) = -\frac{n}{2\sigma^2} + \frac{1}{2\sigma^2} \sum_{j=1}^{n} E[(y_j - \delta_j^T \beta)^2]$$

$$= -\frac{n}{2\sigma^2} + \frac{1}{2\sigma^2} \sum_{j=1}^{n} E[\delta(\eta_j, k)](y_j - \mu_k)^2$$ (7.28)

We now look at the partial derivatives of $L(\theta)$ with respect to the parameters. The observed log likelihood function is

$$L(\theta) = \sum_{j=1}^{n} \ln \sum_{k=1}^{3} \pi_k f_k(y_j) \qquad (7.29)$$

where

$$f_k(y_j) = \frac{1}{\sqrt{2\pi}\sigma} \exp\left[-\frac{1}{2\sigma^2}(y_j - \mu_k)^2\right]. \qquad (7.30)$$

The partial derivatives of $L(\theta)$ with respect to $\beta = [\mu_1 \ \mu_2 \ \mu_3]^T$ are

$$\frac{\partial}{\partial \mu_k} L(\theta) = \sum_{j=1}^{n} \frac{\pi_k}{\sum_{k'=1}^{3} \pi_{k'} f_{k'}(y_j)} \frac{\partial}{\partial \mu_k} f_k(y_j), \qquad (7.31)$$

where

$$\frac{\partial}{\partial \mu_k} f_k(y_j) = f_k(y_j) \left[\frac{1}{\sigma^2}(y_j - \mu_k)\right]. \qquad (7.32)$$

Hence,

$$\frac{\partial}{\partial \mu_k} L(\theta) = \frac{1}{\sigma^2} \sum_{j=1}^{n} \frac{\pi_k f_k(y_j)}{\sum_{k'=1}^{3} \pi_{k'} f_{k'}(y_j)}(y_j - \mu_k). \qquad (7.33)$$

Recall that

$$E[\delta(\eta_j, k)] = \frac{\pi_k f_k(y_j)}{\sum_{k'=1}^{3} \pi_{k'} f_{k'}(y_j)}. \qquad (7.34)$$

Therefore,

$$\frac{\partial}{\partial \mu_k} L(\theta) = \frac{1}{\sigma^2} \sum_{j=1}^{n} E[\delta(\eta_j, k)](y_j - \mu_k), \qquad (7.35)$$

which is exactly the same as $\frac{\partial}{\partial \mu_k} L(\theta|\theta^{(t)})$ given in (7.27). Now, let us look at the partial derivative of $L(\theta)$ with respect to σ^2.

$$\frac{\partial}{\partial \sigma^2} L(\theta) = \sum_{j=1}^{n} \sum_{k=1}^{3} \frac{\pi_k}{\sum_{k'=1}^{3} \pi_{k'} f_{k'}(y_j)} \frac{\partial}{\partial \sigma^2} f_k(y_j), \qquad (7.36)$$

where

$$\frac{\partial}{\partial \sigma^2} f_k(y_j) = -\frac{1}{2\sigma^2} f_k(y_j) + \left[\frac{1}{2\sigma^4}(y_j - \mu_k)^2\right] f_k(y_j). \qquad (7.37)$$

Hence,

$$\frac{\partial}{\partial \sigma^2} L(\theta) = -\frac{1}{2\sigma^2} \sum_{j=1}^{n} \sum_{k=1}^{3} \frac{\pi_k f_k(y_j)}{\sum_{k'=1}^{3} \pi_{k'} f_{k'}(y_j)}$$

$$+ \frac{1}{2\sigma^4} \sum_{j=1}^{n} \left[\sum_{k=1}^{3} \frac{\pi_k f_k(y_j)}{\sum_{k'=1}^{3} \pi_{k'} f_{k'}(y_j)} (y_j - \mu_k)^2 \right] \quad (7.38)$$

Note that

$$\sum_{k=1}^{3} \frac{\pi_k f_k(y_j)}{\sum_{k'=}^{3} \pi_{k'} f_{k'}(y_j)} = \sum_{k=1}^{3} E[\delta(\eta_j, k)] = 1. \quad (7.39)$$

Therefore,

$$\frac{\partial}{\partial \sigma^2} L(\theta) = -\frac{1}{2\sigma^2} \sum_{j=1}^{n} \sum_{k=1}^{3} E[\delta(\eta_j, k)] + \frac{1}{2\sigma^4} \sum_{j=1}^{n} \sum_{k=1}^{3} E[\delta(\eta_j, k)](y_j - \mu_k)^2$$

$$= -\frac{n}{2\sigma^2} + \frac{1}{2\sigma^4} \sum_{j=1}^{n} \sum_{k=1}^{3} E[\delta(\eta_j, k)](y_j - \mu_k)^2 \quad (7.40)$$

which is exactly the same as $\frac{\partial}{\partial \sigma^2} L(\theta|\theta^{(t)})$ given in (7.28). We now have confirmed that

$$\frac{\partial}{\partial \sigma^2} L(\theta|\theta^{(t)}) = \frac{\partial}{\partial \sigma^2} L(\theta) \quad (7.41)$$

and

$$\frac{\partial}{\partial \mu_k} L(\theta|\theta^{(t)}) = \frac{\partial}{\partial \mu_k} L(\theta), \ \forall k = 1, 2, 3. \quad (7.42)$$

This concludes the proof that the EM algorithm does lead to the MLE of the parameters.

7.3   Hypothesis Tests

The overall null hypothesis is "no major gene is segregating" denoted by

$$H_0 : \mu_1 = \mu_2 = \mu_3 = \mu. \quad (7.43)$$

The alternative hypothesis is "at least one of the means is different from others," denoted by

$$H_1 : \mu_1 \neq \mu_3 \text{ or } \mu_2 \neq \frac{1}{2}(\mu_1 + \mu_3). \quad (7.44)$$

The likelihood ratio test statistic is

$$\lambda = -2[L_0(\hat{\theta}) - L_1(\hat{\theta})], \tag{7.45}$$

where $L_1(\hat{\theta})$ is the observed log likelihood function evaluated at the MLE of θ for the full model, and

$$L_0(\hat{\theta}) = -\frac{n}{2}\ln(\hat{\sigma}^2) - \frac{1}{2\hat{\sigma}^2}\sum_{j=1}^{n}(y_j - \hat{\mu})^2 \tag{7.46}$$

is the log likelihood values evaluated at the null model where

$$\hat{\mu} = \frac{1}{n}\sum_{j=1}^{n}y_j \tag{7.47}$$

and

$$\hat{\sigma}^2 = \frac{1}{n}\sum_{j=1}^{n}(y_j - \hat{\mu})^2. \tag{7.48}$$

Under the null hypothesis, λ will follow approximately a chi-square distribution with two degrees of freedom. Therefore, H_0 will be rejected if $\lambda > \chi^2_{2,1-\alpha}$, where $\alpha = 0.05$ may be chosen as the type I error.

7.4   Variances of Estimated Parameters

Unlike other iterative methods of parameter estimation, e.g., Newton–Raphson method, that variance–covariance matrix of the estimated parameters are provided automatically as a by-product of the iteration process, the EM algorithm does not facilitate an easy way for calculating the variance–covariance matrix of the estimated parameters. We now introduce a special method for calculating the variance–covariance matrix. The method was developed by Louis (1982) particularly for calculating the variance–covariance matrix of parameters that are estimated via the EM algorithm. The method requires the first and second partial derivatives of the complete-data log likelihood function (not the observed log likelihood function). The complete-data log likelihood function is

$$L(\theta, \delta) = \sum_{j=1}^{n} L_j(\theta, \delta), \tag{7.49}$$

where

$$L_j(\theta, \delta) = -\frac{1}{2}\ln(\sigma^2) - \frac{1}{2\sigma^2}(y - \delta_j\beta)^2. \tag{7.50}$$

The first partial derivative of this log likelihood with respect to the parameter is called the score function, which is

$$S(\theta, \delta) = \frac{\partial}{\partial \theta} L(\theta, \delta) = \sum_{j=1}^{n} \frac{\partial}{\partial \theta} L_j(\theta, \delta) = \sum_{j=1}^{n} S_j(\theta, \delta), \qquad (7.51)$$

where

$$S_j(\theta, \delta) = \frac{\partial}{\partial \theta} L_j(\theta, \delta) = \begin{bmatrix} \frac{\partial}{\partial \beta} L_j(\theta, \delta) \\ \frac{\partial}{\partial \sigma^2} L_j(\theta, \delta) \end{bmatrix} = \begin{bmatrix} \frac{1}{\sigma^2} \delta_j^T (y_j - \delta_j \beta) \\ -\frac{1}{2\sigma^2} + \frac{1}{2\sigma^4} (y_j - \delta_j \beta)^2 \end{bmatrix}.$$

$$(7.52)$$

The second partial derivative is called the Hessian matrix $H(\theta, \delta)$. The negative value of the Hessian matrix is denoted by $B(\theta, \delta) = -H(\theta, \delta)$,

$$B(\theta, \delta) = -\frac{\partial^2 L(\theta, \delta)}{\partial \theta \, \partial \theta^T} = -\sum_{j=1}^{n} \frac{\partial^2 L_j(\theta, \delta)}{\partial \theta \, \partial \theta^T} = \sum_{j=1}^{n} B_j(\theta, \delta), \qquad (7.53)$$

where

$$B_j(\theta, \delta) = -\frac{\partial^2 L_j(\theta, \delta)}{\partial \theta \, \partial \theta^T} = \begin{bmatrix} -\frac{\partial^2 L_j(\theta, \delta)}{\partial \beta \, \partial \beta^T} & -\frac{\partial^2 L_j(\theta, \delta)}{\partial \beta \, \partial \sigma^2} \\ -\frac{\partial^2 L_j(\theta, \delta)}{\partial \sigma^2 \, \partial \beta^T} & -\frac{\partial^2 L_j(\theta, \delta)}{\partial \sigma^2 \, \partial \sigma^2} \end{bmatrix} \qquad (7.54)$$

Detailed expression of $B_j(\theta, \delta)$ is given below:

$$B_j(\theta, \delta) = \begin{bmatrix} \frac{1}{\sigma^2} \delta_j^T \delta_j & \frac{1}{\sigma^4} \delta_j^T (y_j - \delta_j \beta) \\ \frac{1}{\sigma^4} (y_j - \delta_j \beta)^T \delta_j & \frac{1}{\sigma^6} (y_j - \delta_j \beta)^2 - \frac{1}{2\sigma^2} \end{bmatrix}. \qquad (7.55)$$

Louis (1982) gave the following information matrix:

$$I(\theta) = \mathrm{E}[B(\theta, \delta)] - \mathrm{var}[S(\theta, \delta)]$$

$$= \sum_{j=1}^{n} \mathrm{E}[B_j(\theta, \delta)] - \sum_{j=1}^{n} \mathrm{var}[S_j(\theta, \delta)], \qquad (7.56)$$

where the expectation and variance are taken with respect to the missing variable δ_j using the posterior probability of δ_j. Detailed expressions of $\mathrm{E}[B_j(\theta, \delta)]$ and $\mathrm{var}[S_j(\theta, \delta)]$ are given in the end of this section. Readers may also refer to Han and Xu (2008) and Xu and Hu (2010) for the derivation and the results. Replacing θ by $\hat{\theta}$ and taking the inverse of the information matrix, we get the variance–covariance matrix of the estimated parameters,

$$\mathrm{var}(\hat{\theta}) = I^{-1}(\hat{\theta}) = \{\mathrm{E}[B(\hat{\theta}, \delta)] - \mathrm{var}[S(\hat{\theta}, \delta)]\}^{-1}. \qquad (7.57)$$

This is a 4 × 4 variance–covariance matrix, as shown below:

$$\text{var}(\hat{\theta}) = \begin{bmatrix} \text{var}(\hat{\beta}) & \text{cov}(\hat{\beta}, \hat{\sigma}^2) \\ \text{cov}(\hat{\sigma}^2, \hat{\beta}^T) & \text{var}(\hat{\sigma}^2) \end{bmatrix}, \tag{7.58}$$

where $\text{var}(\hat{\beta})$ is a 3 × 3 variance matrix for the estimated genotypic values.

The additive and dominance effects can be expressed as linear functions of β, as demonstrated below:

$$\begin{bmatrix} a \\ d \end{bmatrix} = \begin{bmatrix} \frac{1}{2} & 0 & -\frac{1}{2} \\ -\frac{1}{2} & 1 & -\frac{1}{2} \end{bmatrix} \begin{bmatrix} \beta_1 \\ \beta_2 \\ \beta_3 \end{bmatrix} = L^T \beta, \tag{7.59}$$

where

$$L = \begin{bmatrix} \frac{1}{2} & 0 & -\frac{1}{2} \\ -\frac{1}{2} & 1 & -\frac{1}{2} \end{bmatrix}^T. \tag{7.60}$$

The variance–covariance matrix for the estimated major gene effects is

$$\text{var}\begin{bmatrix} \hat{a} \\ \hat{d} \end{bmatrix} = L^T \text{var}(\hat{\beta}) L = \begin{bmatrix} \text{var}(\hat{a}) & \text{cov}(\hat{a}, \hat{d}) \\ \text{cov}(\hat{a}, \hat{d}) & \text{var}(\hat{d}) \end{bmatrix}. \tag{7.61}$$

The variance–covariance matrix of the estimated major gene effects also facilitates an alternative method for testing the hypothesis of $H_0 : a = d = 0$. This test is called the Wald-test statistic (Wald 1943),

$$W = \beta^T L[L^T \text{var}(\hat{\beta})L]^{-1} L^T \beta = \begin{bmatrix} \hat{a} & \hat{d} \end{bmatrix} \begin{bmatrix} \text{var}(\hat{a}) & \text{cov}(\hat{a}, \hat{d}) \\ \text{cov}(\hat{a}, \hat{d}) & \text{var}(\hat{d}) \end{bmatrix}^{-1} \begin{bmatrix} \hat{a} \\ \hat{d} \end{bmatrix}. \tag{7.62}$$

The Wald-test statistic is much like the likelihood ratio test statistic. Under the null model, W follows approximately a χ^2 distribution with 2 degrees of freedom. However, Wald test is usually considered inferior compared to the likelihood ratio test statistic, especially when the sample size is small.

Before exiting this section, we now provide the derivation of $E[B_j(\theta, \delta)]$ and $\text{var}[S_j(\theta, \delta)]$. Recall that δ_j is a 1 × 3 multinomial variable with sample size 1 and defined as

$$\delta_j = \begin{bmatrix} \delta(\eta_j, 1) & \delta(\eta_j, 2) & \delta(\eta_j, 3) \end{bmatrix} \tag{7.63}$$

This variable has the following properties:

$$\delta^2(\eta_j, k) = \delta(\eta_j, k) \tag{7.64}$$

and

$$\delta(\eta_j, k)\delta(\eta_j, k') = \begin{cases} \delta(\eta_j, k) & \text{for } k = k' \\ 0 & \text{for } k \neq k' \end{cases} \tag{7.65}$$

Therefore, the expectation of δ_j is

$$E(\delta_j) = \left[E\left[\delta(\eta_j, 1)\right] \ E\left[\delta(\eta_j, 2)\right] \ E\left[\delta(\eta_j, 3)\right] \right] \tag{7.66}$$

The expectation of its quadratic form is

$$E(\delta_j^T \delta_j) = \text{diag}\left[E(\delta_j)\right] = \begin{bmatrix} E\left[\delta(\eta_j, 1)\right] & 0 & 0 \\ 0 & E\left[\delta(\eta_j, 3)\right] & 0 \\ 0 & 0 & E\left[\delta(\eta_j, 3)\right] \end{bmatrix} \tag{7.67}$$

The variance–covariance matrix of δ_j is

$$\text{var}(\delta_j) = E(\delta_j^T \delta_j) - E(\delta_j)E^T(\delta_j) \tag{7.68}$$

To derive the observed information matrix, we need the first and second partial derivatives of the complete-data log likelihood with respect to the parameter vector $\theta = [\beta^T \ \sigma^2]^T$. The score vector is rewritten as

$$S_j(\theta, \delta) = \begin{bmatrix} \frac{1}{\sigma^2}\delta_j^T(y_j - \delta_j\beta) \\ \frac{1}{2\sigma^4}(y_j - \delta_j\beta)^2 \end{bmatrix} + \begin{bmatrix} 0_{3\times 1} \\ -\frac{1}{2\sigma^2} \end{bmatrix} \tag{7.69}$$

where $0_{3\times 1}$ is a 3×1 vector of zeros, and thus, the score is a 4×1 vector. The negative of the second partial derivative is

$$B_j(\theta, \delta) = \begin{bmatrix} \frac{1}{\sigma^2}\delta_j^T\delta_j & \frac{1}{\sigma^4}\delta_j^T(y_j - \delta_j\beta) \\ \frac{1}{\sigma^4}(y_j - \delta_j\beta)^T\delta_j & \frac{1}{\sigma^6}(y_j - \delta_j\beta)^2 \end{bmatrix} + \begin{bmatrix} 0_{3\times 3} & 0_{3\times 1} \\ 0_{1\times 3} & -\frac{1}{2\sigma^2} \end{bmatrix} \tag{7.70}$$

where $0_{3\times 3}$ is a 3×3 matrix of zeros, and thus, $B_j(\theta, \delta)$ is a 4×4 matrix. The expectation of $B_j(\theta, \delta)$ is easy to derive, but derivation of the variance–covariance matrix of the score vector is very difficult. Xu and Xu (2003) used a Monte Carlo approach to approximating the expectation and the variance–covariance matrix. They simulated multiple (e.g., 5,000) samples of δ_j from the posterior distribution and then took the sample mean of $B_j(\theta, \delta)$ and the sample variance–covariance matrix of $S_j(\theta, \delta)$ as the approximations of the corresponding terms. Here, we took a theoretical approach for the derivation and provide explicit expressions for the expectation and variance–covariance matrix. We can express the score vector as a linear function of δ_j and the $B_j(\theta, \delta)$ matrix as a quadratic function of δ_j. By trial and error, we found that

$$
S_j(\theta, \delta) =
\begin{bmatrix}
\frac{1}{\sigma^2}(y_j - \beta_1) & 0 & 0 \\
0 & \frac{1}{\sigma^2}(y_j - \beta_1) & 0 \\
0 & 0 & \frac{1}{\sigma^2}(y_j - \beta_3) \\
\frac{1}{2\sigma^4}(y_j - \beta_1)^2 & \frac{1}{2\sigma^4}(y_j - \beta_2)^2 & \frac{1}{2\sigma^4}(y_j - \beta_3)^2
\end{bmatrix}
\begin{bmatrix}
\delta(\eta_j, 1) \\
\delta(\eta_j, 2) \\
\delta(\eta_j, 3)
\end{bmatrix}
+
\begin{bmatrix}
0 \\
0 \\
0 \\
-\frac{1}{2\sigma^2}
\end{bmatrix}
$$

$$
= A_j^T \delta_j^T + C \tag{7.71}
$$

where A_j^T is the 4×3 coefficient matrix and C is the 4×1 vector of constants. Let us define a 4×1 matrix H_j^T as

$$
H_j^T = T_j^T \delta_j^T =
\begin{bmatrix}
\frac{1}{\sigma^2} & 0 & 0 \\
0 & \frac{1}{\sigma^2} & 0 \\
0 & 0 & \frac{1}{\sigma^2} \\
\frac{1}{\sigma^3}(y_j - \beta_1) & \frac{1}{\sigma^3}(y_j - \beta_2) & \frac{1}{\sigma^3}(y_j - \beta_3)
\end{bmatrix}
\begin{bmatrix}
\delta_j(1) \\
\delta_j(2) \\
\delta_j(3)
\end{bmatrix}
\tag{7.72}
$$

where T_j^T is the 4×3 coefficient matrix. We can now express matrix $B_j(\theta, \delta)$ as

$$
B_j(\theta, \delta) = H_j^T H_j + D = T_j^T \delta_j^T \delta_j T_j + D \tag{7.73}
$$

where

$$
D = \mathrm{diag}(C) =
\begin{bmatrix}
0_{3 \times 3} & 0_{3 \times 1} \\
0_{1 \times 3} & -\frac{1}{2\sigma^2}
\end{bmatrix}
\tag{7.74}
$$

is a 4×4 constant matrix. The expectation of $B_j(\theta, \delta)$ is

$$
E\left[B_j(\theta, \delta)\right] = T_j^T E(\delta_j^T \delta_j) T_j + D \tag{7.75}
$$

The expectation vector and the variance–covariance matrix of $S_j(\theta, \delta)$ are

$$
E\left[S_j(\theta, \delta)\right] = A_j^T E(\delta_j^T) + C \tag{7.76}
$$

and

$$
\mathrm{var}\left[S_j(\theta, \delta)\right] = A_j^T \mathrm{var}(\delta_j) A_j = A_j^T \left[E(\delta_j^T \delta_j) - E(\delta_j^T) E(\delta_j)\right] A_j \tag{7.77}
$$

respectively. Expressing $S_j(\theta, \delta)$ and $B_j(\theta, \delta)$ as linear and quadratic functions of the missing vector δ_j has significantly simplified the derivation of the information matrix.

7.5   Estimation of the Mixing Proportions

We used an F_2 population as an example for segregation analysis. Extension of the segregation analysis to other populations is straightforward and will not be discussed here. For the F_2 population, we assumed that the major gene follows the Mendelian segregation ratio, i.e., $\pi_1 = \pi_3 = \frac{1}{2}\pi_2 = \frac{1}{4}$. Therefore, π_k is a constant, not a parameter for estimation. The method can be extended to a situation where the major gene does not follow the Mendelian segregation ratio. In this case, the values of π_k are also parameters for estimation. This section will introduce a method to estimate the π_k's. These π_k's are called the mixing proportions.

We simply add one more step in the EM algorithm to estimate π_k, $\forall k = 1, 2, 3$. Again, we maximize the expected complete-data log likelihood function. To enforce the restriction that $\sum_{k=1}^{3} \pi_k = 1$, we introduce a Lagrange multiplier ξ. Therefore, the actual function to be maximized is

$$L(\theta|\theta^{(t)}) = -\frac{n}{2}\ln(\sigma^2) - \frac{1}{2\sigma^2}\sum_{j=1}^{n}\sum_{k=1}^{3}\mathrm{E}[\delta(\eta_j, k)](y_j - \mu_k)^2$$

$$+ \sum_{j=1}^{n}\sum_{k=1}^{3}\mathrm{E}[\delta(\eta_j, k)]\ln(\pi_k) + \lambda\left(1 - \sum_{k=1}^{3}\pi_k\right). \tag{7.78}$$

The partial derivatives of $L(\theta|\theta^{(t)})$ with respect to π_k and λ are

$$\frac{\partial}{\partial \pi_k}L(\theta|\theta^{(t)}) = \frac{1}{\pi_k}\sum_{j=1}^{n}\mathrm{E}[\delta(\eta_j, k)] - \lambda, \forall k = 1, 2, 3 \tag{7.79}$$

and

$$\frac{\partial}{\partial \lambda}L(\theta|\theta^{(t)}) = 1 - \sum_{k=1}^{3}\pi_k, \tag{7.80}$$

respectively. Let $\frac{\partial}{\partial \pi_k}L(\theta|\theta^{(t)}) = \frac{\partial}{\partial \lambda}L(\theta|\theta^{(t)}) = 0$, and solve for π_k's and λ. The solution for π_k is

$$\pi_k = \frac{1}{\lambda}\sum_{j=1}^{n}E[\delta(\eta_j, k)], \forall k = 1, 2, 3. \tag{7.81}$$

The solution for λ is obtained by

$$\sum_{k=1}^{3}\pi_k = \frac{1}{\lambda}\sum_{k=1}^{3}\sum_{j=1}^{n}E[\delta(\eta_j, k)] = \frac{1}{\lambda}\sum_{j=1}^{n}\sum_{k=1}^{3}E[\delta(\eta_j, k)] = \frac{n}{\lambda} = 1. \tag{7.82}$$

This is because $\sum_{k=1}^{3}E[\delta(\eta_j, k)] = 1$ and $\sum_{j=1}^{n} = n$. As a result, $\lambda = n$ and thus

$$\pi_k^{(t+1)} = \frac{1}{n}\sum_{j=1}^{n}E[\delta(\eta_j, k)], \forall k = 1, 2, 3. \tag{7.83}$$

7.3 Estimation of the Moving Proportions

Chapter 8
Genome Scanning for Quantitative Trait Loci

In the previous chapters, we learned the basic concept of quantitative genetics, the quantitative genetics model, the method for major gene detection (genotypes of the major gene are observed), and the algorithm for segregation analysis (genotypes of the major gene are not observed). We also learned some analytical techniques to analyze a molecular marker linked to a major gene. The real focus of statistical genomics, however, is to identify functional genes that are responsible for the genetic variation of quantitative traits or complex traits if they are not normally distributed. These chapters provide the necessary technology (knowledge preparation) for gene identification, which is the theme of this book.

Molecular markers are not genes but they are inherited following Mendel's laws, and their genotypes are observable. The functional genes also follow Mendel's laws of inheritance, but their genotypes are not observable. Since both markers and genes are carried by a limited number of chromosomes in the genome, some genes must be physically linked with some markers. If a marker sits in the neighborhood of a gene, the segregation pattern of the marker must be associated with the phenotypic variation of a trait that is controlled by the gene due to linkage. Therefore, we can study marker and trait association and hope to identify important markers that are closely linked to the gene. Since a quantitative trait is often controlled by the segregation of more than one gene, more markers are needed to identify all genes for a quantitative trait. These multiple genes are called quantitative trait loci (QTL). This chapter deals with marker-trait association study in line crosses. The association study using line crosses is different from the association study using randomly sampled populations. The former takes advantage of linkage disequilibrium, while the latter assumes no linkage disequilibrium. As a result, markers associated with the trait of interest in line crosses are not equivalent to the genes, while markers associated with the traits in randomly sampled populations are most likely the actual genes. The statistical methods for association study, however, are the same, regardless whether the populations are derived from line crosses or not.

S. Xu, *Principles of Statistical Genomics*, DOI 10.1007/978-0-387-70807-2_8,

8.1   The Mouse Data

A dataset from an F_2 mouse population consisting of 110 individuals was used as an example for the genetic analysis. The data were published by Lan et al. (2006) and are freely available from the Internet. The parents of the F_2 population were B6 (29 males) and BTBR (31 females). The F_2 mice used in this study were measured for various clinical traits related to obesity and diabetes. The framework map consists of 194 microsatellite markers, with an average marker spacing of about 10 cM. The mouse genome has 19 chromosomes (excluding the sex chromosome). The data analyzed in this chapter contain 110 F_2 mice and 193 markers. The second marker (D7Mit76) on chromosome 7 was excluded from the analysis because it overlaps with the first marker (D7Mit56). The 193 markers cover about 1,800 cM of the entire mouse genome. The trait of interest was the 10th-week body weight. The marker map, the genotypes of the 110 mice for the 193 markers, and the 10th-week body weights of the F_2 mice are also provided in the author's personal website (www.statgen.ucr.edu). The files stored in our website are not the original data but preprocessed by our laboratory members, and thus, they are ready for analysis using QTL mapping software packages such as the QTL procedure in SAS (Hu and Xu 2009).

8.2   Genome Scanning

In major gene identification, we used an F-test statistic to test the significance of a major gene. In genome scanning, we simply treat each marker as a major gene and analyze every single marker. The test statistics of all markers across the genome are plotted against the genome location of the markers, forming a test statistic profile. Some regions of the genome may show peaks, while majority of the genome may be flat. The regions with peaks may suggest QTL nearby the peaks. If the marker density is sufficiently high, some markers may actually overlap with the QTL. The genome scanning is also called individual marker analysis. We scan all markers across the genome but with one marker at a time. The entire genome scanning requires many repeated single-marker analyses. The genetic model and test statistic in genome scanning are of no difference from the major gene detection except that we now deal with multiple markers. Sometimes, investigators already have prior knowledge about the functions of some genes. The functions may be related to the development of the quantitative trait of interest. These genes are called candidate genes for the trait of interest. Genome scanning may also include these candidate genes. Figure 8.1 shows the LOD score profile of the mouse genome for the trait of 10th-week body weight (wt10week). Note that the LOD (log of odds) score is often used in human genetics. The Wald-test statistic is often converted into the LOD score using (see a later section for the definition of LOD)

$$\text{LOD} = \frac{W}{2 \times \ln(10)} \tag{8.1}$$

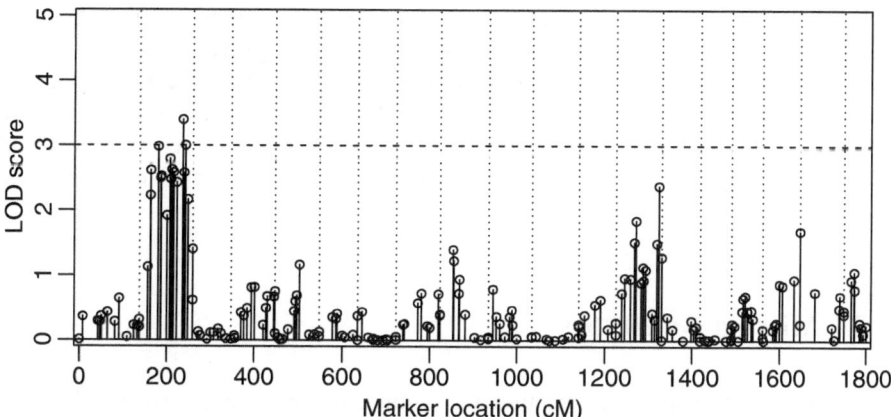

Fig. 8.1 LOD score profile of the entire genome (19 chromosomes) for the 10th-week body weight of F$_2$ mice derived from the cross of two inbred lines. The 19 chromosomes are separated by the dotted reference lines

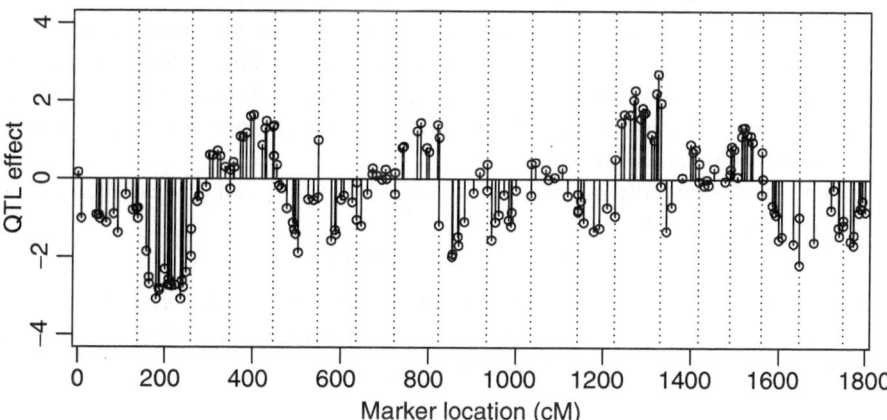

Fig. 8.2 QTL effect profile of the entire genome (19 chromosomes) for the 10th-week body weight of F$_2$ mice derived from the cross of two inbred lines. The 19 chromosomes are separated by the dotted reference lines

There are many peaks in the LOD score profile, but the peaks in chromosome 2 appear to be too high (LOD > 3) to be explained by chance. Therefore, one or more QTL may exist in chromosome two for this trait.

The model used for the analysis is called the additive genetic model because the dominance effect has been ignored. Figure 8.2 shows the additive effects plotted against markers, the so-called QTL effect profile. We can see that QTL effects in some regions of the genome are positive, while in other regions, they are negative. The way we coded the genotypes determined the signs of the QTL effects. Assume that the original genotypic data were coded as "A" for line B6, "B" for line BTBR, and "H" for heterozygote. We numerically recoded the genotype as 1 for "A," 0 for "H," and -1 for "B." Based on this coding system, a negative QTL effect means that the B6 allele is "low" and the BTBR allele is "high." Therefore, the QTL allele carried by B6 in the second chromosome is the "low" allele, that is, it is responsible for the low body weight. Of course, if "A" and "B" alleles represent BTBR and B6, respectively, the negative and positive signs should be explained in just the opposite way.

8.3 Missing Genotypes

In the section of major gene detection, we assumed that the genotype of a major gene is observed for every individual. In the section of segregation analysis, the genotype of the major gene is missing for every individual. This section deals with marker analysis. Although most individuals in the mapping population should be genotyped for all markers, still some individuals may not be genotyped for some markers, either due to technical errors or human errors. If an individual is not genotyped for all markers, this individual should be eliminated from the analysis. However, most individuals may just have a few missing genotypes. These individuals must be included in the analysis; otherwise, we may not have enough sample size to perform genetic mapping. We now use the F_2 population as an example to show how to deal with the missing marker problem.

Let y_j be the phenotypic value of individual j, and it can be described by the following linear model:

$$y_j = b_0 + X_j b_1 + e_j, \tag{8.2}$$

where b_0 is the intercept, b_1 is the additive genetic effect, i.e., a, and e_j is the residual error. The genotype indicator variable X_j depends on the genotype of the marker under consideration. Let us define X_j as

$$X_j = \begin{cases} +1 & \text{for} A_1 A_1 \\ 0 & \text{for} A_1 A_2 \\ -1 & \text{for} A_2 A_2 \end{cases} . \tag{8.3}$$

Let G_j be the genotype of the marker under consideration and $p_j(1) = \Pr(G_j = A_1 A_1 | \text{marker})$ be the probability of $G_j = A_1 A_1$ given the genotypes of the two markers flanking the marker of interest. Similarly, let $p_j(0) = \Pr(G_j = A_1 A_2 | \text{marker})$ and $p_j(-1) = \Pr(G_j = A_2 A_2 | \text{marker})$. The conditional expectation of X_j given the flanking marker genotypes is $E(X_j) = p_j(1) - p_j(-1)$. The model for missing markers is the same as (8.2) except that X_j is replaced by $E(X_j)$, i.e.,

$$y_j = b_0 + E(X_j) b_1 + e_j. \tag{8.4}$$

To include the dominance effect, the revised model is

$$y_j = b_0 + E(X_{j1})b_1 + E(X_{j2})b_2 + e_j \tag{8.5}$$

where X_{j1} is the genotype indicator variable for the additive effect, as defined earlier (8.3), and X_{j2} is the genotype indicator variable for the dominance effect,

$$X_{j2} = \begin{cases} 0 & \text{for} A_1 A_1 \\ 1 & \text{for} A_1 A_2 \\ 0 & \text{for} A_2 A_2 \end{cases} . \tag{8.6}$$

The conditional expectation of X_{j2} is simply $E(X_j) = p_j(0)$. The second regression coefficient b_2 is the dominance effect, i.e., $b_2 = d$.

8.4 Test Statistics

There are many different test statistics we can use for genome scanning. The one we learned in the major gene detection is the F-test statistic. We now discuss the test statistics when only a single model effect is subject to test. In genetic analysis, this is equivalent to testing only the additive effect. Let $\hat{b}_1$ be the estimated genetic effect and $\sigma^2_{\hat{b}_1}$ be the variance of the estimate. The F-test statistic for the null hypothesis $H_0 : b_1 = 0$ is

$$F = \frac{\hat{b}_1^2}{\sigma^2_{\hat{b}_1}}. \tag{8.7}$$

This F-test statistic appears to be different from the F-test statistic occurring in the analysis of variances (ANOVA). The latter is defined as the ratio of the between-group mean squares MS_B to the within-group mean squares MS_W. However, the two test statistics are two different forms of the same test statistic (derivation is not shown). As an F-test statistic, it will follow an F-distribution with a numerator degree of freedom 1 and a denominator degree of freedom $n - 2$.

A single genetic effect can also be tested using the t-test statistic. The t-test statistic is simply the square root of the F-test statistic,

$$t = \sqrt{F} = \frac{|\hat{b}_1|}{\sigma_{\hat{b}_1}} \tag{8.8}$$

Under the null hypothesis $H_0 : b_1 = 0$, this test statistic will follow a t-distribution with $n - 2$ degrees of freedom. As the sample size increases, $n - 2$ is not much different from n; therefore, the degrees of freedom in the F-test and the t-test are approximately equal to the sample size.

When $n \to \infty$, the F-test will be identical to the χ^2-test statistic, which follows a χ^2 distribution with one degree of freedom. The corresponding t-test statistic will approach to the Z-test statistic.

The F-test statistic in the form of $F = \frac{\hat{b}_1^2}{\sigma_{\hat{b}_1}^2}$ is actually called the Wald-test statistic or simply W-test statistic (Wald 1943). Although the Wald-test statistic is not called as often as the F- and t-test statistics in genome scanning, it will be used more often here in this textbook due to the fact that the Wald-test statistic is comparable or similar to the likelihood ratio test statistic.

The likelihood ratio test (LRT) statistic is defined as

$$\lambda = -2[L_0(\hat{\theta}_0) - L_1(\hat{\theta}_1)] \qquad (8.9)$$

where $L_0(\hat{\theta}_0)$ is the log likelihood function evaluated under the null model (H_0 : $b_1 = 0$) and $L_1(\hat{\theta}_1)$ is the log likelihood function evaluated under the full model ($H_1 : b_1 \neq 0$). The null model and the full model differ by one parameter, i.e., $\theta_0 = \{b_0, \sigma^2\}$ and $\theta_1 = \{b_0, b_1, \sigma^2\}$. We often call the null model the restricted model or reduced model because it has $b_1 = 0$ as the restriction or simply has one parameter less than the full model. Because $L_1(\hat{\theta}_1)$ is guaranteed to be larger than $L_0(\hat{\theta}_0)$, the log likelihood difference is negative. A negative test statistic looks strange, and thus, we put a minus sign in front of the difference to make the test statistic positive. The constant multiplier 2 is simply to make the likelihood ratio test statistic follow a standard distribution under the null model. This standard distribution happens to be a χ^2 distribution with one degree of freedom. The degree of freedom is one, not any other value, because the null model has one parameter less than the full model.

We now realize that the Wald-test statistic, the F-test statistic, and the likelihood ratio test statistic all approach a χ^2 distribution with one degree of freedom as the sample size is sufficiently large. Therefore, these three test statistics can be used interchangeably with very little difference, although the likelihood ratio test statistic is considered a slightly better test statistic than the others.

The likelihood ratio test statistic is defined using the natural logarithm, i.e., the logarithm with base $e \approx 2.718281828459$. In human genetics, people often use the LOD (log of odds) score as the test statistic. Let $L_0 = L_0(\hat{\theta}_0)$ and $L_1 = L_1(\hat{\theta}_1)$ be short expressions of the natural logarithms of the likelihood functions under the null model and the full model, respectively. The original likelihood functions (before taking the natural log) are $l_0 = e^{L_0}$ and $l_1 = e^{L_1}$, respectively. The LOD score is defined as

$$\text{LOD} = \log_{10}\left(\frac{l_1}{l_0}\right) = \log_{10}\left(\frac{e^{L_1}}{e^{L_0}}\right)$$

$$= \log_{10} e^{L_1} - \log_{10} e^{L_0} = \log_{10} e^{(L_1 - L_0)} \qquad (8.10)$$

It is the log of the likelihood ratio with base 10 rather than base e. The relationship between LOD and the likelihood ratio test statistic (λ) is

$$\text{LOD} = \log_{10} e^{(L_1 - L_0)} = \frac{1}{2} \log_{10} e^{[-2(L_0 - L_1)]}$$

$$= [-2(L_0 - L_1)] \left(\frac{1}{2} \log_{10} e \right) = \lambda \left(\frac{1}{2} \log_{10} e \right) \quad (8.11)$$

The constant $\frac{1}{2} \log_{10} e \approx 0.2171$ and the inverse of the constant are approximately 4.6052. Therefore, we may use the following approximation to convert λ to LOD:

$$\text{LOD} = 0.2171 \, \lambda = \frac{\lambda}{4.6052} \quad (8.12)$$

The LOD score has an intuitive interpretation because of the base 10. An LOD score of x means that the full model is 10^x times more likely than the restricted model. For example, an LOD score 3 means that the full model (with the marker effect) is 1,000 times more likely than the reduced model (without the marker effect).

We now turn our attention to the hypotheses where two or more genetic effects are tested simultaneously. For example, in an F_2 population, we can test both the additive and dominance effects. The null hypothesis is $H_0 : a = d = 0$ or $H_0 : b_1 = b_2 = 0$. In this case, the t-test is not a valid choice, because it is designed for testing only a single effect. The F-test, although can be used, is rarely chosen as the test statistic for genome scanning. The F-test statistic for testing two effects is defined as

$$F = \frac{1}{2} [\hat{b}_1 \ \hat{b}_2] \begin{bmatrix} \text{var}(\hat{b}_1) & \text{cov}(\hat{b}_1, \hat{b}_2) \\ \text{cov}(\hat{b}_1, \hat{b}_2) & \text{var}(\hat{b}_2) \end{bmatrix}^{-1} \begin{bmatrix} \hat{b}_1 \\ \hat{b}_2 \end{bmatrix} \quad (8.13)$$

The $\frac{1}{2}$ multiplier appears because we are testing two effects. If we test k effects simultaneously, the multiplier will be $\frac{1}{k}$, and the dimensionality of the effect vector and the variance matrix will be changed to $k \times 1$ and $k \times k$ accordingly. The F-test statistic follows an F-distribution with degrees of freedom k and $n - (k + 1)$ or simply k and n when n is sufficiently large.

In contrast to the test for a single genetic effect where the F-test statistic is equivalent to the W-test statistic, when testing two or more effects, the W-test statistic is

$$W = [\hat{b}_1 \ \hat{b}_2] \begin{bmatrix} \text{var}(\hat{b}_1) & \text{cov}(\hat{b}_1, \hat{b}_2) \\ \text{cov}(\hat{b}_1, \hat{b}_2) & \text{var}(\hat{b}_2) \end{bmatrix}^{-1} \begin{bmatrix} \hat{b}_1 \\ \hat{b}_2 \end{bmatrix} \quad (8.14)$$

The relationship between the W-test and the F-test is $W = kF$. When the sample size is sufficiently large, the W-test statistic will approach a χ^2 distribution with k degrees of freedom ($k = 2$ in this case).

The corresponding likelihood ratio test statistic for two or more effects has the same form as that for testing a single effect except that $\theta_0 = \{b_0, \sigma^2\}$ under the null model has two parameters less than the $\theta_1 = \{b_0, b_1, b_2, \sigma^2\}$ under the full model.

As a result, the λ test statistic follows a χ^2 distribution with $k = 2$ degrees of freedom. Both the W-test and the likelihood ratio test statistics follow the same χ^2 distribution, and thus, they can be used interchangeably with very little difference.

The W-test statistic requires calculation of the variance–covariance matrix of the estimated parameters and its inverse. However, some of the algorithms for parameter estimation, e.g., the EM algorithm, do not have an automatic way to calculate this matrix. For these methods, the likelihood ratio test statistic may be preferred because of the ease of calculating the test statistic. When both the W-test and the λ-test statistics are available, which one is better? The answer is that the λ-test statistic is more desirable if the sample size is small. For large samples sizes, these two test statistics are virtually the same.

In summary, the W-test and the λ-test statistics are preferable for genome scanning because they can be used for testing both a single effect and multiple effects (compared to the t-test and the Z-test which are only useful for testing a single effect). The LOD score test statistic is simply a rescaled likelihood ratio test statistic, and thus, they are used interchangeably without any difference at all.

8.5   Bonferroni Correction

Genome scanning involves multiple tests. Sometimes the number of tests may reach hundreds or even thousands. For a single test, the critical value for any test statistic simply takes the 95 % or 99 % quantile of the distribution that the test statistic follows under the null hypothesis. For example, the F-test statistic follows an F-distribution, the likelihood ratio test statistic follows a chi-square distribution, and the W-test statistic also follows a chi-square distribution. When multiple tests are involved, the critical value used for a single test must be adjusted to make the experiment-wise type I error at a desired level, say 0.05.

The Bonferroni correction is a multiple-test correction used when multiple statistical tests are being performed in a single experiment (Dunn 1961). While a given alpha value α may be appropriate for each individual test, it is not for the set of all tests involved in a single experiment. In order to avoid spurious positives, the alpha value needs to be lowered to account for the number of tests being performed. The Bonferroni correction sets the type I error for the entire set of k tests equal to β by taking the alpha value for each test equal to α. The β is now called the experiment-wise type I error rate, and α is called the test-wise type I error rate or nominal type I error rate. The Bonferroni correction states that, in an experiment involving k tests, if you want to control the experiment-wise type I error rate at β, the nominal type I error rate for a single test should be

$$\alpha = \frac{\beta}{k} \tag{8.15}$$

For example, if an experiment involves 100 tests and the investigator wants to control the experiment-wise type I error at $\beta = 0.05$, for each of the individual tests, the nominal type I error rate should be $\alpha = \frac{\beta}{k} = \frac{0.05}{100} = 0.0005$. In other words, for any individual test, the p-value should be less than 0.0005 in order to declare significance for that test. The Bonferroni correction does not require independence of the multiple tests.

When the multiple tests are independent, there is an alternative correction for the type I error, which is called the Šidák correction (Abdi 2007). This correction is often confused with the Bonferroni correction. If a test-wise type I error is α, the probability of nonsignificance is $1 - \alpha$ for this particular test. For k independent tests and none of them is significant, the probability is $(1 - \alpha)^k$. The experiment-wise type I error is defined as the probability that at least one of the k tests is significant. This probability is

$$\beta = 1 - (1 - \alpha)^k \tag{8.16}$$

To find the nominal α value given the experiment-wise value β, we use the reverse function

$$\alpha = 1 - (1 - \beta)^{1/k} \tag{8.17}$$

This correction is the Šidák correction. The two corrections are approximately the same when β is small because $(1 - \beta)^{1/k} \approx 1 - \frac{\beta}{k}$, and thus,

$$\alpha \approx \frac{\beta}{k} \tag{8.18}$$

Therefore, the Bonferroni correction is an approximation of the Šidák correction for multiple independent tests for small β.

8.6 Permutation Test

When the number of tests is large, the Bonferroni and Šidák corrections tend to be overconservative. In addition, if a test statistic does not follow any standard distribution under the null model, calculation of the p-value may be difficult for each individual test. In this case, we can adopt the permutation test to draw an empirical critical value. This method was developed by Churchill and Doerge (1994) for QTL mapping. The idea is simple, but implementation can be time consuming. When the sample size n is small, we can evaluate all $n!$ different permuted samples of the original phenotypic values while keeping the marker genotype data intact. In other words, we only reshuffle the phenotypes, not the marker genotypes. For each permuted sample, we apply any method of genome scanning to calculate the test statistical values for all markers. In each of the permuted sample, the association of the phenotype and genotypes of markers has been (purposely) destroyed so that the distribution of the test statistics will mimic the actual distribution under the null model, from which a desirable critical value can be drawn from the empirical null distribution.

Table 8.1 Phenotypic values of trait y and the genotypes of five markers from ten plants (the original dataset)

Plant	y	M_1	M_2	M_3	M_4	M_5
1	55.0	H	H	H	H	A
2	54.2	H	H	H	H	U
3	61.6	H	H	U	A	A
4	66.6	H	H	H	H	U
5	67.4	H	H	H	B	U
6	64.3	H	H	H	H	H
7	54.0	H	A	B	B	B
8	57.2	H	B	H	H	H
9	63.7	H	H	H	H	H
10	55.0	H	H	A	H	U

The number of permuted samples can be extremely large if the sample size is large. In this case, we can randomly reshuffle the data to purposely destroy the association between the phenotype and the marker genotype. By random reshuffling the phenotypes, individual j may take the phenotypic value of individual i for $i \neq j$, while the marker genotype of individual j remains unchanged. After reshuffling the phenotypes, we analyze the data and scan the entire genome. By chance, we may find some peaks in the test statistic profile. We know that these peaks are false because we have already destroyed the association between markers and phenotypes. We record the value of the test statistic at the highest peak of the profile and denote it by λ_1. We then reshuffle the data and scan the genome again. We may find some false peaks again. We then record the highest peak and write down the value, λ_2, and put it in the dataset. We repeat the reshuffling process many times to form a large sample of λ's, denoted by $\{\lambda_1, \ldots, \lambda_M\}$, where M is a large number, say 1,000. These λ values will form a distribution, called the null distribution. The 95 % or 99 % quantile of the null distribution is the empirical critical value for our test statistic. We then compare our test statistic for each marker (from the original data analysis) against this empirical critical value. If the test statistic of a marker is larger than this critical value, we can declare this marker as being significant. Note that permutation test is time consuming, but it is realistic with the advanced computing system currently available in most laboratories.

We now provide an example to show how to use the permutation test to draw the critical value of a test statistic. Table 8.1 gives a small sample of ten plants (the original dataset).

Ten randomly reshuffled samples are demonstrated in Table 8.2. We can see that the first observation of sample 1 (S_1) takes the phenotype of plant number 6 while the genotypes of the five markers remain unchanged. Another example is that the second observation of sample 2 (S_2) takes the phenotype of plant number 8 while the genotypes of the five markers are still the genotypes for plant number 2.

The phenotypic values corresponding to the ten reshuffled samples are given in Table 8.3. Each sample is subject to genome scanning, i.e., five F-test statistics are calculated, one for each marker. The maximum F-test statistic value for each reshuffled sample is given in the last row of Table 8.3. For example, the maximum

Table 8.2 Plant IDs of ten randomly reshuffled samples, denoted by $S_1, S_2, \ldots, S_{10}$, respectively	S_1	S_2	S_3	S_4	S_5	S_6	S_7	S_8	S_9	S_{10}
	6	5	9	10	5	6	2	10	1	10
	2	8	6	8	1	2	7	6	4	5
	4	1	5	7	8	4	1	9	3	9
	3	10	1	4	10	3	10	8	5	3
	8	3	2	1	7	8	4	5	8	1
	9	4	3	5	2	9	3	3	7	8
	10	7	10	2	4	10	8	2	9	2
	1	9	7	6	3	1	9	1	10	6
	7	6	8	3	6	7	6	7	6	7
	5	2	4	9	9	5	5	4	2	4

Table 8.3 The corresponding phenotypic values and the maximum F-test statistical values (last row) in the ten randomly reshuffled samples

	S_1	S_2	S_3	S_4	S_5	S_6	S_7	S_8	S_9	S_{10}
	64.3	67.4	63.7	55.0	67.4	64.3	54.2	55.0	55.0	55.0
	54.2	57.2	64.3	57.2	55.0	54.2	54.0	64.3	66.6	67.4
	66.6	55.0	67.4	54.0	57.2	66.6	55.0	63.7	61.6	63.7
	61.6	55.0	55.0	66.6	55.0	61.6	55.0	57.2	67.4	61.6
	57.2	61.6	54.2	55.0	54.0	57.2	66.6	67.4	57.2	55.0
	63.7	66.6	61.6	67.4	54.2	63.7	61.6	61.6	54.0	57.2
	55.0	54.0	55.0	54.2	66.6	55.0	57.2	54.2	63.7	54.2
	55.0	63.7	54.0	64.3	61.6	55.0	63.7	55.0	55.0	64.3
	54.0	64.3	57.2	61.6	64.3	54.0	64.3	54.0	64.3	54.0
	67.4	54.2	66.6	63.7	63.7	67.4	67.4	66.6	54.2	66.6
F-test	2.12	1.28	3.51	1.85	2.33	4.51	3.21	3.95	1.11	0.95

F-value in the first sample (S_1) is 2.12, the maximum F-value for S_2 is 1.28, and so on. We then sorted the ten F-values from the ten samples in descending order as shown in the following sequence:

$$\{4.51, 3.95, 3.51, 3.21, 2.33, 2.12, 1.85, 1.28, 1.11, 0.95\}$$

The ten F-values are assumed to be sampled from the null distribution. The empirical 90 % quantile is 3.95, which can be used as the critical value for the F-test statistic to compare under the type I error of 0.10. The number of reshuffled samples in the example is not sufficiently large to give 95 % quantile for the type I error of $\alpha = 0.05$. In practice, the number of randomly reshuffled samples depends on $\alpha = 0.05$ due to Monte Carlo error. Nettleton and Doerge (2000) recommended that the permutation sample size should be at least $\frac{5}{\alpha}$, where α is the experiment-wise type I error rate. In permutation analysis, there is no such a thing as nominal type I error. In practice, we often choose 1,000 as the permutation sample size.

Permutation test is not a method for genome scanning; rather, it is only a way to draw an empirical critical value of a test statistic for us to decide statistical significance of a marker. It applies to all test statistics, e.g., the F-test, the W-test, and the likelihood ratio test. The phrase "permutation test" can be confusing because it is not a method for significance test. "Permutation analysis" may be a better phrase to describe this empirical approach of critical value calculation.

8.7 Piepho's Approximate Critical Value

Permutation analysis is perhaps the best method for drawing the empirical critical value to control the genome-wise type I error rate. However, it can be time consuming. Piepho (2001) developed an approximate method, which does not require randomly reshuffling of the data. The method simply uses exiting test statistical values of all points across the genome. The test statistic must be the likelihood ratio test statistic. If the test statistic is the LOD score, a simple conversion to the likelihood ratio test statistic is required. The W-test statistic may also be used because it also follows a chi-square distribution under the null model. Let $\beta = 0.05$ be the genome-wise type I error and $C = \chi^2_{k,1-\alpha}$ be the $(1-\alpha) \times 100\%$ quantile of the chi-square distribution, and k (the degrees of freedom of the test statistic) is the number of genetic effects subject to statistical test, where $k = 1$ for a BC design and $k = 2$ for an F_2 design. The following relations provide a way to solve for $C = \chi^2_{k,1-\alpha}$, the critical value for the likelihood ratio test statistic to compare so that the genome-wise type I error is controlled at β.

$$\beta = m \Pr(\chi^2_k > C) + \frac{2^{-\frac{1}{2}k} C^{-\frac{1}{2}(1-k)} e^{-\frac{1}{2}C}}{\Gamma(\frac{k}{2})} \sum_{i=1}^{m} v_i \qquad (8.19)$$

where m is the number of chromosomes and $\Gamma(\frac{k}{2})$ is the gamma function. The v_i for the ith chromosome is defined as

$$v_i = \left| \sqrt{\lambda_1} - \sqrt{\lambda_2} \right| + \left| \sqrt{\lambda_2} - \sqrt{\lambda_3} \right| + \cdots + \left| \sqrt{\lambda_{m_i-1}} - \sqrt{\lambda_{m_i}} \right| \qquad (8.20)$$

where λ_l for $l = 1, \ldots, m_i$ is the likelihood ratio test statistic for marker l in chromosome i and m_i is the total number of markers in chromosome i. Once β is given, the above equation is simply a function of C. A numerical solution can be found using the bisection algorithm. Once C is found, the type I error for an individual marker can be obtained using the inverse function of the chi-square distribution function. The gamma function $\Gamma(\frac{k}{2})$ depends on k, the number of genetic effects. For the common designs of experiments, k only takes 1, 2, or 3. For example, $k = 1$ for BC, DH (double haploid), and RIL (recombinant inbred line) designs. If the additive effect is the only one to be tested in the F_2 design, k also equals 1. If both the additive and dominance effects are tested in the F_2 design, $k = 2$. In a four-way cross design, k equals 3. Therefore, we only need the value of $\Gamma(\frac{1}{2}) = \sqrt{\pi}$, $\Gamma(\frac{2}{2}) = \Gamma(1) = 1$, and $\Gamma(\frac{3}{2}) = \frac{\sqrt{\pi}}{2}$.

8.8 Theoretical Consideration

Genome scanning described in this chapter refers to marker analysis. Since a marker is not a QTL, the estimated marker effect only reflects a fraction of the QTL effect. Take a BC design as an example. Assume that a QTL with effect a is d cM away from a marker. If we use this marker to estimate the QTL effect, the marker effect will not be equal to a; rather, it will be $(1 - 2r)a$, where $r = \frac{1}{2}(1 - e^{-d/(2 \times 100)})$ is the recombination fraction between the marker and the QTL. The marker effect is only a fraction of the QTL effect. This fraction is $(1 - 2r)$, which is the correlation coefficient between the marker and the QTL genotype indicator variables. When $r = 0$, a situation where the marker overlaps with the QTL, the marker effect is identical to the QTL effect. On the other hand, if $r = 0.5$, a situation where the marker is not linked to the QTL, the marker effect equals zero, regardless how large the QTL effect is. When $0 < r < 0.5$, what we estimate for the marker is a confounded effect between the QTL effect and the linkage parameter. A small marker effect may be due to a large QTL effect but weakly linked to the marker or a small QTL effect with a strong linkage. There is no way to tell the actual QTL effect unless more markers are taken into account simultaneously, which is the topic to be addressed in the next chapter when interval mapping is introduced.

We now prove that the correlation between the marker and the QTL is $1 - 2r$. Let X be the indicator variable for the QTL genotype, i.e., $X = 1$ for $A_1 A_1$ and $X = 0$ for $A_1 A_2$. Let M be the corresponding indicator variable for the marker genotype, i.e., $M = 1$ and $M = 0$, respectively, for the two genotypes of the marker. The joint distribution of X and M is given in Table 8.4. This joint distribution table is symmetrical, meaning that both M and X have the same marginal distribution. First, let us look at the marginal distribution of variable M. From the joint distribution table, we get $\Pr(M = 1) = \frac{1-r}{2} + \frac{r}{2} = \frac{1}{2}$ and $\Pr(M = 0) = \frac{r}{2} + \frac{1-r}{2} = \frac{1}{2}$. Therefore, the variance of M is

$$\text{var}(M) = E(M^2) - E^2(M)$$

$$= \left(\frac{1}{2} \times 1^2 + \frac{1}{2} \times 0^2 \right) - \left(\frac{1}{2} \times 1 + \frac{1}{2} \times 0 \right)^2$$

$$= \frac{1}{4} \tag{8.21}$$

Table 8.4 Joint distribution of X (QTL genotype) and M (marker genotype)

		M		
		1	0	
X	1	$(1-r)/2$	$r/2$	$1/2$
	0	$r/2$	$(1-r)/2$	$1/2$
		$1/2$	$1/2$	

Similarly, $\mathrm{var}(X) = \frac{1}{4}$, due to the symmetrical nature. We now evaluate the covariance between M and X.

$$
\begin{aligned}
\mathrm{cov}(M, X) &= E(MX) - E(M)E(X) \\
&= \frac{1-r}{2} \times 1 \times 1 - \frac{1}{2} \times \frac{1}{2} = \frac{1}{4}[2(1-r) - 1] \\
&= \frac{1}{4}(1 - 2r)
\end{aligned}
\tag{8.22}
$$

The correlation coefficient between the two variables is

$$
\rho_{MX} = \frac{\mathrm{cov}(M, X)}{\sqrt{\mathrm{var}(M)\mathrm{var}(X)}} = \frac{\frac{1}{4}(1 - 2r)}{\frac{1}{4}} = 1 - 2r
\tag{8.23}
$$

Because of the symmetry of X and M, this correlation is also equal to the regression coefficient, i.e.,

$$
\beta_{XM} = \frac{\mathrm{cov}(M, X)}{\mathrm{var}(M)} = \frac{\frac{1}{4}(1 - 2r)}{\frac{1}{4}} = 1 - 2r
\tag{8.24}
$$

Recall that when a marker is used to estimate the effect of a linked QTL, the QTL effect will be biased by a factor $(1 - 2r)$. This fraction is the correlation between X and M. In fact, it is the regression coefficient of X on M. Because $\rho_{MX} = \beta_{XM}$, we say $(1 - 2r)$ is the correlation coefficient. We now show why the factor of reduction is β_{XM}. Recall that the QTL model is

$$
y_j = b_0 + E(X_j|M_j)b_1 + e_j
\tag{8.25}
$$

where $E(X_j|M)$ is the conditional mean of X_j given M_j. We use marker genotype M_j to infer the QTL genotype X_j. The conditional mean can be expressed as the predicted value of X from M using the following regression equation:

$$
E(X_j|M_j) = E(X_j) + M_j\beta_{XM} = \frac{1}{2} + M_j(1 - 2r)
\tag{8.26}
$$

Substituting this (8.26) into the above model (8.25), we get

$$
\begin{aligned}
y_j &= \left(b_0 + \frac{1}{2}\right) + M_j(\beta_{XM}b_1) + e_j \\
&= b_0^* + M_jb_1^* + e_j
\end{aligned}
\tag{8.27}
$$

where $b_0^* = b_0 + \frac{1}{2}$ and $b_1^* = \beta_{XM}b_1 = (1 - 2r)a$. Note that $b_1 = a$ is the genetic effect and $\beta_{XM} = 1 - 2r$ as given in (8.24).

Chapter 9
Interval Mapping

Interval mapping is an extension of the individual marker analysis so that two markers are analyzed at a time. In the marker analysis (Chap. 8), we cannot estimate the exact position of a QTL. With interval mapping, we use two markers to determine an interval, within which a putative QTL position is proposed. The genotype of the putative QTL is not observable but can be inferred with a certain probability using the three-point or multipoint method introduced in Chap. 4. Once the genotype of the QTL is inferred, we can estimate and test the QTL effect at that particular position. We divide the interval into many putative positions of QTL with one or two cM apart and investigate every putative position within the interval. Once we have searched the current interval, we move on to the next interval and so on until all intervals have been searched. The putative QTL position (not necessarily at a marker) that has the maximum test statistical value is the estimated QTL position. Figure 9.1 demonstrates the process of genome scanning for markers only (panel a), for markers and virtual markers (panel b), and for every point of the chromosome (panel c).

Interval mapping was originally developed by Lander and Botstein (1989) and further modified by numerous authors. Interval mapping has revolutionized genetic mapping because we can really pinpoint the exact location of a QTL. In each of the four sections that follow, we will introduce one specific statistical method of interval mapping based on the F_2 design. Methods of interval mapping for a BC design are straightforward and thus will not be discussed in this chapter. Maximum likelihood (ML) method of interval mapping (Lander and Botstein 1989) is the optimal method for interval mapping. Least-squares (LS) method (Haley and Knott 1992) is a simplified approximation of Lander and Botstein method. The iteratively reweighted least-squares (IRLS) method (Xu 1998a,b) is a further improved method over the least-squares method. Recently Feenstra et al. (2006) developed an estimating equation (EE) method for QTL mapping, which is an extension of the IRLS with improved performance. Han and Xu (2008) developed a Fisher scoring algorithm (FISHER) for QTL mapping. Both the EE and FISHER algorithms maximize the same likelihood function, and thus, they generate identical result. In this chapter, we introduce the methods based on their simplicity rather than their chronological

S. Xu, *Principles of Statistical Genomics*, DOI 10.1007/978-0-387-70807-2_9,
© Springer Science+Business Media, LLC 2013

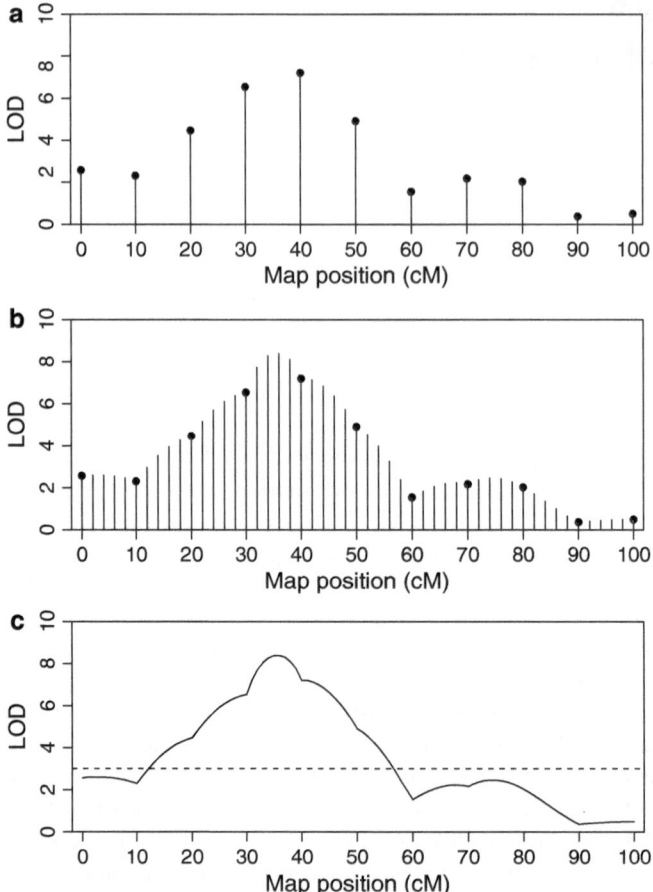

Fig. 9.1 The LOD test statistics for (**a**) marker effects (*top panel*), (**b**) virtual marker effects (panel in the *middle*), and (**c**) every point of a simulated chromosome (*bottom panel*)

orders of development. Therefore, the methods will be introduced in the following order: LS, IRLS, FISHER, and ML. Bayesian method will be discussed in a later chapter where multiple QTL mapping is addressed.

9.1   Least-Squares Method

The LS method was introduced by Haley and Knott (1992) aiming to improving the computational speed. The statistical model for the phenotypic value of the jth individual is

$$y_j = X_j \beta + Z_j \gamma + \varepsilon_j \tag{9.1}$$

where β is a $p \times 1$ vector for some model effects that are irrelevant to QTL effects, X_j is a $1 \times p$ known design vector, $\gamma = \{a, d\}$ is a 2×1 vector for QTL effects of a putative locus (a for additive effect and d for dominance effect), and Z_j is a 1×2 vector for the genotype indicator variable defined as

$$Z_j = \begin{cases} H_1 & \text{for } A_1 A_1 \\ H_2 & \text{for } A_1 A_2 \\ H_3 & \text{for } A_2 A_2 \end{cases} \tag{9.2}$$

where H_k for $k = 1, 2, 3$ is the kth row of matrix

$$H = \begin{bmatrix} +1 & 0 \\ 0 & 1 \\ -1 & 0 \end{bmatrix} \tag{9.3}$$

The residual error ε_j is assumed to be a $N(0, \sigma^2)$ variable. Although normal distribution for ε_j is not a required assumption for the LS method, it is required for the ML method. It is important to include non-QTL effects β in the model to control the residual error variance as small as possible. For example, location and year effects are common in replicated experiments. These effects are not related to QTL but will contribute to the residual error if not included in the model. If there is no such a non-QTL effect to consider in a nice designed experiment, β will be a single parameter (intercept) and X_j will be unity across all $j = 1, \ldots, n$.

With interval mapping, the QTL genotype is never known unless the putative QTL position overlaps with a fully informative marker. Therefore, Haley and Knott (1992) suggested to replace the unknown Z_j by the expectation of Z_j conditional on flanking marker genotype. Let $p_j(1)$, $p_j(0)$, and $p_j(-1)$ be the conditional probabilities for the three genotypes given flanking marker information (see Chap. 4 for the method of calculating conditional probability). The LS model of Haley and Knott (1992) is

$$y_j = X_j \beta + U_j \gamma + e_j \tag{9.4}$$

where

$$U_j = E(Z_j) = p_j(+1)H_1 + p_j(0)H_2 + p_j(-1)H_3 \tag{9.5}$$

is the conditional expectation of Z_j. The residual error e_j (different from ε_j) remains normal with mean zero and variance σ^2, although this assumption has been violated (see next section). The least-squares estimate of β and γ is

$$\begin{bmatrix} \hat{\beta} \\ \hat{\gamma} \end{bmatrix} = \begin{bmatrix} \sum_{j=1}^{n} X_j^T X_j & \sum_{j=1}^{n} X_j^T U_j \\ \sum_{j=1}^{n} U_j^T X_j & \sum_{j=1}^{n} U_j^T U_j \end{bmatrix}^{-1} \begin{bmatrix} \sum_{j=1}^{n} X_j^T y_j \\ \sum_{j=1}^{n} U_j^T y_j \end{bmatrix} \tag{9.6}$$

and the estimated residual error variance is

$$\hat{\sigma}^2 = \frac{1}{n-p-2} \sum_{j=1}^{n} (y_j - X_j\hat{\beta} - U_j\hat{\gamma})^2 \qquad (9.7)$$

The variance–covariance matrix of the estimated parameters is

$$\text{var}\begin{bmatrix} \hat{\beta} \\ \hat{\gamma} \end{bmatrix} = \begin{bmatrix} \sum_{j=1}^{n} X_j^T X_j & \sum_{j=1}^{n} X_j^T U_j \\ \sum_{j=1}^{n} U_j^T X_j & \sum_{j=1}^{n} U_j^T U_j \end{bmatrix}^{-1} \hat{\sigma}^2 \qquad (9.8)$$

which is a $(p+2) \times (p+2)$ matrix. Let

$$\text{var}(\hat{\gamma}) = V = \begin{bmatrix} \text{var}(\hat{a}) & \text{cov}(\hat{a},\hat{d}) \\ \text{cov}(\hat{a},\hat{d}) & \text{var}(\hat{d}) \end{bmatrix} \qquad (9.9)$$

be the 2×2 lower diagonal bock of matrix (9.8). The standard errors of the estimated additive and dominance effects are the square roots of the diagonal elements of matrix (9.9).

We can use either the F-test or the W-test statistic to test the hypothesis of $H_0 : \gamma = 0$. The W-test statistic is

$$W = \hat{\gamma}^T V^{-1} \hat{\gamma} = \begin{bmatrix} \hat{a} & \hat{d} \end{bmatrix} \begin{bmatrix} \text{var}(\hat{a}) & \text{cov}(\hat{a},\hat{d}) \\ \text{cov}(\hat{a},\hat{d}) & \text{var}(\hat{d}) \end{bmatrix}^{-1} \begin{bmatrix} \hat{a} \\ \hat{d} \end{bmatrix} \qquad (9.10)$$

The likelihood ratio test statistic can also be applied if we assume that $e_j \sim N(0,\sigma^2)$ for all $j = 1,\ldots,n$. The log likelihood function for the full model is

$$L_1 = -\frac{n}{2}\ln(\hat{\sigma}^2) - \frac{1}{2\hat{\sigma}^2} \sum_{j=1}^{n} (y - X_j\hat{\beta} - U_j\hat{\gamma})^2$$

$$\approx -\frac{n}{2}\left[\ln(\hat{\sigma}^2) + 1\right] \qquad (9.11)$$

The reduced model under $H_0 : \gamma = 0$ is

$$L_0 = -\frac{n}{2}\ln(\hat{\sigma}^2) - \frac{1}{2\hat{\sigma}^2} \sum_{j=1}^{n} (y - X_j\hat{\beta})^2$$

$$\approx -\frac{n}{2}\left[\ln(\hat{\sigma}^2) + 1\right] \qquad (9.12)$$

where

$$\hat{\hat{\beta}} = \left[\sum_{j=1}^{n} X_j^T X_j \right]^{-1} \left[\sum_{j=1}^{n} X_j^T y_j \right] \tag{9.13}$$

and

$$\hat{\sigma}^2 = \frac{1}{n-p} \sum_{j=1}^{n} (y_j - X_j \hat{\hat{\beta}})^2 \tag{9.14}$$

The likelihood ratio test statistic is

$$\lambda = -2(L_0 - L_1) \tag{9.15}$$

9.2 Weighted Least Squares

Xu (1995) realized that the LS method is flawed because the residual variance is heterogeneous after replacing X_j by its conditional expectation U_j. The conditional variance of X_j given marker information varies from one individual to another, and it will contribute to the residual variance. Xu (1998a,b) modified the exact model

$$y_j = X_j \beta + Z_j \gamma + \varepsilon_j \tag{9.16}$$

by

$$y_j = X_j \beta + U_j \gamma + (Z_j - U_j)\gamma + \varepsilon_j \tag{9.17}$$

which differs from the Haley and Knott's (1992) model by $(Z_j - U_j)\gamma$. Since Z_j is not observable, this additional term is merged into the residual error if ignored. Let

$$e_j = (Z_j - U_j)\gamma + \varepsilon_j \tag{9.18}$$

be the new residual error. The Haley and Knott's (1992) model can be rewritten as

$$y_j = X_j \beta + U_j \gamma + e_j \tag{9.19}$$

Although we assume $\varepsilon_j \sim N(0, \sigma^2)$, this does not validate the normal assumption of e_j. The expectation for e_j is

$$E(e_j) = [E(Z_j) - U_j]\gamma + E(\varepsilon_j) = 0 \tag{9.20}$$

The variance of e_j is

$$\mathrm{var}(e_j) = \sigma_j^2 = \gamma^T \mathrm{var}(Z_j)\gamma + \sigma^2 = \left(\frac{1}{\sigma^2} \gamma^T \Sigma_j \gamma + 1 \right) \sigma^2 \tag{9.21}$$

where $\Sigma_j = \text{var}(Z_j)$, which is defined as a conditional variance–covariance matrix given flanking marker information. The explicit forms of Σ_j are

$$\Sigma_j = E(Z_j^T Z_j) - E(Z_j^T)E(Z_j), \tag{9.22}$$

where

$$E(Z_j^T Z_j) = p_j(1)H_1^T H_1 + p_j(0)H_2^T H_2 + p_j(-1)H_3^T H_3 \tag{9.23}$$

and

$$E(Z_j) = U_j = p_j(1)H_1 + p_j(0)H_2 + p_j(-1)H_3. \tag{9.24}$$

Let

$$\sigma_j^2 = \left(\frac{1}{\sigma^2}\gamma^T \Sigma_j \gamma + 1\right)\sigma^2 = \frac{1}{W_j}\sigma^2 \tag{9.25}$$

where

$$W_j = \left(\frac{1}{\sigma^2}\gamma^T \Sigma_j \gamma + 1\right)^{-1} \tag{9.26}$$

is the weight variable for the jth individual. The weighted least-squares estimate of the parameters is

$$\begin{bmatrix} \hat{\beta} \\ \hat{\gamma} \end{bmatrix} = \begin{bmatrix} \sum\limits_{j=1}^{n} X_j^T W_j X_j & \sum\limits_{j=1}^{n} X_j^T W_j U_j \\ \sum\limits_{j=1}^{n} U_j^T W_j X_j & \sum\limits_{j=1}^{n} U_j^T W_j U_j \end{bmatrix}^{-1} \begin{bmatrix} \sum\limits_{j=1}^{n} X_j^T W_j y_j \\ \sum\limits_{j=1}^{n} U_j^T W_j y_j \end{bmatrix} \tag{9.27}$$

and

$$\hat{\sigma}^2 = \frac{1}{n-p-2}\sum_{j=1}^{n} W_j(y_j - X_j\hat{\beta} - U_j\hat{\gamma})^2 \tag{9.28}$$

Since W_j is a function of σ^2, iterations are required. The iteration process is demonstrated as below:

1. Initialize γ and σ^2.
2. Update β and γ using 9.27.
3. Update σ^2 using 9.28.
4. Repeat Step 2 to Step 3 until a certain criterion of convergence is satisfied.

The iteration process is very fast, usually taking less than 5 iterations to converge. Since the weight is not a constant (it is a function of the parameters), repeatedly updating the weight is required. Therefore, the weighted least-squares method is also called iteratively reweighted least squares (IRLS). The few cycles of iterations make the results of IRLS very close to that of the maximum likelihood method (to be introduced later). A nice property of the IRLS is that the variance–covariance matrix of the estimated parameters is automatically given as a by-product of the iteration process. This matrix is

$$\text{var}\begin{bmatrix} \hat{\beta} \\ \hat{\gamma} \end{bmatrix} = \begin{bmatrix} \sum_{j=1}^{n} X_j^T W_j X_j & \sum_{j=1}^{n} X_j^T W_j U_j \\ \sum_{j=1}^{n} U_j^T W_j X_j & \sum_{j=1}^{n} U_j^T W_j U_j \end{bmatrix}^{-1} \hat{\sigma}^2 \qquad (9.29)$$

As a result, the F- or W-test statistic can be used for significance test. Like the least-squares method, a likelihood ratio test statistic can also be established for significance test. The L_0 under the null model is the same as that described in the section of least-squares method. The L_1 under the alternative model is

$$L_1 = -\frac{n}{2}\ln(\hat{\sigma}^2) + \frac{1}{2}\sum_{j=1}^{n}\ln(W_j) - \frac{1}{2\hat{\sigma}^2}\sum_{j=1}^{n} W_j(y - X_j\hat{\beta} - U_j\hat{\gamma})^2$$

$$\approx -\frac{n}{2}\left[\ln(\hat{\sigma}^2) + 1\right] + \frac{1}{2}\sum_{j=1}^{n}\ln(W_j) \qquad (9.30)$$

9.3 Fisher Scoring

The weighted least-squares solution described in the previous section does not maximize the log likelihood function (9.30). We can prove that it actually maximizes (9.30) if W_j is treated as a constant. The fact that W_j is a function of parameters makes the above weighted least-squares estimates suboptimal. The optimal solution should be obtained by maximizing (9.30) fully without assuming W_j being a constant.

Recall that the linear model for y_j is

$$y_j = X_j\beta + U_j\gamma + e_j \qquad (9.31)$$

where the residual error $e_j = (Z_j - U_j)\gamma + \varepsilon_j$ has a zero mean and variance

$$\sigma_j^2 = \left(\frac{1}{\sigma^2}\gamma^T\Sigma_j\gamma + 1\right)\sigma^2 = \frac{1}{W_j}\sigma^2 \qquad (9.32)$$

If we assume that $e_j \sim N(0, \sigma_j^2)$, we can construct the following log likelihood function:

$$L(\theta) = -\frac{n}{2}\ln(\sigma^2) + \frac{1}{2}\sum_{j=1}^{n}\ln(W_j) - \frac{1}{2\sigma^2}\sum_{j=1}^{n} W_j(y - X_j\beta - U_j\gamma)^2 \qquad (9.33)$$

where $\theta = \{\beta, \gamma, \sigma^2\}$ is the vector of parameters. The maximum likelihood solution for the above likelihood function is hard to obtain because W_j is not a constant but a function of the parameters. The Newton–Raphson algorithm may be adopted, but it requires the second partial derivative of the log likelihood function with respect to the parameter, which is very complicated. In addition, the Newton–Raphson algorithm often misbehaves when the dimensionality of θ is high. We now introduce the Fisher scoring algorithm for finding the MLE of θ. The method requires the first partial derivative of $L(\theta)$ with respect to the parameters, called the score vector and denoted by $S(\theta)$, and the information matrix, denoted by $I(\theta)$. The score vector has the following form:

$$
S(\theta) = \begin{bmatrix} \frac{1}{\sigma^2} \sum\limits_{j=1}^{n} X_j^T W_j (y_j - \mu_j) \\[2mm] \frac{1}{\sigma^2} \sum\limits_{j=1}^{n} U_j^T W_j (y_j - \mu_j) - \frac{1}{\sigma^2} \sum\limits_{j=1}^{n} W_j \Sigma_j \gamma + \frac{1}{\sigma^4} \sum\limits_{j=1}^{n} (y_j - \mu_j)^2 W_j^2 \Sigma_j \gamma \\[2mm] \frac{1}{2\sigma^4} \sum\limits_{j=1}^{n} W_j^2 (y_j - \mu_j)^2 - \frac{1}{2\sigma^2} \sum\limits_{j=1}^{n} W_j \end{bmatrix} \quad (9.34)
$$

where

$$
\mu_j = X_j \beta + U_j \gamma \tag{9.35}
$$

The information matrix is given below

$$
I(\theta) = \begin{bmatrix} \frac{1}{\sigma^2} \sum\limits_{j=1}^{n} X_j^T W_j X_j & \frac{1}{\sigma^2} \sum\limits_{j=1}^{n} X_j^T W_j U_j & 0 \\[2mm] \frac{1}{\sigma^2} \sum\limits_{j=1}^{n} U_j W_j X_j, & \frac{1}{\sigma^2} \sum\limits_{j=1}^{n} U_j^T W_j U_j + \frac{2}{\sigma^4} \sum\limits_{j=1}^{n} W_j^2 \Sigma_j \gamma \gamma^T \Sigma_j, & \frac{1}{\sigma^4} \sum\limits_{j=1}^{n} W_j^2 \Sigma_j \gamma \\[2mm] 0 & \frac{1}{\sigma^4} \sum\limits_{j=1}^{n} W_j^2 \gamma^T \Sigma_j & \frac{1}{2\sigma^4} \sum\limits_{j=1}^{n} W_j^2 \end{bmatrix}
$$

$$(9.36)$$

The Fisher scoring algorithm is implemented using the following iteration equation:

$$
\theta^{(t+1)} = \theta^{(t)} + I^{-1}(\theta^{(t)}) S(\theta^{(t)}) \tag{9.37}
$$

where $\theta^{(t)}$ is the parameter value at iteration t and $\theta^{(t+1)}$ is the updated value. Once the iteration process converges, the variance–covariance matrix of the estimated parameters is automatically given, which is

$$
\text{var}(\hat{\theta}) = I^{-1}(\hat{\theta}) \tag{9.38}
$$

The detailed expression of this matrix is

$$
\operatorname{var}\begin{bmatrix} \hat{\beta} \\ \hat{\gamma} \\ \hat{\sigma}^2 \end{bmatrix} = \begin{bmatrix} \sum_{j=1}^{n} X_j^T W_j X_j & \sum_{j=1}^{n} X_j^T W_j U_j & 0 \\ \sum_{j=1}^{n} U_j W_j X_j, & \sum_{j=1}^{n} U_j^T W_j U_j + \frac{2}{\hat{\sigma}^2} \sum_{j=1}^{n} W_j^2 \Sigma_j \hat{\gamma} \hat{\gamma}^T \Sigma_j, & \frac{1}{\hat{\sigma}^2} \sum_{j=1}^{n} W_j^2 \Sigma_j \hat{\gamma} \\ 0 & \frac{1}{\hat{\sigma}^2} \sum_{j=1}^{n} W_j^2 \hat{\gamma}^T \Sigma_j & \frac{1}{2\hat{\sigma}^2} \sum_{j=1}^{n} W_j^2 \end{bmatrix}^{-1} \hat{\sigma}^2
$$

$$(9.39)$$

which can be compared with the variance–covariance matrix of the iteratively reweighted least-squares estimate given in the previous section (9.29).

We now give the derivation of the score vector and the information matrix. We can write the log likelihood function as

$$
L(\theta) = \sum_{j=1}^{n} L_j(\theta) \tag{9.40}
$$

where

$$
L_j(\theta) = -\frac{1}{2}\ln(\sigma^2) + \frac{1}{2}\ln W_j - \frac{1}{2\sigma^2}W_j(y_j - \mu_j)^2 \tag{9.41}
$$

and

$$
\mu_j = X_j\beta + U_j\gamma \tag{9.42}
$$

The score vector is a vector of the first partial derivatives, as shown below:

$$
S(\theta) = \sum_{j=1}^{n} S_j(\theta) \tag{9.43}
$$

where

$$
S_j(\theta) = \begin{bmatrix} \frac{\partial}{\partial\beta} L_j(\theta) \\ \frac{\partial}{\partial\gamma} L_j(\theta) \\ \frac{\partial}{\partial\sigma^2} L_j(\theta) \end{bmatrix} = \begin{bmatrix} \frac{1}{\sigma^2} X_j^T W_j(y_j - \mu_j) \\ \frac{1}{\sigma^2} U_j^T W_j(y_j - \mu_j) - \frac{1}{\sigma^2} W_j \Sigma_j \gamma + \frac{1}{\sigma^4}(y_j - \mu_j)^2 W_j^2 \Sigma_j \gamma \\ \frac{1}{2\sigma^4} W_j^2(y_j - \mu_j)^2 - \frac{1}{2\sigma^2} W_j \end{bmatrix}
$$

$$(9.44)$$

Therefore, we only need to take the sum of the first partial derivatives across individuals to get the score vector. Note that when deriving $S_j(\theta)$, we need the following derivatives:

$$\frac{\partial W_j}{\partial \theta} = \begin{bmatrix} \frac{\partial W_j}{\partial \beta} \\ \frac{\partial W_j}{\partial \gamma} \\ \frac{\partial W_j}{\partial \sigma^2} \end{bmatrix} = \begin{bmatrix} 0 \\ -\frac{2}{\sigma^2} W_j^2 \Sigma_j \gamma \\ \frac{1}{\sigma^2} W_j (1 - W_j) \end{bmatrix} \tag{9.45}$$

and

$$\frac{\partial \mu_j{'}}{\partial \theta} = \begin{bmatrix} \frac{\partial \mu_j}{\partial \beta} \\ \frac{\partial \mu_j}{\partial \gamma} \\ \frac{\partial \mu_j}{\partial \sigma^2} \end{bmatrix} = \begin{bmatrix} X_j^T \\ U_j^T \\ 0 \end{bmatrix} \tag{9.46}$$

The information matrix is

$$I(\theta) = \sum_{j=1}^{n} I_j(\theta) = \sum_{j=1}^{n} -E[H_j(\theta)] \tag{9.47}$$

where

$$H_j(\theta) = \frac{\partial^2 L_j(\theta)}{\partial \theta \partial \theta^T} = \begin{bmatrix} \frac{\partial^2 L_j(\theta)}{\partial \beta \partial \beta^T} & \frac{\partial^2 L_j(\theta)}{\partial \beta \partial \gamma^T} & \frac{\partial^2 L_j(\theta)}{\partial \beta \partial \sigma^2} \\ \frac{\partial^2 L_j(\theta)}{\partial \gamma \partial \beta^T} & \frac{\partial^2 L_j(\theta)}{\partial \gamma \partial \gamma^T} & \frac{\partial^2 L_j(\theta)}{\partial \gamma \partial \sigma^2} \\ \frac{\partial^2 L_j(\theta)}{\partial \sigma^2 \partial \beta^T} & \frac{\partial^2 L_j(\theta)}{\partial \sigma^2 \partial \gamma^T} & \frac{\partial^2 L_j(\theta)}{\partial \sigma^2 \partial \sigma^2} \end{bmatrix} \tag{9.48}$$

is the second partial derivative of $L_j(\theta)$ with respect to the parameters and called the Hessian matrix. Derivation of this matrix is very tedious, but the negative expectation of the Hessian matrix is identical to the expectation of the product of the score vector (Wedderburn 1974),

$$-E[H_j(\theta)] = E[S_j(\theta) S_j^T(\theta)] \tag{9.49}$$

Using this identity, we can avoid the Hessian matrix. Therefore, the information matrix is

$$I(\theta) = \sum_{j=1}^{n} I_j(\theta) = \sum_{j=1}^{n} E[S_j(\theta) S_j^T(\theta)] \tag{9.50}$$

where

$$E[S_j(\theta) S_j^T(\theta)] = \begin{bmatrix} E\left(\frac{\partial L_j(\theta)}{\partial \beta} \frac{\partial L_j(\theta)}{\partial \beta^T}\right) & E\left(\frac{\partial L_j(\theta)}{\partial \beta} \frac{\partial L_j(\theta)}{\partial \gamma^T}\right) & E\left(\frac{\partial L_j(\theta)}{\partial \beta} \frac{\partial L_j(\theta)}{\partial \sigma^2}\right) \\ E\left(\frac{\partial L_j(\theta)}{\partial \gamma} \frac{\partial L_j(\theta)}{\partial \beta^T}\right) & E\left(\frac{\partial L_j(\theta)}{\partial \gamma} \frac{\partial L_j(\theta)}{\partial \gamma^T}\right) & E\left(\frac{\partial L_j(\theta)}{\partial \gamma} \frac{\partial L_j(\theta)}{\partial \sigma^2}\right) \\ E\left(\frac{\partial L_j(\theta)}{\partial \sigma^2} \frac{\partial L_j(\theta)}{\partial \beta^T}\right) & E\left(\frac{\partial L_j(\theta)}{\partial \sigma^2} \frac{\partial L_j(\theta)}{\partial \gamma^T}\right) & E\left(\frac{\partial L_j(\theta)}{\partial \sigma^2} \frac{\partial L_j(\theta)}{\partial \sigma^2}\right) \end{bmatrix}$$

$$\tag{9.51}$$

Note that the expectation is taken with respect to the phenotypic value y_j. In other words, after taking the expectation, variable y_j will disappear from the expressions. There are six different blocks in the above matrix. We will only provide the derivation for one block as an example. The derivations of the remaining five blocks are left to students for practice. The result can be found in Han and Xu (2008). We now show the derivation of the first block of the matrix. The product (before taking the expectation) is

$$
\frac{\partial L_j(\theta)}{\partial \beta} \frac{\partial L_j(\theta)}{\partial \beta^T} = \left[\frac{1}{\sigma^2} X_j^T W_j(y_j - \mu_j) \right] \left[\frac{1}{\sigma^2} X_j^T W_j(y_j - \mu_j) \right]^T
$$

$$
= \frac{1}{\sigma^4} X_j^T W_j^2 X_j^T (y_j - \mu_j)^2 \tag{9.52}
$$

The expectation of it is

$$
E\left(\frac{\partial L_j(\theta)}{\partial \beta} \frac{\partial L_j(\theta)}{\partial \beta^T} \right) = \frac{1}{\sigma^4} X_j^T W_j^2 X_j^T E\left[(y_j - \mu_j)^2 \right]
$$

$$
= \frac{1}{\sigma^2} X_j^T W_j X_j^T \tag{9.53}
$$

The second line of the above equation requires the following identity:

$$
E\left[(y_j - \mu_j)^2 \right] = \frac{1}{W_j} \sigma^2 \tag{9.54}
$$

Taking the sum of (9.53) across individuals, we get

$$
I_{11}(\theta) = \frac{1}{\sigma^2} \sum_{j=1}^{n} X_j^T W_j X_j \tag{9.55}
$$

which is the first block of the information matrix. When deriving the expectations for the remaining five blocks, we need the following expectations:

$$
E[(y_j - \mu_j)^k] = \begin{cases} 0 & \text{for odd } k \\ W_j^{-1} \sigma^2 & \text{for } k = 2 \\ 3W_j^{-2} \sigma^4 & \text{for } k = 4 \end{cases} \tag{9.56}
$$

The above expectations requires the assumption of $y_j \sim N(\mu_j, \sigma_j^2)$ where $\sigma_j^2 = W_j^{-1} \sigma^2$.

9.4   Maximum Likelihood Method

The maximum likelihood method (Lander and Botstein 1989) is the optimal one compared to all other methods described in this chapter. The linear model for the phenotypic value of y_j is

$$y_j = X_j \beta + Z_j \gamma + \varepsilon_j \tag{9.57}$$

where $\varepsilon_j \sim N(0, \sigma^2)$ is assumed. The genotype indicator variable Z_j is a missing value because we cannot observe the genotype of a putative QTL. Rather than replacing Z_j by U_j as done in the least-squares and the weighted least-squares methods, the maximum likelihood method takes into consideration the mixture distribution of y_j. We have learned the mixture distribution in Chap. 7 when we deal with segregation analysis of quantitative traits. We now extend the mixture model to interval mapping. When the genotype of the putative QTL is observed, the probability density of y_j is

$$f_k(y_j) = \Pr(y_j | Z_j = H_k)$$

$$= \frac{1}{\sqrt{2\pi}\sigma} \exp\left[-\frac{1}{2\sigma^2}(y_j - X_j\beta + H_k\gamma)^2\right] \tag{9.58}$$

When flanking marker information is used, the conditional probability that $Z_j = H_k$ is

$$p_j(k) = \Pr(Z_j = H_k), \forall k = 1, 2, 3 \tag{9.59}$$

for the three genotypes, $A_1 A_1$, $A_1 A_2$, and $A_2 A_2$. These probabilities are different from the Mendelian segregation ratio $(\frac{1}{4}, \frac{1}{2}, \frac{1}{4})$ as described in the segregation analysis. They are the conditional probabilities given marker information and thus vary from one individual to another because different individuals may have different marker genotypes. Using the conditional probabilities as weights, we get the mixture distribution

$$f(y_j) = \sum_{k=1}^{3} p_j(2-k) f_k(y_j) \tag{9.60}$$

where

$$p_j(2-k) = \begin{cases} p_j(-1) & \text{for } k = 1 \\ p_j(0) & \text{for } k = 2 \\ p_j(+1) & \text{for } k = 3 \end{cases} \tag{9.61}$$

is a special notation for the conditional probability and should not be interpreted as p_j times $(2-k)$. The log likelihood function is

$$L(\theta) = \sum_{j=1}^{n} L_j(\theta) \tag{9.62}$$

where $L_j(\theta) = \ln f(y_j)$.

9.4.1 EM Algorithm

The MLE of θ can be obtained using any numerical algorithms but the EM algorithm is generally more preferable than others because we can take advantage of the mixture distribution. Derivation of the EM algorithm has been given in Chap. 7 when segregation analysis was introduced. Here we simply give the result of the EM algorithm. Assuming that the genotypes of all individuals are observed, the maximum likelihood estimates of parameters would be

$$\begin{bmatrix} \beta \\ \gamma \end{bmatrix} = \begin{bmatrix} \sum_{j=1}^{n} X_j^T X_j & \sum_{j=1}^{n} X_j^T Z_j \\ \sum_{j=1}^{n} Z_j^T X_j & \sum_{j=1}^{n} Z_j^T Z_j \end{bmatrix}^{-1} \begin{bmatrix} \sum_{j=1}^{n} X_j^T y_j \\ \sum_{j=1}^{n} Z_j^T y_j \end{bmatrix} \tag{9.63}$$

and

$$\sigma^2 = \frac{1}{n} \sum_{j=1}^{n} (y_j - X_j \beta - Z_j \gamma)^2 \tag{9.64}$$

The EM algorithm takes advantage of the above explicit solutions of the parameters by substituting all entities containing the missing value Z_j by their posterior expectations, i.e.,

$$\begin{bmatrix} \beta \\ \gamma \end{bmatrix} = \begin{bmatrix} \sum_{j=1}^{n} X_j^T X_j & \sum_{j=1}^{n} X_j^T E(Z_j) \\ \sum_{j=1}^{n} E(Z_j^T) X_j & \sum_{j=1}^{n} E(Z_j^T Z_j) \end{bmatrix}^{-1} \begin{bmatrix} \sum_{j=1}^{n} X_j^T y_j \\ \sum_{j=1}^{n} E(Z_j^T) y_j \end{bmatrix} \tag{9.65}$$

and

$$\sigma^2 = \frac{1}{n} \sum_{j=1}^{n} E\left[(y_j - X_j \beta - Z_j \gamma)^2\right] \tag{9.66}$$

where the expectations are taken using the posterior probabilities of QTL genotypes, which is defined as

$$p_j^*(2-k) = \frac{p_j(2-k) f_k(y_j)}{\sum_{k'=1}^{3} p_j(2-k') f_{k'}(y_j)}, \forall k = 1, 2, 3 \tag{9.67}$$

The posterior expectations are

$$E(Z_j) = \sum_{k=1}^{3} p_j^*(2-k)H_k$$

$$E(Z_j^T Z_j) = \sum_{k=1}^{3} p_j^*(2-k)H_k^T H_k$$

$$E\left[(y_j - X_j\beta - Z_j\gamma)^2\right] = \sum_{k=1}^{3} p_j^*(2-k)(y_j - X_j\beta - H_k\gamma)^2 \qquad (9.68)$$

Since $f_k(y_j)$ is a function of parameters, thus $p_j^*(2-k)$ is also a function of the parameters. However, the parameters are unknown, and they are the very quantities we want to find out. Therefore, iterations are required. Here is the iteration process:

1. Initialize $\theta = \theta^{(t)}$ for $t = 0$.
2. Calculate the posterior expectations using (9.67) and (9.68).
3. Update parameters using (9.65) and (9.66).
4. Increment t by 1, and repeat Step 2 to Step 3 until a certain criterion of convergence is satisfied.

Once the iteration converges, the MLE of the parameters is $\hat{\theta} = \theta^{(t)}$, where t is the number of iterations required for convergence.

9.4.2   Variance–Covariance Matrix of $\hat{\theta}$

Unlike the weighted least-squares and the Fisher scoring algorithms where the variance–covariance matrix of the estimated parameters is automatically given as a by-product of the iteration process, the EM algorithm requires an additional step to calculate this matrix. The method was developed by Louis (1982), and it requires the score vectors and the Hessian matrix for the complete-data log likelihood function rather than the actual observed log likelihood function. The complete-data log likelihood function is the log likelihood function as if Z_j were observed, which is

$$L(\theta, Z) = \sum_{j=1}^{n} L_j(\theta, Z) \qquad (9.69)$$

wheres

$$L_j(\theta, Z) = -\frac{1}{2}\ln(\sigma^2) - \frac{1}{2\sigma^2}(y_j - X_j\beta - Z_j\gamma)^2 \qquad (9.70)$$

The score vector is

$$S(\theta, Z) = \sum_{j=1}^{n} S_j(\theta, Z) \tag{9.71}$$

where

$$S_j(\theta, Z) = \begin{bmatrix} \frac{\partial}{\partial \beta} L_j(\theta, Z) \\ \frac{\partial}{\partial \gamma} L_j(\theta, Z) \\ \frac{\partial}{\partial \sigma^2} L_j(\theta, Z) \end{bmatrix} = \begin{bmatrix} \frac{1}{\sigma^2} X_j^T (y_j - X_j \beta - Z_j \gamma) \\ \frac{1}{\sigma^2} Z_j^T (y_j - X_j \beta - Z_j \gamma) \\ -\frac{1}{2\sigma^2} + \frac{1}{2\sigma^4}(y_j - X_j \beta - Z_j \gamma)^2 \end{bmatrix} \tag{9.72}$$

The second partial derivative (Hessian matrix) is

$$H(\theta, Z) = \sum_{j=1}^{n} H_j(\theta, Z) \tag{9.73}$$

where

$$H_j(\theta, Z) = \begin{bmatrix} \frac{\partial^2 L_j(\theta,Z)}{\partial \beta \partial \beta^T} & \frac{\partial^2 L_j(\theta,Z)}{\partial \beta \partial \gamma^T} & \frac{\partial^2 L_j(\theta,Z)}{\partial \beta \partial \sigma^2} \\ \frac{\partial^2 L_j(\theta,Z)}{\partial \gamma \partial \beta^T} & \frac{\partial^2 L_j(\theta,Z)}{\partial \gamma \partial \gamma^T} & \frac{\partial^2 L_j(\theta,Z)}{\partial \gamma \partial \sigma^2} \\ \frac{L_j(\theta,Z)}{\partial \sigma^2 \partial \beta^T} & \frac{L_j(\theta,Z)}{\partial \sigma^2 \partial \gamma^T} & \frac{L_j(\theta,Z)}{\partial \sigma^2 \partial \sigma^2} \end{bmatrix} \tag{9.74}$$

The six different blocks of the above matrix are

$$\frac{\partial^2 L_j(\theta)}{\partial \beta \partial \beta^T} = -\frac{1}{\sigma^2} X_j^T X_j$$

$$\frac{\partial^2 L_j(\theta)}{\partial \beta \partial \gamma^T} = -\frac{1}{\sigma^2} X_j^T Z_j$$

$$\frac{\partial^2 L_j(\theta)}{\partial \beta \partial \sigma^2} = -\frac{1}{\sigma^4} X_j^T (y_j - X_j \beta - Z_j \gamma)$$

$$\frac{\partial^2 L_j(\theta)}{\partial \gamma \partial \gamma^T} = -\frac{1}{\sigma^2} Z_j^T Z_j$$

$$\frac{\partial^2 L_j(\theta)}{\partial \gamma \partial \sigma^2} = -\frac{1}{\sigma^4} Z_j^T (y_j - X_j \beta - Z_j \gamma)$$

$$\frac{\partial^2 L_j(\theta)}{\partial \sigma^2 \partial \sigma^2} = \frac{1}{2\sigma^4} - \frac{1}{\sigma^6}(y_j - X_j \beta - Z_j \gamma)^2 \tag{9.75}$$

We now have the score vector and the Hessian matrix available for the complete-data log likelihood function. The Louis information matrix is

$$I(\theta) = -E[H(\theta, Z)] - E[S(\theta, Z)S^T(\theta, Z)] \tag{9.76}$$

where the expectations are taken with respect to the missing value (Z_j) using the posterior probabilities of QTL genotypes. At the MLE of parameters, $E[S(\hat{\theta}, Z)] = 0$. Therefore,

$$E[S(\theta, Z)S^T(\theta, Z)] = \text{var}[S(\theta, Z)] + E[S(\theta, Z)]E[S^T(\theta, Z)]$$

$$= \text{var}[S(\theta, Z)] \tag{9.77}$$

As a result, an alternative expression of the Louis information matrix is

$$I(\theta) = -E[H(\theta, Z)] - \text{var}[S(\theta, Z)] \tag{9.78}$$

$$= -\sum_{j=1}^{n} E[H_j(\theta, Z)] - \sum_{j=1}^{n} \text{var}[S_j(\theta, Z)] \tag{9.79}$$

The expectations are

$$E[H_j(\theta, Z)] = \begin{bmatrix} E\left(\frac{\partial^2 L_j(\theta,Z)}{\partial\beta\partial\beta^T}\right) & E\left(\frac{\partial^2 L_j(\theta,Z)}{\partial\beta\partial\gamma^T}\right) & E\left(\frac{\partial^2 L_j(\theta,Z)}{\partial\beta\partial\sigma^2}\right) \\ E\left(\frac{\partial^2 L_j(\theta,Z)}{\partial\gamma\partial\beta^T}\right) & E\left(\frac{\partial^2 L_j(\theta,Z)}{\partial\gamma\partial\gamma^T}\right) & E\left(\frac{\partial^2 L_j(\theta,Z)}{\partial\gamma\partial\sigma^2}\right) \\ E\left(\frac{L_j(\theta,Z)}{\partial\sigma^2\partial\beta^T}\right) & E\left(\frac{L_j(\theta,Z)}{\partial\sigma^2\partial\gamma^T}\right) & E\left(\frac{L_j(\theta,Z)}{\partial\sigma^2\partial\sigma^2}\right) \end{bmatrix} \tag{9.80}$$

The six different blocks of the above matrix are

$$E\left(\frac{\partial^2 L_j(\theta)}{\partial\beta\partial\beta^T}\right) = -\frac{1}{\sigma^2} X_j^T X_j$$

$$E\left(\frac{\partial^2 L_j(\theta)}{\partial\beta\partial\gamma^T}\right) = -\frac{1}{\sigma^2} X_j^T E(Z_j)$$

$$E\left(\frac{\partial^2 L_j(\theta)}{\partial\beta\partial\sigma^2}\right) = -\frac{1}{\sigma^4} X_j^T \left[y_j - X_j\beta - E(Z_j)\gamma\right]$$

$$E\left(\frac{\partial^2 L_j(\theta)}{\partial\gamma\partial\gamma^T}\right) = -\frac{1}{\sigma^2} E(Z_j^T Z_j)$$

$$E\left(\frac{\partial^2 L_j(\theta)}{\partial\gamma\partial\sigma^2}\right) = -\frac{1}{\sigma^4} E\left[Z_j^T(y_j - X_j\beta - Z_j\gamma)\right]$$

$$E\left(\frac{\partial^2 L_j(\theta)}{\partial\sigma^2\partial\sigma^2}\right) = \frac{1}{2\sigma^4} - \frac{1}{\sigma^6} E\left[(y_j - X_j\beta - Z_j\gamma)^2\right] \tag{9.81}$$

Again, all the expectations are taken with respect to the missing value Z_j, not the observed phenotype y_j. This is very different from the information matrix of the Fisher scoring algorithm. The variance–covariance matrix of the score vector is

$$\text{var}[S(\theta, Z)] - \sum_{j=1}^{n} \text{var}[S_j(\theta, Z)] \tag{9.82}$$

where $\text{var}[S_j(\theta, Z)]$ is a symmetric matrix as shown below:

$$\begin{bmatrix} \text{var}\left(\frac{\partial L_j(\theta,Z)}{\partial \beta}\right) & \text{cov}\left(\frac{\partial L_j(\theta,Z)}{\partial \beta}, \frac{\partial L_j(\theta,Z)}{\partial \gamma^T}\right) & \text{cov}\left(\frac{\partial L_j(\theta,Z)}{\partial \beta}, \frac{\partial L_j(\theta,Z)}{\partial \sigma^2}\right) \\ \text{cov}\left(\frac{\partial L_j(\theta,Z)}{\partial \gamma}, \frac{\partial L_j(\theta,Z)}{\partial \beta^T}\right) & \text{var}\left(\frac{\partial L_j(\theta,Z)}{\partial \gamma}\right) & \text{cov}\left(\frac{\partial L_j(\theta,Z)}{\partial \gamma}, \frac{\partial L_j(\theta,Z)}{\partial \sigma^2}\right) \\ \text{cov}\left(\frac{\partial L_j(\theta,Z)}{\partial \sigma^2}, \frac{\partial L_j(\theta,Z)}{\partial \beta^T}\right) & \text{cov}\left(\frac{\partial L_j(\theta,Z)}{\partial \sigma^2}, \frac{\partial L_j(\theta,Z)}{\partial \gamma^T}\right) & \text{var}\left(\frac{\partial L_j(\theta,Z)}{\partial \sigma^2}\right) \end{bmatrix}$$

$$\tag{9.83}$$

The variances are calculated with respect to the missing value Z_j using the posterior probabilities of QTL genotypes. We only provide the detailed expression of one block of the above matrix. The remaining blocks are left to students for practice. The block that is used as an example is the (1,2) block.

$$\text{cov}\left(\frac{\partial L_j(\theta, Z)}{\partial \beta}, \frac{\partial L_j(\theta, Z)}{\partial \gamma^T}\right) = E\left[\frac{\partial L_j(\theta, Z)}{\partial \beta} \frac{\partial L_j(\theta, Z)}{\partial \gamma^T}\right]$$
$$-E\left[\frac{\partial L_j(\theta, Z)}{\partial \beta}\right] E\left[\frac{\partial L_j(\theta, Z)}{\partial \gamma^T}\right] \tag{9.84}$$

where

$$E\left[\frac{\partial L_j(\theta, Z)}{\partial \beta}\right] = \frac{1}{\sigma^2} X_j^T [y_j - X_j \beta - E(Z_j)\gamma]$$

$$E\left[\frac{\partial L_j(\theta, Z)}{\partial \gamma^T}\right] = \frac{1}{\sigma^2} E[(y_j - X_j\beta - Z_j\gamma)Z_j]$$

$$E\left[\frac{\partial L_j(\theta, Z)}{\partial \beta} \frac{\partial L_j(\theta, Z)}{\partial \gamma^T}\right] = \frac{1}{\sigma^4} E\left[X_j^T(y_j - X_j\beta - Z_j\gamma)^2 Z_j\right] \tag{9.85}$$

We already learned how to calculate $E(Z_j)$ using the posterior probability of QTL genotype. The other expectations are

$$E[(y_j - X_j\beta - Z_j\gamma)Z_j] = \sum_{k=1}^{3} p_j^*(2 - k)(y_j - X_j\beta - H_k\gamma)H_k$$

$$E\left[X_j^T(y_j - X_j\beta - Z_j\gamma)^2 Z_j^T\right] = \sum_{k=1}^{3} p_j^*(2 - k)X_j^T(y_j - X_j\beta - H_k\gamma)^2 H_k^T$$

$$\tag{9.86}$$

When calculating the information matrix, the parameter θ is substituted by $\hat{\theta}$, the MLE of θ. Therefore, the observed information matrix is

$$I(\hat{\theta}) = -E[H(\hat{\theta}, Z)] - \mathrm{var}[S(\hat{\theta}, Z)] \tag{9.87}$$

and the variance–covariance matrix of the estimated parameters is $\mathrm{var}(\hat{\theta}) = I^{-1}(\hat{\theta})$.

9.4.3   Hypothesis Test

The hypothesis that $H_0 : \gamma = 0$ can be tested using several different ways. If $\mathrm{var}(\hat{\theta})$ is already calculated, we can use the F- or W-test statistic, which requires $\mathrm{var}(\hat{\gamma})$, the variance–covariance matrix of the estimated QTL effects. It is a submatrix of $\mathrm{var}(\hat{\theta})$. The W-test statistic is

$$W = \hat{\gamma}^T \mathrm{var}^{-1}(\hat{\gamma})\hat{\gamma} \tag{9.88}$$

Alternatively, the likelihood ratio test statistic can be applied to test H_0. We have presented two log likelihood functions; one is the complete-data log likelihood function, denoted by $L(\theta, Z)$, and the other is the observed log likelihood function, denoted by $L(\theta)$. The log likelihood function used to construct the likelihood ratio test statistic is $L(\theta)$, not $L(\theta, Z)$. This complete-data log likelihood function, $L(\theta, Z)$, is only used to derive the EM algorithm and the observed information matrix. The likelihood ratio test statistic is

$$\lambda = -2(L_0 - L_1)$$

where $L_1 = L(\hat{\theta})$ is the observed log likelihood function evaluated at $\hat{\theta} = \{\hat{\beta}, \hat{\gamma}, \hat{\sigma}^2\}$ and L_0 is the log likelihood function evaluated at $\hat{\hat{\theta}} = \{\hat{\hat{\beta}}, 0, \hat{\hat{\sigma}}^2\}$ under the restricted model. The estimated parameter $\hat{\hat{\theta}}$ under the restricted model and L_0 are the same as those given in the section of the least-squares method.

9.5   Remarks on the Four Methods of Interval Mapping

The LS method (Haley and Knott 1992) is an approximation of the ML method, aiming to improve the computational speed. The method has been extended substantially to many other situations, e.g., multiple-trait QTL mapping (Knott and Haley 2000) and QTL mapping for binary traits (Visscher et al. 1996). When used for binary and other nonnormal traits, the method is no longer called LS. Because of the fast speed, the method remains a popular method, even though the computer power has increased by many orders of magnitude since the LS was developed. In some literature (e.g., Feenstra et al. 2006), the LS method is also called the H–K method in honor of the authors, Haley and Knott (1992). Xu (1995) noticed that the LS method, although a good approximation to ML in terms of estimates of QTL effects and test statistic, may lead to a biased (inflated) estimate for the residual error variance. Based on this work, Xu (1998a,b) eventually developed the iteratively reweighted least-squares (IRLS) method. In these works (Xu 1998a,b), the iteratively reweighted least squares was abbreviated IRWLS. Xu (1998b) compared LS, IRLS, and ML in a variety of situations and concluded that IRLS is always better than LS and as efficient as ML. When the residual error does not have a normal distribution, which is required by the ML method, LS and IRLS can be better than ML. In other words, LS and IRLS are more robust than ML to the departure from normality. Kao (2000) and Feenstra et al. (2006) conducted more comprehensive investigation on LS, IRLS, and ML and found that when epistatic effects exist, LS can generate unsatisfactory results, but IRLS and ML usually map QTL better than LS. In addition, Feenstra et al. (2006) modified the weighted least-squares method by using the estimating equations (EE) algorithm. This algorithm further improved the efficiency of the weighted least squares by maximizing an approximate likelihood function. Most recently, Han and Xu (2008) developed a Fisher scoring (FISHER) algorithm to maximize the approximate likelihood function. Both the EE and Fisher algorithm maximize the same likelihood function, and thus, they produce identical results.

The LS method ignores the uncertainty of the QTL genotype. The IRLS, FISHER (or EE), and ML methods use different ways to extract information from the uncertainty of QTL genotype. If the putative location of QTL overlaps with a fully informative marker, all four methods produce identical result. Therefore, if the marker density is sufficiently high, there is virtually no difference for the four methods. For low marker density, when the putative position is far away from either flanking marker, the four methods will show some difference. This difference will be magnified by large QTL. Han and Xu (2008) compared the four methods in a simulation experiment and showed that when the putative QTL position is fixed in the middle of a 10-cM interval, the four methods generated almost identical results. However, when the interval expands to 20 cM, the differences among the four methods become noticeable.

Interval mapping with a 1-cM increment for the mouse 10th-week body weight data was conducted using all the four methods by Han and Xu (2008). The LOD test statistic profiles are shown in Fig. 9.2 for the four methods of interval mapping (LS, IRLS, FISHER, and ML). There is virtually no difference for the four methods. The difference in LOD profiles is noticeable when the marker density is low. Comparisons for the estimated QTL effects were also conducted for the mouse data. Figure 9.3 shows the estimated QTL effect profiles along the genome for the four methods. Again the difference is barely noticeable.

A final remark on interval mapping is the way to infer the QTL genotype using flanking markers. If only flanking markers are used to infer the genotype of a putative position bracketed by the two markers, the method is called interval

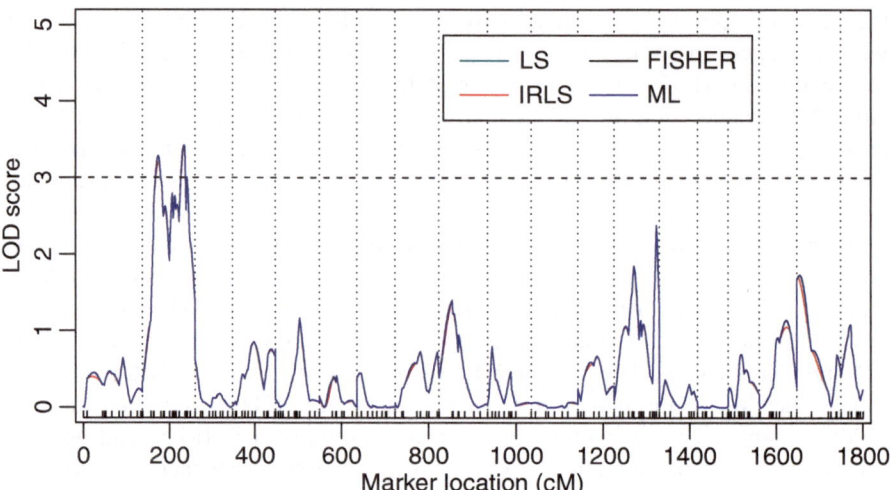

Fig. 9.2 The LOD test statistic profiles for four methods of interval mapping (LS, least square; IRLS, iteratively reweighted least square; FISHER, Fisher scoring; and ML, maximum likelihood). The mouse data were obtained from Lan et al. (2006). The trait investigated is the 10th-week body weight. The 19 chromosomes (excluding the sex chromosome) are separated by the *vertical dotted lines*. The unevenly distributed *black ticks* on the horizontal axis indicate the marker locations

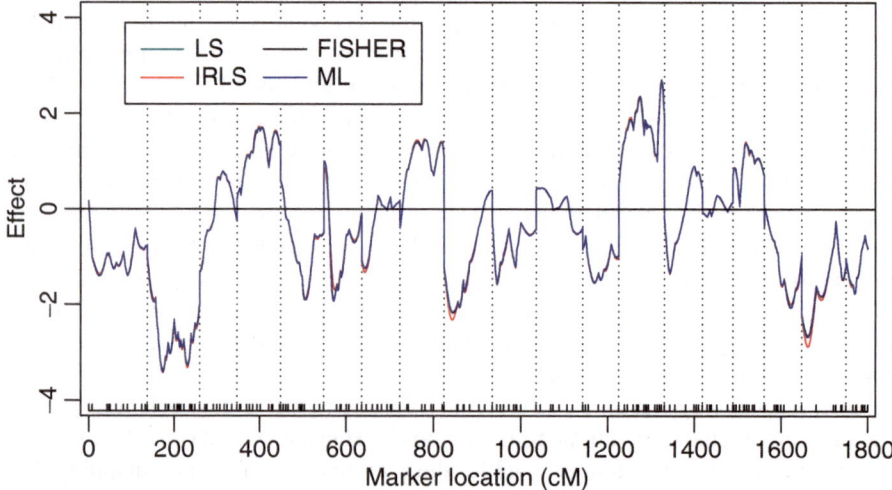

Fig. 9.3 The QTL effect profiles for four methods of interval mapping (LS, least square; IRLS, iteratively reweighted least square; FISHER, Fisher scoring; and ML, maximum likelihood). The mouse data were obtained from Lan et al. (2006). The trait investigated is the 10th-week body weight. The 19 chromosomes (excluding the sex chromosome) are separated by the *vertical dotted lines*. The unevenly distributed *black ticks* on the horizontal axis indicate the marker locations

mapping. Strictly speaking, interval mapping only applies to fully informative markers because we always use flanking markers to infer the QTL genotype. However, almost all datasets obtained from real-life experiments contain missing, uninformative, or partially informative markers. To extract maximum information from markers, people always use the multipoint method (Jiang and Zeng 1997) to infer a QTL genotype. The multipoint method uses more markers or even all markers of the entire chromosome (not just flanking markers) to infer the genotype of a putative position. With the multipoint analysis, we no longer have the notion of interval, and thus, interval mapping is no longer an appropriate phrase to describe QTL mapping. Unfortunately, a more appropriate phrase has not been proposed, and people are used to the phrase of interval mapping. Therefore, the so-called interval mapping in the current literature means QTL mapping under a single QTL model, regardless of whether the genotype of a putative QTL position is inferred from flanking markers or all markers.

Chapter 10
Interval Mapping for Ordinal Traits

Many disease resistance traits in agricultural crops are measured in ordered categories. The generalized linear model (GLM) methodology (Nelder and Wedderburn 1972; Wedderburn 1974; McCullagh and Nelder 1999) is an ideal tool to analyze these traits. Ordinal traits are usually controlled by the segregation of multiple QTL and environmental factors. The genetic architecture of such traits can be studied using linkage analysis. One can analyze the association of each marker with the disease phenotype. If the marker information is fully observable, i.e., marker genotypes can be observed, the standard GLM methodology can be directly applied to the association study by screening markers of the entire genome for their association with the disease trait. Many statistical software packages, e.g., SAS (SAS Institute 2008b), have built-in functions or procedures to perform the standard GLM analysis. One can simply execute the built-in procedures many times, one for each marker, to scan the entire genome without developing a new computer program. In any genetic experiments, missing marker genotypes are unavoidable. In addition, interval mapping requires detection of association between the trait phenotype and loci that are not necessarily located at marker positions. Genotypes of these additional loci are never observed. Therefore, GLM with missing values must be applied. There is a rich literature on the missing value GLM analysis (Ibrahim 1990; Horton and Laird 1999; Ibrahim et al. 2002, 2005). The most popular method is the maximum likelihood (ML) method implemented via the EM algorithm (Horton and Laird 1999). Other methods are also available, such as multiple imputation (MI, Rubin (1987)), fully Bayesian (FB, Ibrahim et al. (2002)) and weighted estimating equations (WEE, Ibrahim et al. (2005)). A complete review on the methods can be found in Ibrahim et al. (2005). Hackett and Weller (1995) first applied the ML method to mapping ordinal trait QTL. They took advantage of an existing software package named GeneStat for the standard GLM analysis (without missing covariates) and modified the software by incorporating a weight variable. The modified GLM for missing data duplicates the data by the number of genotypes per locus, e.g., two for a backcross population and three for an F_2 population. The weight variable is simply the posterior probabilities of the missing genotypes. The weight variable is updated iteratively until the iteration

S. Xu, *Principles of Statistical Genomics*, DOI 10.1007/978-0-387-70807-2_10,
© Springer Science+Business Media, LLC 2013

converges. The modified GLM program is not necessarily simpler than a program written anew. Furthermore, the variance–covariance matrix of estimated parameters is not available for the modified GML algorithm. Xu et al. (2003) developed an explicit EM algorithm using the posterior probability of missing covariates as the weight variable and further provided the variance–covariance matrix of the estimated parameters by using the Louis' (1982) adjustment for the information matrix. Standard deviations (square roots of the variances) of estimated parameters represent the precisions of the estimates, which are required in the final report for publication. The variance–covariance matrix of the estimated QTL effects can also be used to calculate the Wald-test statistic (Wald 1943), which is an alternative test that can replace the likelihood ratio test statistic. Although using the large sample distribution for the likelihood ratio test gives more accurate approximation for small and moderate-sized samples, the latter has a computational advantage since it does not require calculation of the likelihood function under the null model (McCulloch and Searle 2001). A missing QTL genotype usually has partial information, which can be extracted from linked markers. This information can be used to infer the QTL genotypes using several different ways (McCulloch and Searle 2001). In QTL mapping for continuously distributed traits, mixture model (Lander and Botstein 1989) is the most efficient way to take advantage of marker information. The least-squares method of Haley and Knott (1992) is the simplest way to incorporate linked markers. Performances of the weighted least-squares method of Xu (1998a,b) and estimating equations (EE) algorithm of Feenstra et al. (2006) are usually between the least-squares and mixture model methods. These methods have been successfully applied to QTL mapping for continuous traits, but they have not been investigated for ordinal trait QTL mapping. This chapter will introduce several alternative GLM methods for mapping quantitative trait loci of ordinal traits.

10.1   Generalized Linear Model

Suppose that a disease phenotype of individual j ($j = 1, \ldots, n$) is measured by an ordinal variable denoted by $S_j = 1, \ldots, p + 1$, where $p + 1$ is the total number of disease classes and n is the sample size. Let $Y_j = \{Y_{jk}\}, \forall k = 1, \ldots, p + 1$, be a $(p + 1) \times 1$ vector to indicate the disease status of individual j. The kth element of Y_j is defined as

$$Y_{jk} = \begin{cases} 1 & \text{if } S_j = k \\ 0 & \text{if } S_j \neq k \end{cases} \tag{10.1}$$

Using the probit link function, the expectation of Y_{jk} is defined as

$$\mu_{jk} = E(Y_{jk}) = \Phi(\alpha_k + X_j\beta + Z_j\gamma) - \Phi(\alpha_{k-1} + X_j\beta + Z_j\gamma) \tag{10.2}$$

where α_k ($\alpha_0 = -\infty$ and $\alpha_{p+1} = +\infty$) is the intercept, β is a $q \times 1$ vector for some systematic effects (not related to the effects of quantitative trait loci), and γ is an $r \times 1$ vector for the effects of a quantitative trait locus. The symbol $\Phi(.)$ is the standardized cumulative normal function. The design matrix X_j is assumed to be known, but Z_j may not be fully observable because it is determined by the genotype of j for the locus of interest. Because the link function is probit, this type of analysis is called probit analysis. Let $\mu_j = \{\mu_{jk}\}$ be a $(p+1) \times 1$ vector. The expectation for vector Y_j is $E(Y_j) = \mu_j$, and the variance matrix of Y_j is

$$V_j = \text{var}(Y_j) = \psi_j + \mu_j \mu_j^T \tag{10.3}$$

where $\psi_j = \text{diag}(\mu_j)$. The method to be developed requires the inverse of matrix V_j. However, V_j is not of full rank. We can use a generalized inverse of V_j, such as $V_j^- = \psi_j^{-1}$, in place of V_j^{-1}. The parameter vector is $\theta = \{\alpha, \beta, \gamma\}$ with a dimensionality of $(p + q + r) \times 1$. Binary data is a special case of ordinal data in that $p = 1$ so that there are only two categories, $S_j = \{1, 2\}$. The expectation of Y_{jk} is

$$\mu_{jk} = \begin{cases} \Phi(\alpha_1 + X_j\beta + Z_j\gamma) - \Phi(\alpha_0 + X_j\beta + Z_j\gamma) & \text{for } k = 1 \\ \Phi(\alpha_2 + X_j\beta + Z_j\gamma) - \Phi(\alpha_1 + X_j\beta + Z_j\gamma) & \text{for } k = 2 \end{cases} \tag{10.4}$$

Because $\alpha_0 = -\infty$ and $\alpha_2 = +\infty$ in the binary case, we have

$$\mu_{jk} = \begin{cases} \Phi(\alpha_1 + X_j\beta + Z_j\gamma) & \text{for } k = 1 \\ 1 - \Phi(\alpha_1 + X_j\beta + Z_j\gamma) & \text{for } k = 2 \end{cases} \tag{10.5}$$

We can see that $\mu_{j2} = 1 - \mu_{j1}$ and

$$\Phi^{-1}(\mu_{j1}) = \alpha_1 + X_j\beta + Z_j\gamma \tag{10.6}$$

The link function is $\Phi^{-1}(.)$, and thus, it is called the probit link function. Once we take the probit transformation, the model becomes a linear model. Therefore, this type of model is called a generalized linear model (GLM). The ordinary linear model we learned before for continuous traits is a special case of the GLM because the link function is simply the identity, i.e.,

$$I^{-1}(\mu_{j1}) = \alpha_1 + X_j\beta + Z_j\gamma \tag{10.7}$$

or simply

$$\mu_{j1} = \alpha_1 + X_j\beta + Z_j\gamma \tag{10.8}$$

Most techniques we learned for the linear model apply to the generalized linear model.

10.2   ML Under Homogeneous Variance

Let us first assume that the genotypes of the QTL are observed for all individuals. In this case, variable Z_j is not missing. The log likelihood function under the probit model is

$$L(\theta) = \sum_{j=1}^{n} L_j(\theta) \tag{10.9}$$

where

$$L_j(\theta) = \sum_{k=1}^{p+1} Y_{jk} \ln[\Phi(\alpha_k + X_j\beta + Z_j\gamma) - \Phi(\alpha_{k-1} + X_j\beta + Z_j\gamma)] \tag{10.10}$$

and $\theta = \{\alpha, \beta, \gamma\}$ is the vector of parameters. This is the simplest GLM problem, and the classical iteratively reweighted least-squares approach for GLM (Nelder and Wedderburn 1972; Wedderburn 1974) can be used without any modification. The iterative equation under the classical GLM is given below:

$$\theta^{(t+1)} = \theta^{(t)} + I^{-1}(\theta^{(t)})S(\theta^{(t)}) \tag{10.11}$$

where $\theta^{(t)}$ is the parameter value in the current iteration, $I(\theta^{(t)})$ is the information matrix, and $S(\theta^{(t)})$ is the score vector, both evaluated at $\theta^{(t)}$. We can interpret

$$\Delta\theta = I^{-1}(\theta^{(t)})S(\theta^{(t)}) \tag{10.12}$$

in (10.11) as the adjustment for $\theta^{(t)}$ to improve the solution in the direction that leads to the ultimate maximum likelihood estimate of θ. Equation (10.3) shows that the variance of Y_j is a function of the expectation of Y_j. This special relationship leads to a convenient way to calculate the information matrix and the score vector, as given by Wedderburn (1974),

$$I(\theta) = \sum_{j=1}^{n} D_j^T W_j D_j \tag{10.13}$$

and

$$S(\theta) = \sum_{j=1}^{n} D_j^T W_j (Y_j - \mu_j) \tag{10.14}$$

where $W_j = \psi_j^{-1}$. Therefore, the increment (adjustment) of the parameter can be estimated using the following iteratively reweighted least-squares approach:

$$\Delta\theta = \left[\sum_{j=1}^{n} D_j^T W_j D_j\right]^{-1} \left[\sum_{j=1}^{n} D_j^T W_j (Y_j - \mu_j)\right] \tag{10.15}$$

where D_j is a $(p+1) \times (p+q+r)$ matrix for the first partial derivatives of μ_j with respect to the parameters and $W_j = V_j^- = \psi_j^{-1}$ is the weight matrix. Matrix D_j can be partitioned into three blocks,

$$D_j = \frac{\partial \mu_j}{\partial \theta^T} = \left[\frac{\partial \mu_j}{\partial \alpha^T} \frac{\partial \mu_j}{\partial \beta^T} \frac{\partial \mu_j}{\partial \gamma^T} \right] \tag{10.16}$$

The first block $\partial \mu_j / \partial \alpha^T = \{\partial \mu_{jk} / \partial \alpha_l\}$ is a $(p+1) \times p$ matrix with

$$\frac{\partial \mu_{jk}}{\partial \alpha_{k-1}} = -\phi(\alpha_{k-1} + X_j \beta + Z_j \gamma)$$

$$\frac{\partial \mu_{jk}}{\partial \alpha_k} = \phi(\alpha_k + X_j \beta + Z_j \gamma)$$

$$\frac{\partial \mu_{jk}}{\partial \alpha_l} = 0, \ \forall l \neq \{k-1, k\} \tag{10.17}$$

The second block $\partial \mu_j / \partial \beta^T = \{\partial \mu_{jk} / \partial \beta\}$ is a $(p+1) \times q$ matrix with

$$\frac{\partial \mu_{jk}}{\partial \beta} = X_j^T[\phi(\alpha_k + X_j \beta + Z_j \gamma) - \phi(\alpha_{k-1} + X_j \beta + Z_j \gamma)] \tag{10.18}$$

The third block $\partial \mu_j / \partial \gamma^T = \{\partial \mu_{jk} / \partial \gamma\}$ is a $(p+1) \times r$ matrix with

$$\frac{\partial \mu_{jk}}{\partial \gamma} = Z_j^T[\phi(\alpha_k + X_j \beta + Z_j \gamma) - \phi(\alpha_{k-1} + X_j \beta + Z_j \gamma)] \tag{10.19}$$

In all the above partial derivatives, the range of k is $k = 1, \ldots, p+1$. The sequence of parameter values during the iteration process converges to a local maximum likelihood estimate, denoted by $\hat{\theta}$. The variance–covariance matrix of $\hat{\theta}$ is approximately equal to $\mathrm{var}(\hat{\theta}) = I^{-1}(\hat{\theta})$, which is a by-product of the iteration process. Here, we are actually dealing with a situation where the QTL overlaps with a fully informative marker because observed marker genotypes represent the genotypes of the disease locus. If the QTL of interest does not overlap with any markers, the genotype of the QTL is not observable, i.e., Z_j is missing. The classical GLM does not apply directly to such a situation. The missing value Z_j still has some information due to linkage with some markers. Again, we use an F_2 population as an example to show how to handle the missing value of Z_j. The ML estimation of parameters under the homogeneous variance model is obtained simply by substituting Z_j with the conditional expectation of Z_j given flanking marker information. Let

$$p_j(2-g) = \Pr(Z_j = H_g | \text{marker}), \forall g = 1, 2, 3 \tag{10.20}$$

be the conditional probability of the QTL genotype given marker information, where the marker information can be either drawn from two flanking markers (interval mapping, Lander and Botstein 1989) or multiple markers (multipoint analysis, Jiang and Zeng 1997). Note that $p_j(2-g)$ is not p_j multiplied by $(2-g)$; rather, it is a notation for the probabilities of the three genotypes. For $g = 1, 2, 3$, we have $p_j(-1)$, $p_j(0)$, and $p_j(+1)$, respectively, where $p_j(-1)$, etc., are defined early in Chap. 9. Vector H_g for $g = 1, 2, 3$ is also defined in Chap. 9 as genotype indicator variables.

Using marker information, we can calculate the expectation of Z_j, which is

$$U_j = E(Z_j) = \sum_{g=1}^{3} p_j (2 - g) H_g \tag{10.21}$$

The method is called ML under the homogeneous residual variance because when we substitute Z_j by U_j, the residual error variance is no longer equal to unity; rather it is inflated, and the inflation varies across individuals. However, the homogeneous variance model here assumed the residual variance is constant across individuals. This method is the ordinal trait analogy of the Haley and Knott's (1992) method of QTL mapping.

10.3   ML Under Heterogeneous Variance

The homogeneous variance model is only a first moment approximation because the uncertainty of the estimated Z_j has been ignored. Let

$$\Sigma_j = \text{var}(Z_j) = \sum_{g=1}^{3} p_j (2 - g) H_g^T H_g - U_j^T U_j \tag{10.22}$$

be the conditional covariance matrix for Z_j. Note that model (10.2) with Z_j substituted by U_j is

$$\mu_{jk} = E(Y_{jk}) = \Phi(\alpha_k + X_j \beta + U_j \gamma) - \Phi(\alpha_{k-1} + X_j \beta + U_j \gamma) \tag{10.23}$$

An underlying assumption for this probit model is that the residual error variance for the "underlying liability" of the disease trait is unity across individuals. Once U_j is used in place of Z_j, the residual error variance becomes

$$\sigma_j^2 = \gamma^T \Sigma_j \gamma + 1 \tag{10.24}$$

This is an inflated variance, and it is heterogeneous across individuals. In order to apply the probit model, we need to rescale the model effects as follows (Xu and Hu 2010):

$$\mu_{jk} = \Phi\left[\frac{1}{\sigma_j}(\alpha_k + X_j \beta + U_j \gamma)\right] - \Phi\left[\frac{1}{\sigma_j}(\alpha_{k-1} + X_j \beta + U_j \gamma)\right] \tag{10.25}$$

This modification leads to a change in the partial derivatives of μ_j with respect to the parameters. Corresponding changes in the derivatives are given below.

$$\frac{\partial \mu_{jk}}{\partial \alpha_{k-1}} = -\frac{1}{\sigma_j}\phi\left[\frac{1}{\sigma_j}(\alpha_{k-1} + X_j\beta + U_j\gamma)\right]$$

$$\frac{\partial \mu_{jk}}{\partial \alpha_k} = \frac{1}{\sigma_j}\phi\left[\frac{1}{\sigma_j}(\alpha_k + X_j\beta + U_j\gamma)\right]$$

$$\frac{\partial \mu_{jk}}{\partial \alpha_l} = 0, \ \forall l \neq \{k-1, k\} \tag{10.26}$$

$$\frac{\partial \mu_{jk}}{\partial \beta} = \frac{1}{\sigma_j}\phi\left[\frac{1}{\sigma_j}(\alpha_k + X_j\beta + U_j\gamma)\right]X_j^T$$

$$-\frac{1}{\sigma_j}\phi\left[\frac{1}{\sigma_j}(\alpha_{k-1} + X_j\beta + U_j\gamma)\right]X_j^T \tag{10.27}$$

and

$$\frac{\partial \mu_{jk}}{\partial \gamma} = \frac{1}{\sigma_j}\phi\left[\frac{1}{\sigma_j}(\alpha_k + X_j\beta + U_j\gamma)\right]\left[U_j^T - \frac{1}{\sigma_j^2}(\alpha_k + X_j\beta + U_j\gamma)\Sigma_j\gamma\right]$$

$$-\frac{1}{\sigma_j}\phi\left[\frac{1}{\sigma_j}(\alpha_{k-1} + X_j\beta + U_j\gamma)\right]\left[U_j^T - \frac{1}{\sigma_j^2}(\alpha_{k-1} + X_j\beta + U_j\gamma)\Sigma_j\gamma\right]$$

$$\tag{10.28}$$

The iteration formula remains the same as (10.11) except that the modified weight and partial derivatives are used under the heterogeneous residual variance model.

10.4   ML Under Mixture Distribution

The mixture model approach defines genotype-specific expectation, variance matrix, and all derivatives for each individual. Let

$$\mu_{jk}(g) = E(Y_{jk}) = \Phi(\alpha_k + X_j\beta + H_g\gamma) - \Phi(\alpha_{k-1} + X_j\beta + H_g\gamma) \tag{10.29}$$

be the expectation of Y_{jk} if j takes the gth genotype for $g = 1, 2, 3$. The corresponding variance–covariance matrix is

$$V_j(g) = \psi_j(g) - \mu_j(g)\mu_j^T(g) \tag{10.30}$$

where $\psi_j(g) = \text{diag}[\mu_j(g)]$. Let $D_j(g)$ be the partial derivatives of the expectation with respect to the parameters. The corresponding values of $D_j(g)$ are

$$\frac{\partial \mu_{jk}(g)}{\partial \alpha_{k-1}} = -\phi(\alpha_{k-1} + X_j\beta + H_g\gamma)$$

$$\frac{\partial \mu_{jk}(g)}{\partial \alpha_k} = \phi(\alpha_k + X_j\beta + H_g\gamma)$$

$$\frac{\partial \mu_{jk}(g)}{\partial \alpha_l} = 0, \ \forall l \neq \{k-1, k\} \tag{10.31}$$

$$\frac{\partial \mu_{jk}(g)}{\partial \beta} = X_j^T[\phi(\alpha_k + X_j\beta + H_g\gamma) - \phi(\alpha_{k-1} + X_j\beta + H_g\gamma)] \tag{10.32}$$

and

$$\frac{\partial \mu_{jk}(g)}{\partial \gamma} = H_g^T[\phi(\alpha_k + X_j\beta + H_g\gamma) - \phi(\alpha_{k-1} + X_j\beta + H_g\gamma)] \tag{10.33}$$

Let us define the posterior probability of QTL genotype after incorporating the disease phenotype for individual j as

$$p_j^*(2-g) = \frac{p_j(2-g)Y_j^T\mu_j(g)}{\sum_{g'=1}^3 p_j(2-g')Y_j^T\mu_j(g')} \tag{10.34}$$

The increment for parameter updating under the mixture model is

$$\Delta\theta = \left[\sum_{j=1}^n E\left(D_j^T W_j D_j\right)\right]^{-1} \left[\sum_{j=1}^n E\left(D_j^T W_j(Y_j - \mu_j)\right)\right] \tag{10.35}$$

where

$$E\left(D_j^T W_j D_j\right) = \sum_{g=1}^3 p_j^*(g)D_j^T(g)W_j(g)D_j(g) \tag{10.36}$$

$$E\left(D_j^T W_j(Y_j - \mu_j)\right) = \sum_{g=1}^3 p_j^*(2-g)D_j^T(g)W_j(g)(Y_j - \mu_j(g)) \tag{10.37}$$

and

$$W_j(g) = \psi_j^{-1}(g) \tag{10.38}$$

This is actually an EM algorithm where calculating the posterior probabilities of QTL genotype and using the posterior probabilities to calculate $E\left(D_j^T W_j D_j\right)$ and $E\left(D_j^T W_j (Y_j - \mu_j)\right)$ constitute the E-step and calculating the increment of the parameter using the weighted least-squares formula makes up the M-step. A problem with this EM algorithm is that $\text{var}(\hat{\theta})$ is not a by-product of the iteration process. For simplicity, if the markers are sufficiently close to the trait locus of interest, we can use

$$\text{var}(\hat{\theta}) \approx \left[\sum_{j=1}^{n} E\left(D_j^T W_j D_j\right)\right]^{-1} \tag{10.39}$$

to approximate the covariance matrix of estimated parameters. This is an underestimated variance matrix. A more precise method to calculate $\text{var}(\hat{\theta})$ is to adjust the above equation by the information loss due to uncertainty of the QTL genotype. Let

$$S(\hat{\theta}|Z) = \sum_{j=1}^{n} D_j^T W_j (Y_j - \mu_j) \tag{10.40}$$

be the score vector as if Z were observed. Louis (1982) showed that the information loss is due to the variance–covariance matrix of the score vector, which is

$$\text{var}[S(\hat{\theta}|Z)] = \sum_{j=1}^{n} \text{var}\left[D_j^T W_j (Y_j - \mu_j)\right] \tag{10.41}$$

The variance is taken with respect to the missing value Z using the posterior probability of QTL genotype. The information matrix after adjusting for the information loss is

$$I(\hat{\theta}) = \sum_{j=1}^{n} E\left(D_j^T W_j D_j\right) - \sum_{j=1}^{n} \text{var}\left[D_j^T W_j (Y_j - \mu_j)\right] \tag{10.42}$$

The variance–covariance matrix for the estimated parameters is then approximated by $\text{var}(\hat{\theta}) = I^{-1}(\hat{\theta})$. Details of $\text{var}[D_j^T W_j (Y_j - \mu_j)]$ are given by Xu and Hu (2010).

10.5 ML via the EM Algorithm

The EM algorithm to be introduced here is different from the EM under the mixture model described in the previous section. We now use a liability model (Xu et al. 2003) to derive the EM algorithm. Xu et al. (2003) hypothesizes that there is an underlying liability that controls the observed phenotype. The liability is a continuous variable and has exactly the same behavior as a quantitative trait.

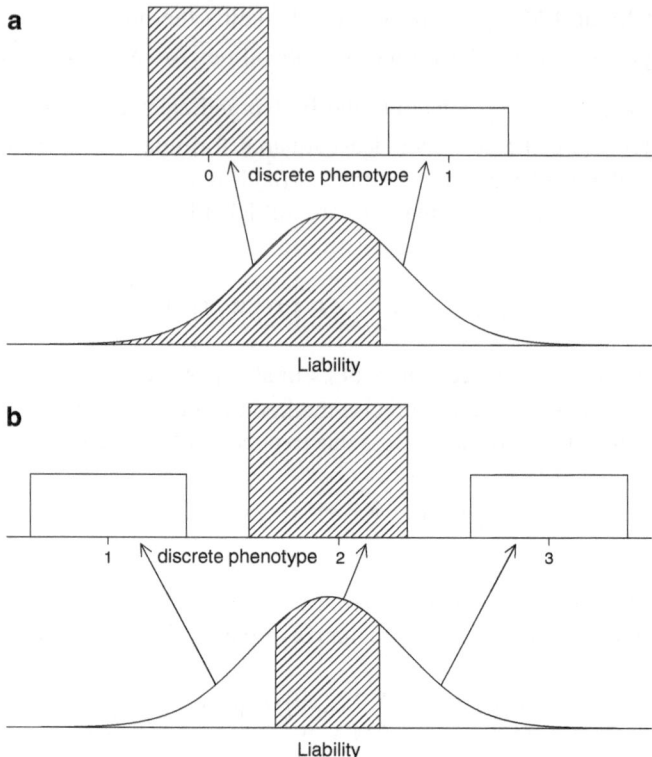

Fig. 10.1 Connection between the unobserved continuous liability and the observed discrete phenotype. The *top panel* shows the connection for an ordinal trait with two categories, and the *bottom panel* shows the connection for an ordinal trait with three categories

The only difference is that the liability is not observable while the quantitative trait can be measured in experiments. The observed ordinal trait phenotype is connected with the liability by a series of thresholds, as demonstrated in Fig. 10.1. In the generalized linear model under the mixture distribution, the EM algorithm treats the QTL genotype as missing value. Here, we treat the liability as missing value as well. Let y_j be the liability for the jth individual. This is different from $Y_j = \{Y_{jk}\}$, the multivariate representation of the ordered categorical phenotype in the generalized linear model. The liability can be described by the following linear model:

$$y_j = X_j \beta + Z_j \gamma + \varepsilon_j \tag{10.43}$$

where $\varepsilon_j \sim N(0, \sigma^2)$ is assumed. Under the liability model, σ^2 cannot be estimated, and thus, we set $\sigma^2 = 1$. This arbitrary scale will not affect the significance test because the estimated parameters $\theta = \{\alpha, \beta, \gamma\}$ are defined relative to σ^2. The connection between y_j and the observed phenotype is

$$S_j = k \text{ , for } \alpha_{k-1} < y_j \le \alpha_k \tag{10.44}$$

where $k = 1, \dots, p + 1$. The thresholds α do not appear in the linear model explicitly but serve as converters from y_j to S_j. Xu et al. (2003) developed an EM algorithm for ordinal trait QTL mapping by using this liability model. They used a three-step approach, where the first step is to estimate the non-QTL effects (β), the second step is to estimate the QTL effects (γ), and the third step is to estimate the thresholds (α). The method does not have a simple way to calculate the variance–covariance matrix of the estimated parameters. Xu and Xu (2006) extended the method using a multivariate version of the GLM. This method gives a way to calculate the variance–covariance matrix of the estimated parameters. Both methods (Xu et al. 2003; Xu and Xu 2006) are quite complicated in the E-step. When the number of categories is two (the binary case), both methods can be simplified. This section will deal with the simplified binary trait QTL mapping where only one threshold is applied. In this case, the single threshold is set to zero so that it is not a parameter for estimation, and thus, we only estimate β and γ. In the binary situation, $S_j = \{1, 2\}$ and

$$Y_{j1} = \begin{cases} 1 & \text{for } S_j = 1 \\ 0 & \text{for } S_j = 2 \end{cases} \tag{10.45}$$

and

$$Y_{j2} = \begin{cases} 0 & \text{for } S_j = 1 \\ 1 & \text{for } S_j = 2 \end{cases} \tag{10.46}$$

The liability model remains the same as that given in (10.43). The derivation of the EM algorithm starts with the complete-data situation. If both Z_j and y_j were observed, the ML estimates of β and γ would be

$$\begin{bmatrix} \beta \\ \gamma \end{bmatrix} = \begin{bmatrix} \sum_{j=1}^{n} X_j^T X_j & \sum_{j=1}^{n} X_j^T Z_j \\ \sum_{j=1}^{n} Z_j^T X_j & \sum_{j=1}^{n} Z_j^T Z_j \end{bmatrix}^{-1} \begin{bmatrix} \sum_{j=1}^{n} X_j^T y_j \\ \sum_{j=1}^{n} Z_j^T y_j \end{bmatrix} \tag{10.47}$$

This is simply the ordinary least-squares estimates of the parameters. The EM algorithm takes advantage of this explicit solution in the maximization step. If we had observed y_j but still not been able to estimate Z_j, the maximization step of the EM algorithm would be

$$\begin{bmatrix} \beta \\ \gamma \end{bmatrix} = \begin{bmatrix} \sum_{j=1}^{n} X_j^T X_j & \sum_{j=1}^{n} X_j^T E(Z_j) \\ \sum_{j=1}^{n} E(Z_j^T) X_j & \sum_{j=1}^{n} E(Z_j^T Z_j) \end{bmatrix}^{-1} \begin{bmatrix} \sum_{j=1}^{n} X_j^T y_j \\ \sum_{j=1}^{n} E(Z_j^T) y_j \end{bmatrix} \tag{10.48}$$

The problem here is that we observe neither Z_j nor y_j. Intuitively, the maximization step of the EM should be

$$
\begin{bmatrix} \beta \\ \gamma \end{bmatrix} = \begin{bmatrix} \sum_{j=1}^{n} X_j^T X_j & \sum_{j=1}^{n} X_j^T E(Z_j) \\ \sum_{j=1}^{n} E(Z_j^T) X_j & \sum_{j=1}^{n} E(Z_j^T Z_j) \end{bmatrix}^{-1} \begin{bmatrix} \sum_{j=1}^{n} X_j^T E(y_j) \\ \sum_{j=1}^{n} E(Z_j^T y_j) \end{bmatrix}
\tag{10.49}
$$

where the expectations are taken with respect to both Z_j and y_j using the posterior probabilities of QTL genotypes. We now present the method for calculating these expectation terms. We first address $E(Z_j)$ and $E(Z_j^T Z_j)$ using the posterior probabilities of the QTL genotypes.

$$
p_j^*(2-g) = \frac{p_j(2-g)\left[\Phi(X_j\beta + H_g\gamma)\right]^{Y_{j1}}\left[1 - \Phi(X_j\beta + H_g\gamma)\right]^{Y_{j2}}}{\sum_{g'=1}^{3} p_j(2-g')\left[\Phi(X_j\beta + H_{g'}\gamma)\right]^{Y_{j1}}\left[1 - \Phi(X_j\beta + H_{g'}\gamma)\right]^{Y_{j2}}}
\tag{10.50}
$$

Given the posterior probabilities, we have

$$
E(Z_j) = \sum_{g=1}^{3} p_j^*(2-g) H_g
\tag{10.51}
$$

and

$$
E(Z_j^T Z_j) = \sum_{g=1}^{3} p_j^*(2-g) H_g^T H_g
\tag{10.52}
$$

The expectations for terms that involve y_j can be expressed as

$$
E(y_j) = \underset{Z}{E}\left[\underset{y}{E}(y_j|Z_j)\right] = \sum_{g=1}^{3} p_j^*(2-g)\, \underset{y}{E}(y_j|H_g)
\tag{10.53}
$$

and

$$
E(Z_j^T y_j) = \underset{Z}{E}\left[Z_j^T \underset{y}{E}(y_j|Z_j)\right] = \sum_{g=1}^{3} p_j^*(2-g) H_g^T \underset{y}{E}(y_j|H_g)
\tag{10.54}
$$

where

$$
\underset{y}{E}(y_j|H_g) = X_j\beta + H_g\gamma + \frac{(Y_{j2} - Y_{j1})\phi(X_j\beta + H_g\gamma)}{\left[\Phi(X_j\beta + H_g\gamma)\right]^{Y_{j1}}\left[1 - \Phi(X_j\beta + H_g\gamma)\right]^{Y_{j2}}}
\tag{10.55}
$$

Therefore, the EM algorithm can be summarized as:

1. Initialize parameters $\theta^{(0)} = \{\beta^{(0)}, \gamma^{(0)}\}$.
2. Calculate $E(Z_j)$, $E(Z_j^T Z_j)$, $E(y_j)$, and $E(Z_j^T y_j)$.
3. Update β and γ using (10.49).
4. Repeat Step 2 to Step 3 until convergence is reached.

Once the EM algorithm converges, we obtain the estimated parameters and are ready to calculate the Louis (1982) information matrix. The variance–covariance matrix of the estimated parameters simply takes the inverse of the information matrix. Let

$$
H(\theta, Z, y) = - \begin{bmatrix} \sum_{j=1}^{n} X_j^T X_j & \sum_{j=1}^{n} X_j^T Z_j \\ \sum_{j=1}^{n} Z_j^T X_j & \sum_{j=1}^{n} Z_j^T Z_j \end{bmatrix} \tag{10.56}
$$

be the Hessian matrix of the complete-data log likelihood function and

$$
S(\theta, Z, y) = \begin{bmatrix} \sum_{j=1}^{n} X_j^T (y_j - X_j \beta - Z_j \gamma) \\ \sum_{j=1}^{n} Z_j^T (y_j - X_j \beta - Z_j \gamma) \end{bmatrix} \tag{10.57}
$$

be the score vector of the complete-data log likelihood function. The Louis information matrix is

$$
I(\theta) = -E\left[H(\theta, Z, y)\right] - E\left[S(\theta, Z, y)S^T(\theta, Z, y)\right] \tag{10.58}
$$

where the expectations are taken with respect to the missing values of Z and y. Note that

$$
\mathrm{var}\left[S(\theta, Z, y)\right] = E\left[S(\theta, Z, y)S^T(\theta, Z, y)\right] - E\left[S(\theta, Z, y)\right] E\left[S^T(\theta, Z, y)\right]
$$

$$\tag{10.59}$$

and $E[S(\theta, Z, y)] = 0$ at $\theta = \hat{\theta}$. This leads to

$$
E\left[S(\theta, Z, y)S^T(\theta, Z, y)\right] = \mathrm{var}\left[S(\theta, Z, y)\right] \tag{10.60}
$$

Therefore, the Louis information matrix is also expressed as

$$
I(\theta) = -E\left[H(\theta, Z, y)\right] - \mathrm{var}\left[S(\theta, Z, y)\right] \tag{10.61}
$$

The first term is easy to obtain, as shown below:

$$- E\left[H(\theta, Z, y)\right] = \begin{bmatrix} \sum\limits_{j=1}^{n} X_j^T X_j & \sum\limits_{j=1}^{n} X_j^T E(Z_j) \\ \sum\limits_{j=1}^{n} E(Z_j^T) X_j & \sum\limits_{j=1}^{n} E(Z_j^T Z_j) \end{bmatrix} \tag{10.62}$$

The second term can be expressed as

$$\mathrm{var}\left[S(\theta, Z, y)\right] = \sum_{j=1}^{n} \mathrm{var}\left[S_j(\theta, Z, y)\right] \tag{10.63}$$

where

$$S_j(\theta, Z, y) = \begin{bmatrix} X_j^T (y_j - X_j \beta - Z_j \gamma) \\ Z_j^T (y_j - X_j \beta - Z_j \gamma) \end{bmatrix} \tag{10.64}$$

Explicit form of $\mathrm{var}\left[S_j(\theta, Z, y)\right]$ can be derived. This matrix is a 2×2 block matrix, denoted by

$$\mathrm{var}\left[S_j(\theta, Z, y)\right] = \begin{bmatrix} \Sigma_{11} & \Sigma_{12} \\ \Sigma_{21} & \Sigma_{22} \end{bmatrix} \tag{10.65}$$

we now provide detailed expressions of the blocks.

$$\begin{aligned} \Sigma_{11} &= E\left[X_j^T \mathrm{var}(y_j - X_j \beta - Z_j \gamma) X_j\right] \\ \Sigma_{22} &= E\left[Z_j^T \mathrm{var}(y_j - X_j \beta - Z_j \gamma) Z_j\right] \\ \Sigma_{12} &= E\left[X_j^T \mathrm{var}(y_j - X_j \beta - Z_j \gamma) Z_j\right] \end{aligned} \tag{10.66}$$

where $\mathrm{var}(y_j - X_j \beta - Z_j \gamma)$ is the variance of a truncated normal variable (the truncation point being zero) conditional on $Y_j = \{Y_{j1}, Y_{j2}\}$ and Z_j. Let

$$\varphi(Z_j) = \mathrm{var}(y_j - X_j \beta - Z_j \gamma) \tag{10.67}$$

be the short notation for the variance of the truncated normal variable. With some manipulation on Cohen (1991) formula, we get

$$\varphi(Z_j) = 1 - \psi(X_j \beta + Z_j \gamma)\left[\psi(X_j \beta + Z_j \gamma) - (Y_{j1} - Y_{j2})(X_j \beta + Z_j \gamma)\right] \tag{10.68}$$

where

$$\psi(X_j \beta + Z_j \gamma) = \frac{\phi(X_j \beta + Z_j \gamma)}{\left[1 - \Phi(X_j \beta + Z_j \gamma)\right]^{Y_{j1}} \left[\Phi(X_j \beta + Z_j \gamma)\right]^{Y_{j2}}} \tag{10.69}$$

Therefore,

$$\Sigma_{11} = \sum_{g=1}^{3} p_j^*(2-g)X_j^T \varphi(H_g)X_j$$

$$\Sigma_{12} = \sum_{g=1}^{3} p_j^*(2-g)X_j^T \varphi(H_g)H_g$$

$$\Sigma_{22} = \sum_{g=1}^{3} p_j^*(2-g)H_g^T \varphi(H_g)H_g \qquad (10.70)$$

Further manipulation on the information matrix, we get

$$I(\theta) = \begin{bmatrix} \sum_{j=1}^{n} E\left[X_j^T\left(1-\varphi(Z_j)\right)X_j\right], & \sum_{j=1}^{n} E\left[X_j^T\left(1-\varphi(Z_j)\right)Z_j\right] \\ \sum_{j=1}^{n} E\left[Z_j^T\left(1-\varphi(Z_j)\right)X_j\right], & \sum_{j=1}^{n} E\left[Z_j^T\left(1-\varphi(Z_j)\right)Z_j\right] \end{bmatrix} \qquad (10.71)$$

which is a 2×2 matrix.

Xu and Xu (2003) proposed an alternative method to calculate the Louis information matrix via Monte Carlo simulations. The method does not involve the above complicated derivation; instead, it simply simulates the QTL genotype (Z_j) using the posterior distribution for each individual and the liability (y_j) conditional on the genotype using the truncated normal distribution for the individual. The method directly uses the following information matrix:

$$I(\theta) = -E\left[H(\theta, Z, y)\right] - E\left[S(\theta, Z, y)S^T(\theta, Z, y)\right] \qquad (10.72)$$

with $E[S(\theta, Z, y)S^T(\theta, Z, y)]$ obtained via Monte Carlo simulations. Let $Z^{(t)}$ and $y^{(t)}$ be simulated Z and y at the tth sample so that $S(\theta, Z^{(t)}, y^{(t)})$ is the score vector given $Z^{(t)}$, $y^{(t)}$, and $\theta = \hat{\theta}$. The Monte Carlo approximation of $E[S(\theta, Z, y)S^T(\theta, Z, y)]$ is

$$E\left[S(\theta, Z, y)S^T(\theta, Z, y)\right] \approx \frac{1}{T}\sum_{t=1}^{T} S(\theta, Z^{(t)}, y^{(t)})S^T(\theta, Z^{(t)}, y^{(t)}) \qquad (10.73)$$

where T is a large number, say 10,000. The liability for the jth individual, y_j, is simulated from a truncated normal distribution. We adopt the inverse transformation method that has an acceptance rate of 100 % (Rubinstein 1981). With this method, we first defined

$$v = 1 - \Phi(X_j\beta + Z_j\gamma) \qquad (10.74)$$

and then simulated a variable u from $U(0, 1)$. Finally, we took the inverse function of the standardized normal distribution to obtain

$$y_j = Y_{j1}\Phi^{-1}(u\,v) + Y_{j2}\Phi^{-1}[v + u(1 - v)] \tag{10.75}$$

Intrinsic functions for both $\Phi(.)$ and $\Phi^{-1}(.)$ are available in many computer software packages. For example, in the SAS package (SAS Institute 2008a), $\Phi(x)$ is coded as $\Phi(x) = \text{probnorm}(x)$ and $\Phi^{-1}(u)$ is coded as $\Phi^{-1}(u) = \text{probit}(u)$. The Monte Carlo approximation is time consuming so that we cannot calculate the information matrix for every point of the genome scanned. Instead, we only calculate the information matrix at the points where evidences of QTL are strong.

10.6   Logistic Analysis

Similar to the probit link function, we may also use the logit link function to perform the generalized linear model analysis. Let

$$\zeta_{jk} = \frac{\exp(\alpha_k + X_j\beta + Z_j\gamma)}{1 + \exp(\alpha_k + X_j\beta + Z_j\gamma)} \tag{10.76}$$

be the cumulative distribution function of $\alpha_k + X_j\beta + Z_j\gamma$. Under the logistic model, the mean of Y_{jk} is modeled by

$$\mu_{jk} = E(Y_{jk}) = \zeta_{jk} - \zeta_{j(k-1)} \tag{10.77}$$

The logistic model for the binary data is

$$\mu_{jk} = \begin{cases} \zeta_{j1} & \text{for } k = 1 \\ 1 - \zeta_{j1} & \text{for } k = 2 \end{cases} \tag{10.78}$$

From $\mu_{j1} = \zeta_{j1}$, we obtain

$$\text{logit}(\mu_{j1}) = \ln\left(\frac{\mu_{j1}}{1 - \mu_{j1}}\right) = \alpha_1 + X_j\beta + Z_j\gamma \tag{10.79}$$

Both the probit and logit transformations of the expectation of Y_{j1} lead to a linear model. Note that the linear model obtained here only shows the property of the transformation. In the actual theory development and data analysis, the linear transformations in (10.6) and (10.79) are never used. Showing the linear transformations may potentially cause confusion to students because, by intuition, they may try to transform the ordinal data (Y_{jk}) first and then conduct the usual linear regression on the transformed data, which is not appropriate and certainly not the intention of the GLM developers. The maximum likelihood analysis under the homogeneous variance, heterogeneous variance, and mixture model and the EM algorithm described previously in the probit analysis apply to the logistic analysis. We only show the logistic analysis under the homogeneous variance model as an example. Note that under this model, we only need to substitute Z_j by U_j to define the expectation, i.e.,

$$\zeta_{jk} = \frac{\exp(\alpha_k + X_j\beta + U_j\gamma)}{1 + \exp(\alpha_k + X_j\beta + U_j\gamma)} \tag{10.80}$$

and

$$\mu_{jk} = E(Y_{jk}) = \zeta_{jk} - \zeta_{j(k-1)} \tag{10.81}$$

Once μ_j is defined, the weight W_j is also defined. The only item left is D_j, which is

$$D_j = \frac{\partial \mu_j}{\partial \theta^T} = \left[\frac{\partial \mu_j}{\partial \alpha^T} \frac{\partial \mu_j}{\partial \beta^T} \frac{\partial \mu_j}{\partial \gamma^T} \right] \tag{10.82}$$

The first block $\partial \mu_j / \partial \alpha^T = \{\partial \mu_{jk} / \partial \alpha_l\}$ is a $(p+1) \times p$ matrix with

$$\frac{\partial \mu_{jk}}{\partial \alpha_{k-1}} = -\zeta_{j(k-1)}(1 - \zeta_{j(k-1)})$$

$$\frac{\partial \mu_{jk}}{\partial \alpha_k} = \zeta_{jk}(1 - \zeta_{jk})$$

$$\frac{\partial \mu_{jk}}{\partial \alpha_l} = 0, \ \forall l \neq \{k-1, k\} \tag{10.83}$$

The second block $\partial \mu_j / \partial \beta^T = \{\partial \mu_{jk} / \partial \beta\}$ is a $(p+1) \times q$ matrix with

$$\frac{\partial \mu_{jk}}{\partial \beta} = X_j^T \zeta_{jk}(1 - \zeta_{jk}) - X_j^T \zeta_{j(k-1)}(1 - \zeta_{j(k-1)}) \tag{10.84}$$

The third block $\partial \mu_j / \partial \gamma^T = \{\partial \mu_{jk} / \partial \gamma\}$ is a $(p+1) \times r$ matrix with

$$\frac{\partial \mu_{jk}}{\partial \gamma} = U_j^T \zeta_{jk}(1 - \zeta_{jk}) - U_j^T \zeta_{j(k-1)}(1 - \zeta_{j(k-1)}) \tag{10.85}$$

In the above partial derivatives, the range of k is $k = 1, \ldots, p+1$.

10.7 Example

The experiment was conducted by Dou et al. (2009). A female sterile line of wheat XND126 and an elite wheat cultivar Gaocheng 8901 with normal fertility were crossed for genetic analysis of female sterility measured as the number of

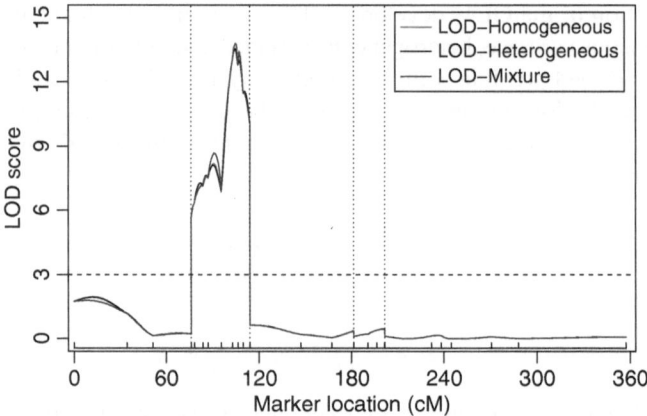

Fig. 10.2 The LOD test statistic profiles for three methods of interval mapping (HOMOGE-NEOUS, HETEROGENEOUS, and MIXTURE). The data were obtained from Dou et al. (2009). The trait investigated is the female fertility of wheat measured as a binary trait (seed presence and absence). The five chromosomes (part of the wheat genome) are separated by the *vertical dotted lines*. The unevenly distributed *black ticks* on the horizontal axis indicate the marker locations

seeded spikelets per plant. The parents, their F_1 and F_2 progeny, were planted at the Huaian experimental station in China for the 2006–2007 growing season under the normal autumn sowing condition. The mapping population was an F_2 family consisting of 243 individual plants. About 84 % of the F_2 progeny had seeded spikelets, and the remaining 16 % plants did not have any seeds at all. Among the plants with seeded spikelets, the number of seeded spikelets varied from one to as many as 31. The phenotype is the count data point and can be modeled using the Poisson distribution. The phenotype can also be treated as a binary data point and analyzed using the Bernoulli distribution. In this example, we treated the phenotype as a binary data (seed presence and absence) and analyzed it using the Bernoulli distribution. A total of 28 SSR markers were used in this experiment. These markers covered five chromosomes of the wheat genome with an average genome marker density of 15.5 cM per marker interval. The five chromosomes are only part of the wheat genome. These chromosomes were scanned for QTL of the binary data. Let A_1 and A_2 be the alleles carried by Gaocheng 8901 and XDN128, respectively. Let A_1A_1, A_1A_2, and A_2A_2 be the three genotypes for the QTL of interest. The genotype is numerically coded as 1, 0, and -1, respectively, for the three genotypes. The genome was scanned with 1-cM increment. All the three methods described in this chapter were used for the interval mapping. They are the homogeneous variance model (HOMOGENEOUS), the heterogeneous variance model (HETEROGENEOUS), and the mixture model (MIXTURE). The LOD score profiles are depicted in Fig. 10.2. When LOD = 3 is used as the threshold value, all three methods detected two major QTL on chromosome 2. The LOD score for the mixture model appears to be higher than the other two models, but the difference is very small and can be safely ignored.

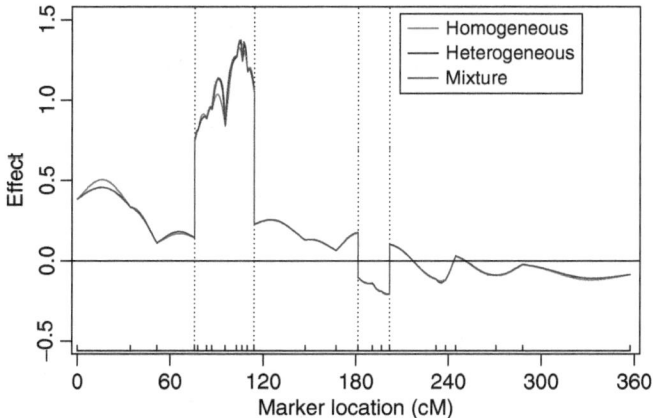

Fig. 10.3 The QTL effect profiles for three methods of interval mapping (HOMOGENEOUS, HETEROGENEOUS, and MIXTURE). The data were obtained from Dou et al. (2009). The trait investigated is the female fertility of wheat measured as a binary trait (seed presence and absence). The five chromosomes (part of the wheat genome) are separated by the *vertical dotted lines*. The unevenly distributed *black ticks* on the horizontal axis indicate the marker locations

The estimated QTL effect profiles are given in Fig. 10.3. Again the three methods are almost the same for the estimated QTL effects except that the mixture model and the heterogeneous model give slightly higher estimates than the homogeneous model. In practice, we recommend the heterogeneous model because it produces almost the same result as the mixture model but with much less computing time than the mixture model.

Chapter 11
Mapping Segregation Distortion Loci

A basic assumption in QTL mapping is that genomic loci (QTL and markers) follow the Mendelian segregation ratio. The Mendelian ratio depends on the population under investigation. For example, in a BC population, the Mendelian ratio is $1 : 1$ for the two genotypes ($A_1 A_1$, $A_1 A_2$). In an F_2 population, the Mendelian ratio is $1 : 2 : 1$ for the three genotypes ($A_1 A_1$, $A_1 A_2$, and $A_2 A_2$). If the segregation ratio of a locus deviates from the Mendelian segregation ratio, we say that the locus is a non-Mendelian locus or segregation distortion locus (SDL). In fact, a marker whose segregation deviates from the Mendelian ratio is not necessarily an SDL. It is most likely that a true SDL sits nearby the marker and the observed segregation distortion of the marker is caused by the SDL because of linkage. Sometimes we may see markers in several regions of the genome that show segregation distortion. This may be caused by several SDL across the genome. The SDL themselves may be caused by viability selection. In other words, different genotypes of the SDL may have different viabilities. Genotypes that are favored by the viability selection are overrepresented, while genotypes that are against by the viability selection are underrepresented. Therefore, an SDL may also be called viability locus (VL). Viability selection may happen in the gametic level or zygotic level or both. But it is hard to tell the difference between gametic selection and zygotic selection unless we can directly observe the gametes. Like quantitative trait loci, segregation distortion loci can be mapped using marker information. Evolutionary biologists may be more interested in SDL, while agricultural scientists may be more interested in QTL. In a single experiment of genetic mapping, we may simultaneously investigate both SDL and QTL.

The earliest work in SDL mapping was Fu and Ritland (1994). For the first time, the authors proposed the viability selection hypothesis and tried to map SDL using marker information under the maximum likelihood framework. Mitchell-Olds (1995) also developed a similar ML method to map SDL in an F_2 population. A more systematic treatment of SDL mapping was made by Lorieux et al. (1995a,b) using genome-wide markers. Vogl and Xu (2000) took a Bayesian approach to mapping viability selection loci. Luo and Xu (2003) developed an EM algorithm to estimate the segregation ratio under the ML framework. Luo et al. (2005) eventually

S. Xu, *Principles of Statistical Genomics*, DOI 10.1007/978-0-387-70807-2_11,
© Springer Science+Business Media, LLC 2013

developed a quantitative genetics model to estimate the genetic effects of viability
loci using a four-way cross design. Some of the methods have been applied to SDL
mapping in rice (Wang et al. 2005a).

This chapter will introduce methods for mapping SDL under two different
models. One is called the probabilistic model (Luo and Xu 2003), and the other
is called the liability model (Luo et al. 2005). Under some special situations, both
models will generate the same result, but in most situations, the liability model is
more efficient. In the last section, we will combine QTL mapping and SDL mapping
together and jointly map QTL and SDL (Xu and Hu 2009).

11.1   Probabilistic Model

Consider an SDL with an arbitrary segregation ratio in an F_2 family derived from
the cross of two inbred lines. Let M and N be the left and right flanking markers
bracketing the SDL (denoted by G for short). The interval of the genome carrying
the three loci is denoted by a segment MGN. The three genotypes of the SDL are
denoted by $G_1 G_1$, $G_1 G_2$, and $G_2 G_2$, respectively. Similar notation also applies to
the genotypes of the flanking markers. The interval defined by markers M and N
is divided into two segments. Let r_1 and r_2 be the recombination fractions for
segment MG and segment GN, respectively. The joint distribution of the marker
genotypes conditional on the SDL genotype can be derived using the Markovian
property under the assumption of no segregation interference between consecutive
loci. Let us order the three genotypes, $G_1 G_1$, $G_1 G_2$, and $G_2 G_2$, as genotypes 1, 2,
and 3, respectively. If individual j takes the κth genotype for the SDL, we denote
the event by $G_j = \kappa$, $\forall \kappa = 1, 2, 3$. The joint probability of the two markers
conditional on the genotype of the SDL is

$$\Pr(M_j = \xi, N_j = \zeta | G_j = \kappa)$$
$$= \Pr(M_j = \xi | G_j = \kappa) \Pr(N_j = \zeta | G_j = \kappa) \qquad (11.1)$$

for all $\kappa, \xi, \zeta = 1, 2, 3$, where $\Pr(M_j = \xi | G_j = \kappa) = T_1(\kappa, \xi)$ and $\Pr(N_j = \zeta | G_j = \kappa) = T_2(\kappa, \zeta)$. We use $T_i(\kappa, \xi)$ to denote the κth row and the ξth column of
the following transition matrix

$$T_i = \begin{bmatrix} (1 - r_i)^2 & 2r_i(1 - r_i) & r_i^2 \\ r_i(1 - r_i) & (1 - r_i)^2 + r_i^2 & r_i(1 - r_i) \\ r_i^2 & 2r_i(1 - r_i) & (1 - r_i)^2 \end{bmatrix}, \forall i = 1, 2 \qquad (11.2)$$

For example,

$$\Pr(M_j = 1, N_j = 2 | G_j = 3)$$
$$= \Pr(M_j = 1 | G_j = 3) \Pr(N_j = 2 | G_j = 3) \qquad (11.3)$$
$$= T_1(3, 1) \, T_2(3, 2) = 2r_1^2 r_2(1 - r_2) \qquad (11.4)$$

Let $\omega_\kappa = \Pr(G = \kappa)$, $\forall \kappa = 1, 2, 3$, be the probability that a randomly sampled individual from the F_2 family takes the κth genotype. Let $\omega = \{\omega_1, \omega_2, \omega_3\}$ be the array of the genotype frequencies, and it is the vector of parameters for estimation and test. Under Mendelian segregation, the three genotype frequencies are denoted by $\phi = \{\frac{1}{4}, \frac{1}{2}, \frac{1}{4}\}$. Therefore, the null hypothesis is that the F_2 population is a Mendelian population, i.e., $\omega = \phi$. We use a generic notation p for probability, so that $p(G_j = \kappa)$ represents $\Pr(G_j = \kappa)$ and $p(M_j, N_j | G_j = \kappa)$ stands for $\Pr(M_j, N_j | G_j = \kappa)$. Given the parameters ω, the data (flanking marker genotypes), and the multinomial probability model, we are ready to construct the log likelihood function, which is

$$L(\omega) = \sum_{j=1}^{n} \ln \left[\sum_{\kappa=1}^{3} p(G_j = \kappa) p(M_j, N_j | G_j = \kappa) \right]$$

$$= \sum_{j=1}^{n} \ln \left[\sum_{\kappa=1}^{3} \omega_\kappa T_1(\kappa, M_j) T_2(\kappa, N_j) \right] \tag{11.5}$$

where the parameters have a restriction $\sum_{\kappa=1}^{3} \omega_\kappa = 1$. Note that without any other information, $p(G_j = \kappa) = \omega_\kappa$, $\forall j = 1, \ldots, n$. Under the assumption of Mendelian segregation, $\omega = \phi$, i.e., $\omega_1 = \omega_3 = \frac{1}{2}\omega_2 = \frac{1}{4}$. However, we treat ω as unknown parameters. We postulate that deviation of ω from the Mendelian ratio will cause a marker linked to locus G to show distorted segregation. This likelihood function has been used by Luo et al. (2005) for mapping SDL.

11.1.1 The EM Algorithm

The MLE of the parameters can be solved via the EM algorithm (Dempster et al. 1977). We need to rewrite the likelihood function in a form of complete-data. Let us define a delta function as

$$\delta(G_j, \kappa) = \begin{cases} 1 & \text{if } G_j = \kappa \\ 0 & \text{if } G_j \neq \kappa \end{cases} \tag{11.6}$$

If the genotypes of the SDL are known for all individuals, i.e., given $\delta(G_j, \kappa)$ for all $j = 1, \ldots, n$ and $\kappa = 1, 2, 3$, the complete-data log likelihood is

$$L(\omega, \delta) = \sum_{j=1}^{n} \ln[p(M_j, N_j | G_j) p(G_j)] \tag{11.7}$$

where

$$p(M_j, N_j | G_j) = \prod_{\kappa=1}^{3} p(M_j, N_j | G_j = \kappa)^{\delta(G_j, \kappa)}$$

$$= \prod_{\kappa=1}^{3} [T_1(\kappa, M_j) T_2(\kappa, N_j)]^{\delta(G_j, \kappa)} \quad (11.8)$$

and

$$p(G_j) = \prod_{\kappa=1}^{3} \omega_{\kappa}^{\delta(G_j, \kappa)} \quad (11.9)$$

Therefore, the complete-data log likelihood function can be rewritten as

$$L(\omega, \delta) = \sum_{j=1}^{n} \sum_{\kappa=1}^{3} \delta(G_j, \kappa) \{ \ln[T_1(\kappa, M_j)] + \ln[T_2(\kappa, N_j)] + \ln(\omega_{\kappa}) \} \quad (11.10)$$

This log likelihood function involves missing value $\delta(G_j, \kappa)$ and thus cannot be used directly. We need to take expectation of this function with respect to $\delta(G_j, \kappa)$. In addition, we introduce a Lagrange multiplier to make sure that the parameters are estimated within their restriction, i.e., $\sum_{\kappa=1}^{3} \omega_{\kappa} = 1$. Therefore, the actual log likelihood function that is maximized in the EM algorithm is

$$E[L(\omega, \delta)] = \sum_{j=1}^{n} \sum_{\kappa=1}^{3} E[\delta(G_j, \kappa)] \{ \ln[T_1(\kappa, M_j)] + \ln[T_2(\kappa, N_j)] + \ln(\omega_{\kappa}) \}$$

$$+ \lambda \left(1 - \sum_{\kappa=1}^{3} \omega_{\kappa} \right) \quad (11.11)$$

where λ is a Lagrange multiplier and is treated as a parameter for estimation. Before we maximize the above expected complete-data log likelihood function, we need to calculate $E[\delta(G_j, \kappa)]$, which is called the posterior expectation of the missing genotype and is calculated using Bayes' theorem,

$$E[\delta(G_j, \kappa)] = \frac{\omega_{\kappa} T_1(\kappa, M_j) T_2(\kappa, N_j)}{\sum_{\kappa'}^{3} \omega_{\kappa'} T_1(\kappa', M_j) T_2(\kappa', N_j)} \quad (11.12)$$

Note that this posterior expectation requires the value of parameter ω, which happens to be what we want to estimate. Therefore, iterations are required. Once an initial value of ω is provided, we can find the posterior expectation of the missing genotype, which further allows us to maximize the expected complete-data log likelihood function, (11.11). To maximize $E[L(\omega, \delta)]$, we take the partial derivatives of $E[L(\omega, \delta)]$ with respect to the parameters and equate the partial derivative to zero and solve for the parameters. The partial derivatives are

$$\frac{\partial}{\partial \omega_\kappa} E[L(\theta, \delta)] = \sum_{j=1}^{n} E[\delta(G_j, \kappa)] \frac{1}{\omega_\kappa} - \lambda, \forall \kappa = 1, 2, 3 \qquad (11.13)$$

and

$$\frac{\partial}{\partial \lambda} E[L(\omega, \delta)] = 1 - \sum_{\kappa=1}^{3} \omega_\kappa \qquad (11.14)$$

Setting (11.13) to zero, we get

$$\omega_\kappa = \frac{1}{\lambda} \sum_{j=1}^{n} E[\delta(G_j, \kappa)], \forall \kappa = 1, 2, 3 \qquad (11.15)$$

Equation (11.14) is just the restriction, which allows us to solve for λ. Note that $\sum_{\kappa=1}^{3} E[\delta(G_j, \kappa)] = 1$, i.e., the sum of the three conditional probabilities is unity. This leads to

$$\sum_{\kappa=1}^{3} \omega_\kappa = \frac{1}{\lambda} \sum_{\kappa=1}^{3} \sum_{j=1}^{n} E[\delta(G_j, \kappa)] = \frac{1}{\lambda} \sum_{j=1}^{n} \sum_{\kappa=1}^{3} E[\delta(G_j, \kappa)] = \frac{n}{\lambda} = 1 \quad (11.16)$$

As a result, we get $\lambda = n$. Substituting $\lambda = n$ into (11.15) leads to

$$\omega_\kappa = \frac{1}{n} \sum_{j=1}^{n} E[\delta(G_j, \kappa)], \forall \kappa = 1, 2, 3 \qquad (11.17)$$

The Lagrange multiplier λ is a nuisance parameter that allows us to find solution of ω_k in a convenient way. The EM algorithm is summarized as:

1. Initialize $\omega = \omega^{(0)}$.
2. Calculate $E[\delta(G_j, \kappa)]$ using (11.12) (the E-step).
3. Update ω using (11.17) (the M-step).
4. Repeat the E-step and the M-step until a certain criterion of convergence is satisfied.

11.1.2 Hypothesis Test

The null hypothesis is H_0: $\omega = \phi$. The alternative hypothesis is H_A: $\omega \neq \phi$. The likelihood ratio test statistic is used to test the null hypothesis. The likelihood ratio test statistic is

$$LRT = -2[L_0(\phi) - L_1(\hat{\omega})] \qquad (11.18)$$

where

$$L_1(\hat{\omega}) = \sum_{j=1}^{n} \ln \left[\sum_{\kappa=1}^{3} \hat{\omega}_\kappa T_1(\kappa, M_j) T_2(\kappa, N_j) \right] \qquad (11.19)$$

is the observed log likelihood function evaluated at $\omega = \hat{\omega}$ and

$$L_0(\phi) = \sum_{j=1}^{n} \ln \left[\sum_{\kappa=1}^{3} \phi_\kappa T_1(\kappa, M_j) T_2(\kappa, N_j) \right] \qquad (11.20)$$

is the log likelihood function evaluated at $\omega = \phi$. Under the null hypothesis, LRT will follow a chi-square distribution with 2 degrees of freedom. The reason for the 2 degrees of freedom is that we only have two (not three) independent parameters to estimate.

11.1.3   Variance Matrix of the Estimated Parameters

There are three parameters in vector $\omega = \{\omega_1, \omega_2, \omega_3\}$, but only two are independent because of the restriction $\sum_{\kappa=1}^{3} \omega_\kappa = 1$. Therefore, we only need to find the variance matrix of two components. Let us express $\omega_3 = 1 - \omega_1 - \omega_2$ so that

$$\text{var}(\hat{\omega}_3) = \text{var}(\hat{\omega}_1) + \text{var}(\hat{\omega}_2) + 2\text{cov}(\hat{\omega}_1, \hat{\omega}_2) \qquad (11.21)$$

$$\text{var}(\hat{\omega}_1, \hat{\omega}_3) = \text{cov}(\hat{\omega}_1, 1 - \hat{\omega}_1 - \hat{\omega}_2) = -\text{var}(\hat{\omega}_1) - \text{cov}(\hat{\omega}_1, \hat{\omega}_2) \qquad (11.22)$$

and

$$\text{var}(\hat{\omega}_2, \hat{\omega}_3) = \text{cov}(\hat{\omega}_2, 1 - \hat{\omega}_1 - \hat{\omega}_2) = -\text{var}(\hat{\omega}_2) - \text{cov}(\hat{\omega}_1, \hat{\omega}_2) \qquad (11.23)$$

We now redefine $\omega = \{\hat{\omega}_1, \hat{\omega}_2\}$ as a vector with two components only. Therefore, we only need to derive the variance–covariance matrix for vector $\omega = \{\hat{\omega}_1, \hat{\omega}_2\}$ because the variance for $\hat{\omega}_3$ and the covariances involving $\hat{\omega}_3$ are all functions of $\text{var}(\omega)$. Let

$$L_j(\omega, \delta) = \sum_{\kappa=1}^{3} \delta(G_j, \kappa)\{\ln[T_1(\kappa, M_j)] + \ln[T_2(\kappa, N_j)] + \ln(\omega_\kappa)\} \qquad (11.24)$$

be the complete-data log likelihood function for individual j so that $L(\omega, \delta) = \sum_{j=1}^{n} L_j(\omega, \delta)$. Note that whenever ω_3 occurs, it is replaced by $\omega_3 = 1 - \omega_1 - \omega_2$. The Louis (1982) information matrix is

$$I(\hat{\omega}) = -E[H(\hat{\omega}, \delta)] - \text{var}[S(\hat{\omega}, \delta)] \qquad (11.25)$$

where

$$
E[H(\hat{\omega}, \delta)] =
\begin{bmatrix}
\sum\limits_{j=1}^{n} E\left(\frac{\partial^2 L_j(\omega,\delta)}{\partial\omega_1^2}\right) & \sum\limits_{j=1}^{n} E\left(\frac{\partial^2 L_j(\omega,\delta)}{\partial\omega_1\partial\omega_2}\right) \\
\sum\limits_{j=1}^{n} E\left(\frac{\partial^2 L_j(\omega,\delta)}{\partial\omega_1\partial\omega_2}\right) & \sum\limits_{j=1}^{n} E\left(\frac{\partial^2 L_j(\omega,\delta)}{\partial\omega_2^2}\right)
\end{bmatrix}
\tag{11.26}
$$

is the expectation of the Hessian matrix of the complete-data log likelihood function and

$$
\text{var}[S(\hat{\omega}, \delta)] =
\begin{bmatrix}
\sum\limits_{j=1}^{n} \text{var}\left(\frac{\partial L_j(\omega,\delta)}{\partial\omega_1}\right) & \sum\limits_{j=1}^{n} \text{cov}\left(\frac{\partial L_j(\omega,\delta)}{\partial\omega_1}, \frac{\partial L_j(\omega,\delta)}{\partial\omega_2}\right) \\
\sum\limits_{j=1}^{n} \text{cov}\left(\frac{\partial L_j(\omega,\delta)}{\partial\omega_1}, \frac{\partial L_j(\omega,\delta)}{\partial\omega_2}\right) & \sum\limits_{j=1}^{n} \text{var}\left(\frac{\partial L_j(\omega,\delta)}{\partial\omega_2}\right)
\end{bmatrix}
\tag{11.27}
$$

is the variance–covariance matrix of the score vector of the complete-data log likelihood function. Both the expectation and the variance are taken with respect to the missing value $\delta(G_j, \kappa)$ using the posterior distribution of the genotype of the SDL. The inverse of the information matrix is used as an approximation of the variance matrix of $\hat{\omega} = \{\hat{\omega}_1, \hat{\omega}_2\}$ as shown below:

$$
\text{var}(\hat{\omega}) =
\begin{bmatrix}
\text{var}(\hat{\omega}_1) & \text{cov}(\hat{\omega}_1, \hat{\omega}_2) \\
\text{cov}(\hat{\omega}_1, \hat{\omega}_2) & \text{var}(\hat{\omega}_2)
\end{bmatrix}
\tag{11.28}
$$

i.e., $\text{var}(\hat{\omega}) \approx I^{-1}(\hat{\omega})$.

We now evaluate each element of $E[H(\hat{\omega}, \delta)]$ and $\text{var}[S(\hat{\omega}, \delta)]$. For the expected Hessian matrix, we have

$$
E\left(\frac{\partial^2 L_j(\omega, \delta)}{\partial\omega_1^2}\right) = -E[\delta(G_j, 1)]\frac{1}{\omega_1^2} - E[\delta(G_j, 3)]\frac{1}{(1 - \omega_1 - \omega_2)^2}
$$

$$
E\left(\frac{\partial^2 L_j(\omega, \delta)}{\partial\omega_2^2}\right) = -E[\delta(G_j, 2)]\frac{1}{\omega_2^2} - E[\delta(G_j, 3)]\frac{1}{(1 - \omega_1 - \omega_2)^2}
$$

$$
E\left(\frac{\partial^2 L_j(\omega, \delta)}{\partial\omega_1\partial\omega_2}\right) = -E[\delta(G_j, 3)]\frac{1}{(1 - \omega_1 - \omega_2)^2}
\tag{11.29}
$$

For the variance matrix of the score vector, we have

$$
\text{var}\left(\frac{\partial L_j(\omega, \delta)}{\partial\omega_1}\right) = \frac{1}{\omega_1^2}\text{var}[\delta(G_j, 1)] + \frac{1}{\omega_3^2}\text{var}[\delta(G_j, 3)]
$$

$$
- \frac{2}{\omega_1\omega_3}\text{cov}[\delta(G_j, 1), \delta(G_j, 3)]
$$

$$\mathrm{var}\left(\frac{\partial L_j(\omega,\delta)}{\partial\omega_2}\right) = \frac{1}{\omega_2^2}\mathrm{var}[\delta(G_j,2)] + \frac{1}{\omega_3^2}\mathrm{var}[\delta(G_j,3)]$$

$$- \frac{2}{\omega_2\omega_3}\mathrm{cov}[\delta(G_j,2),\delta(G_j,3)]$$

$$\mathrm{cov}\left(\frac{\partial L_j(\omega,\delta)}{\partial\omega_1},\frac{\partial L_j(\omega,\delta)}{\partial\omega_2}\right) = \frac{1}{\omega_1\omega_2}\mathrm{cov}[\delta(G_j,1),\delta(G_j,2)]$$

$$- \frac{1}{\omega_2\omega_3}\mathrm{cov}[\delta(G_j,2),\delta(G_j,3)]$$

$$- \frac{1}{\omega_1\omega_3}\mathrm{cov}[\delta(G_j,1),\delta(G_j,3)]$$

$$+ \frac{1}{\omega_3^2}\mathrm{var}[\delta(G_j,3)] \qquad (11.30)$$

Note again that $\omega_3 = 1-\omega_1-\omega_2$ for notational simplicity. The variance–covariance matrix of the score vector requires the variance–covariance matrix of vector $\delta_j = \{\delta(G_j,1),\delta(G_j,2),\delta(G_j,3)\}$. Let $\pi_{j\kappa} = E[\delta(G_j,\kappa)], \forall\kappa = 1,2,3$ be the short notation for the posterior expectation of $\delta(G_j,\kappa)$. The variance–covariance matrix of vector δ_j is

$$\mathrm{var}(\delta_j) = \begin{bmatrix} \pi_{j1}(1-\pi_{j1}) & -\pi_{j1}\pi_{j2} & -\pi_{j1}\pi_{j3} \\ -\pi_{j1}\pi_{j2} & \pi_{j2}(1-\pi_{j2}) & -\pi_{j2}\pi_{j3} \\ -\pi_{j1}\pi_{j3} & -\pi_{j2}\pi_{j3} & \pi_{j3}(1-\pi_{j3}) \end{bmatrix} \qquad (11.31)$$

Elements of the score vector and the Hessian matrix for individual j are the first and second partial derivatives of $L_j(\omega,\delta)$ with respect to ω. These are given as follows:

$$\frac{\partial L_j(\omega,\delta)}{\partial\omega_1} = \delta(G_j,1)\frac{1}{\omega_1} - \delta(G_j,3)\frac{1}{1-\omega_1-\omega_2}$$

$$\frac{\partial L_j(\omega,\delta)}{\partial\omega_2} = \delta(G_j,2)\frac{1}{\omega_2} - \delta(G_j,3)\frac{1}{1-\omega_1-\omega_2} \qquad (11.32)$$

and

$$\frac{\partial^2 L_j(\omega,\delta)}{\partial\omega_1^2} = -\delta(G_j,1)\frac{1}{\omega_1^2} - \delta(G_j,3)\frac{1}{(1-\omega_1-\omega_2)^2}$$

$$\frac{\partial^2 L_j(\omega,\delta)}{\partial\omega_2^2} = -\delta(G_j,2)\frac{1}{\omega_2^2} - \delta(G_j,3)\frac{1}{(1-\omega_1-\omega_2)^2}$$

$$\frac{\partial^2 L_j(\omega,\delta)}{\partial\omega_1\partial\omega_2} = -\delta(G_j,3)\frac{1}{(1-\omega_1-\omega_2)^2} \qquad (11.33)$$

11.1.4 Selection Coefficient and Dominance

In viability selection, we often use selection coefficient and the degree of dominance to express the intensity of selection. There is a unique relationship between segregation distortion and the selection intensity. Let w_{11}, w_{12}, and w_{22} be the relative fitness of the three genotypes ($G_1 G_1, G_1 G_2$, and $G_2 G_2$) of the SDL, respectively. Let s and h be the selection coefficient and degree of dominance. The relative fitness can be expressed as (Hartl and Clark 1997)

$$\begin{aligned}
w_{11} &= 1 \\
w_{12} &= 1 - sh \\
w_{22} &= 1 - s
\end{aligned}$$

(11.34)

In an F_2 population, the average fitness is

$$\bar{w} = \frac{1}{4}w_{11} + \frac{1}{2}w_{12} + \frac{1}{4}w_{22} = \frac{1}{4} + \frac{1}{2}(1 - sh) + \frac{1}{4}(1 - s) \tag{11.35}$$

The segregation ratio after the viability selection is

$$\omega_1 = \frac{\frac{1}{4}w_{11}}{\bar{w}} = \frac{1}{1 + 2(1 - sh) + (1 - s)}$$

$$\omega_2 = \frac{\frac{1}{2}w_{11}}{\bar{w}} = \frac{2(1 - sh)}{1 + 2(1 - sh) + (1 - s)}$$

$$\omega_3 = \frac{\frac{1}{4}w_{11}}{\bar{w}} = \frac{1 - s}{1 + 2(1 - sh) + (1 - s)} \tag{11.36}$$

This equation system represents the relationship between the segregation ratio and the intensity of viability selection. The inverse relationship is given by Luo et al. (2005)

$$s = \frac{\omega_1 - \omega_3}{\omega_1}$$

$$h = \frac{\omega_1 - \frac{1}{2}\omega_2}{\omega_1 - \omega_3} \tag{11.37}$$

which is used to obtain the MLE of s and h given the MLE of $\omega = \{\omega_1, \omega_2, \omega_3\}$.

11.2 Liability Model

Systematic environmental effects may mask the effects of viability loci and cause low power of detection. It is impossible to remove the systematic error from the analysis using the probabilistic model described above. However, the liability model proposed here provides an extremely convenient way to remove such systematic errors. Let y_j be an underlying liability for individual j in the F_2 population. We use the following linear model to describe y_j:

$$y_j = X_j\beta + Z_j\gamma + \varepsilon_j \tag{11.38}$$

where β is a vector of nongenetic effects (systematic error effects), X_j is a design matrix for the systematic errors, $Z_j = \{Z_{j1}, Z_{j2}\}$ represents the genotypes of the SDL and has been defined earlier in QTL mapping, $\gamma = \{a, d\}$ are the genetic effects of QTL as defined earlier, and $\varepsilon_j \sim N(0, 1)$ is the residual error for the liability. We can see that the liability is simply a regular quantitative trait, except that it is not observable. Because the liability is a hypothetical variable, the residual variance cannot be estimated, and thus, we set the variance to unity. We assume that viability selection acts on the liability under the truncation selection scheme, i.e., individual j will survive if $y_j \geq 0$; otherwise, it will be eliminated from the population. Since all individuals observed in the F_2 population are survivors, $y_j \geq 0$ applies to all individuals. The probability that $y_j \geq 0$ is

$$\Pr(y_j \geq 0) = \Phi(X_j\beta + Z_j\gamma) \tag{11.39}$$

where $\Phi(.)$ is the standardized cumulative normal function. This probability may be considered as the relative fitness. Recall that

$$Z_{j1} = \begin{cases} +1 & \text{for } G_1G_1 \\ 0 & \text{for } G_1G_2 \\ -1 & \text{for } G_2G_2 \end{cases} \tag{11.40}$$

and

$$Z_{j2} = \begin{cases} 0 & \text{for } G_1G_1 \\ 1 & \text{for } G_1G_2 \\ 0 & \text{for } G_2G_2 \end{cases} \tag{11.41}$$

are the indicator variables for the QTL genotype. Therefore, given each of the three genotypes, we have

$$\begin{aligned} \Pr(y_j \geq 0|G_1G_1) &= w_j(11) = \Phi(X_j\beta + a) \\ \Pr(y_j \geq 0|G_1G_2) &= w_j(12) = \Phi(X_j\beta + d) \\ \Pr(y_j \geq 0|G_2G_2) &= w_j(22) = \Phi(X_j\beta - a) \end{aligned} \tag{11.42}$$

Let us define the expected relative fitness for individual j by

$$\begin{aligned} \bar{w}_j &= \frac{1}{4}w_j(11) + \frac{1}{2}w_j(12) + \frac{1}{4}w_j(22) \\ &= \frac{1}{4}\Phi(X_j\beta + a) + \frac{1}{2}\Phi(X_j\beta + d) + \frac{1}{4}\Phi(X_j\beta - a) \end{aligned} \tag{11.43}$$

The normalized fitness for individual j is

$$\omega_j(1) = \frac{\frac{1}{4}w_j(11)}{\bar{w}_j} = \frac{\Phi(X_j\beta + a)}{\Phi(X_j\beta + a) + 2\Phi(X_j\beta + d) + \Phi(X_j\beta - a)}$$

$$\omega_j(2) = \frac{\frac{1}{2}w_j(12)}{\bar{w}_j} = \frac{2\Phi(X_j\beta + d)}{\Phi(X_j\beta + a) + 2\Phi(X_j\beta + d) + \Phi(X_j\beta - a)}$$

$$\omega_j(3) = \frac{\frac{1}{4}w_j(22)}{\bar{w}_j} = \frac{\Phi(X_j\beta - a)}{\Phi(X_j\beta + a) + 2\Phi(X_j\beta + d) + \Phi(X_j\beta - a)}$$

$$(11.44)$$

Under the liability model, the parameter vector is $\theta = \{\beta, \gamma\}$. We have formulated the problem of mapping SDL into that of mapping QTL. The log likelihood function is

$$L(\theta) = \sum_{j=1}^{n} \ln\left[\sum_{\kappa=1}^{3} \omega_j(\kappa)T_1(\kappa, M_j)T_2(\kappa, N_j)\right] \qquad (11.45)$$

11.2.1   EM Algorithm

Due to the complexity of the likelihood function, there has been no simple algorithm for the MLE of the parameters. Therefore, Luo et al. (2005) used the simplex algorithm (Nelder and Mead 1965) to search for the MLE of parameters. An EM algorithm does exist except that the maximization step is much more complicated than that under the probabilistic model. Let us look at the log likelihood function used in the complete-data situation, i.e., $\delta(G_j, \kappa)$ is treated as known:

$$L(\theta, \delta) = \sum_{j=1}^{n} L_j(\theta, \delta) \qquad (11.46)$$

where

$$
\begin{aligned}
L_j(\theta, \delta) = {} & + \delta(G_j, 1)[\ln(T_1(1, M_j)) + \ln(T_2(1, N_j)) + \ln \Phi(X_j\beta + a)] \\
& + \delta(G_j, 2)[\ln(T_1(2, M_j)) + \ln(T_2(2, N_j)) + \ln 2 + \ln \Phi(X_j\beta + d)] \\
& + \delta(G_j, 3)[\ln(T_1(3, M_j)) + \ln(T_2(3, N_j)) + \ln \Phi(X_j\beta - a)] \\
& - \ln[\Phi(X_j\beta + a) + 2\Phi(X_j\beta + d) + \Phi(X_j\beta - a)] \qquad (11.47)
\end{aligned}
$$

The first partial derivatives are

$$S(\theta, \delta) = \sum_{j=1}^{n} S_j(\theta, \delta) = \begin{bmatrix} \sum_{j=1}^{n} \frac{\partial L_j(\theta,\delta)}{\partial \beta} \\ \sum_{j=1}^{n} \frac{\partial L_j(\theta,\delta)}{\partial a} \\ \sum_{j=1}^{n} \frac{\partial L_j(\theta,\delta)}{\partial d} \end{bmatrix} \qquad (11.48)$$

where

$$\frac{\partial L_j(\theta, \delta)}{\partial \beta} = + \delta(G_j, 1) \frac{X_j^T \phi(X_j\beta + a)}{\Phi(X_j\beta + a)} + \delta(G_j, 2) \frac{X_j^T \phi(X_j\beta + d)}{\Phi(X_j\beta + d)}$$

$$+ \delta(G_j, 3) \frac{X_j^T \phi(X_j\beta - a)}{\Phi(X_j\beta - a)}$$

$$- \frac{X_j^T \phi(X_j\beta + a) + 2X_j^T \phi(X_j\beta + d) + X_j^T \phi(X_j\beta - a)}{\Phi(X_j\beta + a) + 2\Phi(X_j\beta + d) + \Phi(X_j\beta - a)}$$

$$\frac{\partial L_j(\theta, \delta)}{\partial a} = + \delta(G_j, 1) \frac{\phi(X_j\beta + a)}{\Phi(X_j\beta + a)} - \delta(G_j, 3) \frac{\phi(X_j\beta - a)}{\Phi(X_j\beta - a)}$$

$$- \frac{\phi(X_j\beta + a) - \phi(X_j\beta - a)}{\Phi(X_j\beta + a) + 2\Phi(X_j\beta + d) + \Phi(X_j\beta - a)}$$

$$\frac{\partial L_j(\theta, \delta)}{\partial d} = + \delta(G_j, 2) \frac{\phi(X_j\beta + d)}{\Phi(X_j\beta + d)}$$

$$- \frac{2\phi(X_j\beta + d)}{\Phi(X_j\beta + a) + 2\Phi(X_j\beta + d) + \Phi(X_j\beta - a)} \qquad (11.49)$$

The Fisher information matrix is

$$I(\theta) = \sum_{j=1}^{n} \begin{bmatrix} E\left[\frac{\partial L_j(\theta,\delta)}{\partial \beta} \frac{\partial L_j(\theta,\delta)}{\partial \beta^T}\right] & E\left[\frac{\partial L_j(\theta,\delta)}{\partial \beta} \frac{\partial L_j(\theta,\delta)}{\partial a}\right] & E\left[\frac{\partial L_j(\theta,\delta)}{\partial \beta} \frac{\partial L_j(\theta,\delta)}{\partial d}\right] \\ E\left[\frac{\partial L_j(\theta,\delta)}{\partial a} \frac{\partial L_j(\theta,\delta)}{\partial \beta^T}\right] & E\left[\frac{\partial L_j(\theta,\delta)}{\partial a} \frac{\partial L_j(\theta,\delta)}{\partial a}\right] & E\left[\frac{\partial L_j(\theta,\delta)}{\partial a} \frac{\partial L_j(\theta,\delta)}{\partial d}\right] \\ E\left[\frac{\partial L_j(\theta,\delta)}{\partial d} \frac{\partial L_j(\theta,\delta)}{\partial \beta^T}\right] & E\left[\frac{\partial L_j(\theta,\delta)}{\partial d} \frac{\partial L_j(\theta,\delta)}{\partial a}\right] & E\left[\frac{\partial L_j(\theta,\delta)}{\partial d} \frac{\partial L_j(\theta,\delta)}{\partial d}\right] \end{bmatrix}$$

$$(11.50)$$

Let $S(\theta) = E[S(\theta, \delta)]$ be the expectation of the first partial derivative. We have the following iteration equation, which is the maximization step of the EM algorithm:

$$\theta^{(t+1)} = \theta^{(t)} + I^{-1}(\theta^{(t)})S(\theta^{(t)}) \qquad (11.51)$$

The expectation step is to calculate the expectation of δ_j using

$$E[\delta(G_j, \kappa)] = \frac{\omega_j(\kappa)T_1(\kappa, M_j)T_2(\kappa, N_j)}{\sum_{\kappa'}^{3} \omega_j(\kappa')T_1(\kappa', M_j)T_2(\kappa', N_j)} \tag{11.52}$$

Before we proceed to the next section, let us look at the details of the Fisher information matrix. In a slightly more compact notation, it is rewritten as

$$I(\theta) = \sum_{j=1}^{n} E\left[S_j(\theta, \delta)S_j^T(\theta, \delta) \right] \tag{11.53}$$

where $S_j(\theta, \delta)$ can be expressed as a linear function of vector δ_j, i.e.,

$$S_j(\theta, \delta) = A_j^T \delta_j + C_j \tag{11.54}$$

where A_j is a $3 \times (p+2)$ matrix and C_j is a $(p+2) \times 1$ vector. The expressions of A_j and C_j can be found from (11.49). The dimension of vector β is p. Since $\mathrm{var}(\delta_j)$ and $E(\delta_j)$ are known (given before), we can write

$$I(\theta) = \sum_{j=1}^{n} E\left(A_j^T \delta_j \delta_j^T A_j + A_j^T \delta_j C_j^T + C_j \delta_j^T A_j + C_j C_j^T \right)$$

$$= \sum_{j=1}^{n} A_j^T E(\delta_j \delta_j^T)A_j + A_j^T E(\delta_j)C_j^T + C_j E(\delta_j^T)A_j + C_j C_j^T \tag{11.55}$$

where

$$E(\delta_j \delta_j^T) = \mathrm{var}(\delta_j) + E(\delta_j)E(\delta_j^T) \tag{11.56}$$

Definition of $\mathrm{var}(\delta_j)$ can be found in (11.31).

11.2.2 Variance Matrix of Estimated Parameters

The variance–covariance matrix of the estimated parameters can be approximated by $\mathrm{var}(\hat{\theta}) \approx I^{-1}(\hat{\theta})$. However, a better approximation is to adjust the Fisher information matrix by the variance–covariance matrix of the score vector, i.e.,

$$\mathrm{var}(\hat{\theta}) \approx \left[I(\hat{\theta}) - \sum_{j=1}^{n} \mathrm{var}[S_j(\theta, \delta)] \right]^{-1} \tag{11.57}$$

where

$$\text{var}[S_j(\theta, \delta)] = \text{var}(A_j \delta_j) = A_j^T \text{var}(\delta_j) A_j \tag{11.58}$$

This adjustment gives the Louis (1982) information matrix.

11.2.3   Hypothesis Test

The null hypothesis is that there is no segregation distortion. This has been formulated as $H_0 : a = d = 0$. The log likelihood function evaluated at $\theta = \hat{\theta}$ is

$$L_1(\hat{\theta}) = \sum_{j=1}^{n} \ln \left[\sum_{\kappa=1}^{3} \omega_j(\kappa) T_1(\kappa, M_j) T_2(\kappa, N_j) \right] \tag{11.59}$$

The log likelihood function evaluated under the null model is

$$L_0(\phi) = \sum_{j=1}^{n} \ln \left[\sum_{\kappa=1}^{3} \phi_\kappa T_1(\kappa, M_j) T_2(\kappa, N_j) \right] \tag{11.60}$$

This is because under $H_0 : a = d = 0$, we have $\omega_j(\kappa) = \phi_\kappa, \forall \kappa = 1, 2, 3$. Given L_0 and L_1, the usual likelihood ratio test statistic LRT is used to test the null hypothesis, where $\text{LRT} = -2(L_0 - L_1)$.

The liability model has two advantages over the probabilistic model: (1) Cofactors can be removed from the analysis by fitting a β vector in the model and (2) the Wald (1943) test statistic may be used to test the null hypothesis.

11.3   Mapping QTL Under Segregation Distortion

Segregation distortion has long been treated as an error in the area of QTL mapping. Its impact on the result of QTL mapping is generally considered detrimental. Therefore, QTL mappers usually delete markers with segregation distortion before conducting QTL mapping. However, a recent study (Xu 2008) shows that segregation distortion can help QTL mapping in some circumstances. Rather than deleting markers with segregation distortion, we can take advantage of these markers in QTL mapping. This section will combine QTL mapping and SDL mapping to map QTL and SDL jointly. The method was recently published by Xu and Hu (2009).

11.3.1   Joint Likelihood Function

Consider that a QTL itself is also an SDL, i.e., the QTL is not necessarily a Mendelian locus. We now go back to the probabilistic model for the SDL. The parameter for SDL is $\omega = \{\omega_1, \omega_2, \omega_3\}$. Let y_j be the phenotypic value of a quantitative trait (not the liability) measured from individual j. The probability density of y_j conditional on $G_j = \kappa$ for individual j is normal with mean $\mu_j = X_j\beta + H_\kappa\gamma$ and variance σ^2, i.e.,

$$p(y_j|G_j - \kappa) = f_\kappa(y_j) = \frac{1}{\sqrt{2\pi\sigma^2}} \exp\left[-\frac{1}{2\sigma^2}(y_j - X_j\beta - H_\kappa\gamma)^2\right] \quad (11.61)$$

The conditional probability for the (flanking) markers is

$$p(M_j, N_j|G_j = \kappa) = T_1(\kappa, N_j)T_2(\kappa, N_j) \quad (11.62)$$

The probability that $G_j = \kappa$ is

$$p(G_j = \kappa) = \omega_\kappa \quad (11.63)$$

The joint likelihood function can be obtained by combining the three probabilities,

$$L(\theta) = \sum_{j=1}^{n} \ln\left[\sum_{\kappa=1}^{3} p(G_j = \kappa)p(y_j|G_j = \kappa)p(M_j, N_j|G_j = \kappa)\right] \quad (11.64)$$

which is rewritten as

$$L(\theta) = \sum_{j=1}^{n} \ln\left\{\sum_{\kappa=1}^{3} \omega_\kappa f_\kappa(y_j)T_1(\kappa, M_j)T_2(\kappa, N_j)\right\} \quad (11.65)$$

where the parameter vector is $\theta = \{\beta, \gamma, \omega\}$.

11.3.2   EM Algorithm

Derivation of the EM algorithm is given by Xu and Hu (2009). Here we only provide the final result. The expectation step of the EM algorithm requires computing the expectation of δ_j conditional on the data and θ. Because δ_j is a multivariate Bernoulli variable, the expectation is simply the probability of $\delta(G_j, \kappa) = 1$, i.e.,

$$E[\delta(G_j, \kappa)] = \frac{p(G_j = \kappa)p(y_j|G_j = \kappa)p(M_j, N_j|G_j = \kappa)}{\sum_{\kappa'=1}^{3} p(G_j = \kappa')p(y_j|G_j = \kappa')p(M_j, N_j|G_j = \kappa')}$$

$$= \frac{\omega_\kappa f_\kappa(y_j)T_1(\kappa, M_j)T_2(\kappa, N_j)}{\sum_{\kappa'=1}^{3} \omega_{\kappa'} f_\kappa(y_j)T_1(\kappa', M_j)T_2(\kappa', N_j)} \quad (11.66)$$

The maximization step of the EM algorithm involves the following equations:

$$\beta = \left[\sum_{j=1}^{n} X_j^T X_j^T \right]^{-1} \left[\sum_{j=1}^{n} \sum_{\kappa=1}^{3} E[\delta(G_j,\kappa)](y_j - H_\kappa\gamma) \right]$$

$$\gamma = \left[\sum_{j=1}^{n} \sum_{\kappa=1}^{3} E[\delta(G_j,\kappa)] \left(H_\kappa^T H_\kappa \right) \right]^{-1} \left[\sum_{j=1}^{n} (y_j - X_j\beta) \right]$$

$$\sigma^2 = \frac{1}{n} \sum_{j=1}^{n} \sum_{\kappa=1}^{3} E[\delta(G_j,\kappa)] \left(y_j - X_j\beta - H_\kappa\gamma \right)^2$$

$$\omega_\kappa = \frac{1}{n} \sum_{j=1}^{n} E[\delta(G_j,\kappa)], \ \forall \kappa = 1,2,3 \tag{11.67}$$

11.3.3   Variance–Covariance Matrix of Estimated Parameters

Let us define the complete-data log likelihood function for individual j as

$$L_j(\theta,\delta) = -\frac{1}{2} \ln(\sigma^2) - \frac{1}{2\sigma^2} \sum_{\kappa=1}^{3} \delta(G_j,\kappa)(y_j - X_j\beta - H_\kappa\gamma)^2$$

$$+ \sum_{\kappa=1}^{3} \delta(G_j,\kappa)\{\ln[T_1(\kappa,M_j)] + \ln[T_2(\kappa,N_j)]\}$$

$$+ \sum_{\kappa=1}^{3} \delta(G_j,\kappa) \ln \omega_\kappa \tag{11.68}$$

where $\omega_3 = 1 - \omega_1 - \omega_2$ so that ω_3 is excluded from the parameter vector. Elements of the score vector for individual j are

$$\frac{\partial L_j(\theta,\delta)}{\partial \beta} = \frac{1}{\sigma^2} \sum_{k=1}^{3} \delta(G_j,\kappa) X_j^T (y_j - X_j\beta - H_\kappa\gamma)$$

$$\frac{\partial L_j(\theta,\delta)}{\partial \gamma} = \frac{1}{\sigma^2} \sum_{\kappa=1}^{3} \delta(G_j,\kappa) H_\kappa^T (y_j - X_j\beta - H_\kappa\gamma)$$

$$\frac{\partial L_j(\theta,\delta)}{\partial \sigma^2} = -\frac{1}{2\sigma^2} + \frac{1}{2\sigma^4} \sum_{\kappa=1}^{3} \delta(G_j,\kappa)(y_j - X_j\beta - H_\kappa\gamma)^2$$

$$\frac{\partial L_j(\theta, \delta)}{\partial \omega_1} = \delta(G_j, 1)\frac{1}{\omega_1} - \delta(G_j, 3)\frac{1}{1 - \omega_1 - \omega_2}$$

$$\frac{\partial L_j(\theta, \delta)}{\partial \omega_2} = \delta(G_j, 2)\frac{1}{\omega_2} - \delta(G_j, 3)\frac{1}{1 - \omega_1 - \omega_2} \qquad (11.69)$$

Elements of the Hessian matrix are

$$\frac{\partial^2 L_j(\theta, \delta)}{\partial \beta \partial \beta^T} = -\frac{1}{\sigma^2} X_j^T X_j$$

$$\frac{\partial^2 L_j(\theta, \delta)}{\partial \beta \partial \gamma^T} = -\frac{1}{\sigma^2} \sum_{\kappa=1}^{3} \delta(G_j, \kappa) X_j^T H_\kappa$$

$$\frac{\partial^2 L_j(\theta, \delta)}{\partial \beta \partial \sigma^2} = -\frac{1}{\sigma^4} \sum_{\kappa=1}^{3} \delta(G_j, \kappa) X_j^T (y_j - X_j \beta - H_\kappa \gamma) \qquad (11.70)$$

$$\frac{\partial^2 L_j(\theta, \delta)}{\partial \gamma \partial \gamma^T} = -\frac{1}{\sigma^2} \sum_{\kappa=1}^{3} \delta(G_j, \kappa) H_\kappa^T H_\kappa$$

$$\frac{\partial^2 L_j(\theta, \delta)}{\partial \gamma \partial \beta^T} = -\frac{1}{\sigma^2} \sum_{\kappa=1}^{3} \delta(G_j, \kappa) H_\kappa^T X_j$$

$$\frac{\partial^2 L_j(\theta, \delta)}{\partial \gamma \partial \sigma^2} = -\frac{1}{\sigma^4} \sum_{\kappa=1}^{3} \delta(G_j, \kappa) H_\kappa^T (y_j - X_j \beta - H_\kappa \gamma) \qquad (11.71)$$

$$\frac{\partial^2 L_j(\theta, \delta)}{\partial \sigma^2 \partial \sigma^2} = +\frac{1}{2\sigma^4} - \frac{1}{\sigma^6} \sum_{\kappa=1}^{3} \delta(G_j, \kappa)(y_j - X_j \beta - H_\kappa \gamma)^2$$

$$\frac{\partial^2 L_j(\theta, \delta)}{\partial \sigma^2 \partial \beta^T} = -\frac{1}{\sigma^4} \sum_{\kappa=1}^{3} \delta(G_j, \kappa)(y_j - X_j \beta - H_\kappa \gamma) X_j$$

$$\frac{\partial^2 L_j(\theta, \delta)}{\partial \sigma^2 \partial \gamma^T} = -\frac{1}{\sigma^4} \sum_{\kappa=1}^{3} \delta(G_j, \kappa)(y_j - X_j \beta - H_\kappa \gamma) H_\kappa \qquad (11.72)$$

$$\frac{\partial^2 L_j(\theta, \delta)}{\partial \omega_1 \partial \omega_1} = -\delta(G_j, \kappa)\frac{1}{\omega_1^2} + \delta(G_j, 3)\frac{1}{\omega_3^2}$$

$$\frac{\partial^2 L_j(\theta, \delta)}{\partial \omega_1 \partial \omega_2} = +\delta(G_j, 3)\frac{1}{\omega_3^2}$$

$$\frac{\partial^2 L_j(\theta, \delta)}{\partial \omega_2 \partial \omega_2} = -\delta(G_j, \kappa)\frac{1}{\omega_2^2} + \delta(G_j, 3)\frac{1}{\omega_3^2} \qquad (11.73)$$

The score vector and the Hessian matrix provide the original material from which $-E[H(\theta, \delta)]$ and $\text{var}[S(\theta, \delta)]$ are calculated (see Xu and Hu 2009). The Louis

(1982) information matrix is

$$I(\theta) = -E[H(\theta, \delta)] - \text{var}[S(\theta, \delta)] \tag{11.74}$$

from which, we can get the variance matrix of the estimated parameters using $\text{var}(\hat{\theta}) \approx I^{-1}(\hat{\theta})$.

11.3.4   Hypothesis Tests

Hypothesis 1

There are several different hypotheses we can test. The first null hypothesis is H_0 : $\gamma = 0$, i.e., there is no QTL for the quantitative trait. To test this hypothesis, we need the full-model likelihood value as shown below:

$$L_1(\hat{\theta}) = \sum_{j=1}^{n} \ln \left\{ \sum_{\kappa=1}^{3} \omega_\kappa f_\kappa(y_j) T_1(\kappa, M_j) T_2(\kappa, N_j) \right\} \tag{11.75}$$

where the parameters in the right-hand side of the equation are replaced by the MLE. The reduced-model likelihood value is calculated using

$$L_0(\hat{\hat{\theta}}) = \sum_{j=1}^{n} \ln \left\{ \sum_{\kappa=1}^{3} \omega_\kappa T_1(\kappa, M_j) T_2(\kappa, N_j) \right\}$$

$$- \frac{1}{2\sigma^2} \sum_{j=1}^{n} (y_j - X_j \beta)^2 - \frac{n}{2} \ln(\sigma^2) \tag{11.76}$$

where $\gamma = 0$ is enforced and $\hat{\hat{\theta}}$ is the estimated parameter vector under the reduced model. The usual likelihood ratio test statistic is then constructed using the two likelihood values.

Hypothesis 2

The second hypothesis is H_0 : $\omega = \phi$, i.e., the population is Mendelian. The log likelihood functions under the full model are

$$L_1(\hat{\theta}) = \sum_{j=1}^{n} \ln \left\{ \sum_{\kappa=1}^{3} \omega_\kappa f_\kappa(y_j) T_1(\kappa, M_j) T_2(\kappa, N_j) \right\} \tag{11.77}$$

This is the same as that given in (11.75). The likelihood value under the reduced model is

$$L_1(\hat{\hat{\theta}}) = \sum_{j=1}^{n} \ln \left\{ \sum_{\kappa=1}^{3} \phi_\kappa f_\kappa(y_j) T_1(\kappa, M_j) T_2(\kappa, N_j) \right\} \quad (11.78)$$

where $\omega = \phi$ is enforced and $\hat{\hat{\theta}}$ is the estimated parameter vector under the restricted model. The usual likelihood ratio test statistic is then constructed using the two likelihood values.

Hypothesis 3

The third hypothesis is $H_0 : \gamma = 0 \,\&\, \omega = \phi$, i.e., Mendelian population with no QTL effect for the quantitative trait. The full model remains the same as that given in (11.75) and (11.77). The reduced model is

$$L_0(\hat{\hat{\theta}}) = \sum_{j=1}^{n} \ln \left\{ \sum_{\kappa=1}^{3} \phi_\kappa T_1(\kappa, M_j) T_2(\kappa, N_j) \right\}$$

$$- \frac{1}{2\sigma^2} \sum_{j=1}^{n} (y_j - X_j \beta)^2 - \frac{n}{2} \ln(\sigma^2) \quad (11.79)$$

where $\hat{\hat{\theta}}$ is the estimated parameter vector under the restricted model. This hypothesis will be rejected if $\gamma \neq 0$ or $\omega \neq \phi$ or both inequalities hold. This hypothesis is particularly interesting for QTL mapping under selective genotyping. If the F_2 population is a Mendelian population, i.e., there is no segregation distortion, individuals are only genotyped based on the extremity of the phenotype values. Selective genotyping will lead to $\omega \neq \phi$, even if the original F_2 population is Mendelian.

11.3.5 Example

The mouse data introduced in Sect. 8.1 of Chap. 8 is used again for the joint QTL and SDL analysis. The mouse genome is scanned for QTL of the 10th-week body weight, the segregation distortion locus (SDL), and both QTL and SDL with a 1-cM increment for all the 19 chromosomes (excluding the sex chromosome) of the genome. The LOD scores are depicted in Fig. 11.1. Let LOD = 3 be the criterion of significance for gene detection. Two QTL appear to be significant, and both are on chromosome 2. Three SDL are significant with one on chromosome 6 (LOD $\approx$ 42.5), one on chromosome 14 (LOD $\approx$ 5.5), and one on chromosome 18 (LOD $\approx$ 3.5). The joint test has the highest LOD score across the entire genome.

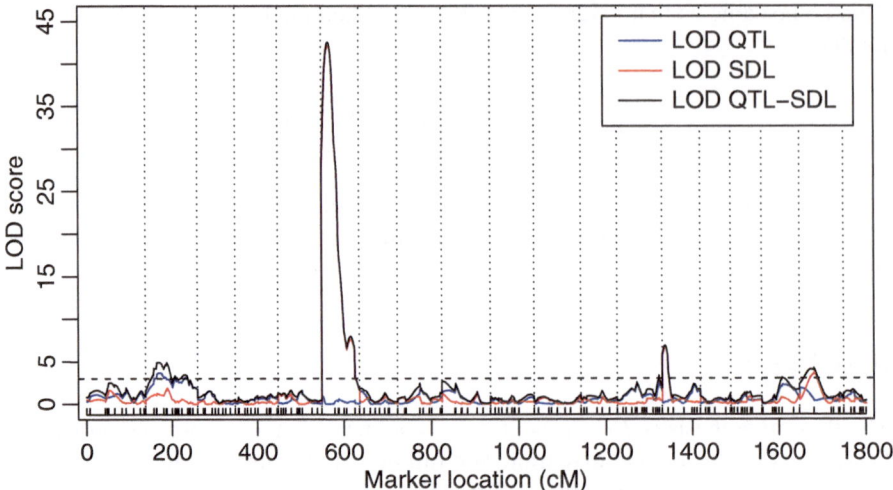

Fig. 11.1 The LOD test statistics profiles for the mouse genome (excluding the sex chromosome). The three LOD score profiles represent (1) the LOD test for QTL of the 10th-week body weight (*blue*), (2) the LOD score for SDL (segregation distortion locus, *red*), and (3) the LOD score for both the QTL and SDL (*black*). The dashed horizontal line indicates the LOD = 3 criterion. The 19 chromosomes are separated by the *vertical reference dotted lines*

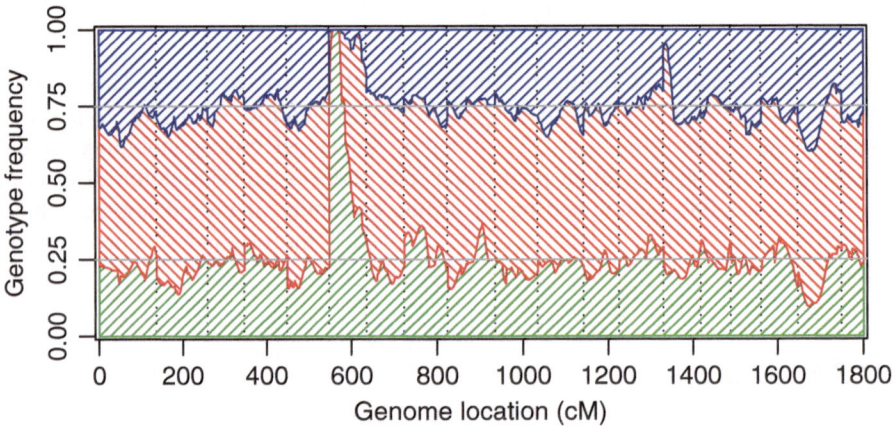

Fig. 11.2 Estimated genotypic frequencies for the mouse genome. Frequencies of the three genotypes are represented by areas with different patterns (A_1A_1 at *top*, A_1A_2 in the *middle*, and A_2A_2 at the *bottom*). The chromosomes are separated by the reference lines on the horizontal axis. The two reference lines on the vertical axis (0.25 and 0.75) divide the area into three parts based on the Mendelian segregation ratio (0.25, 0.5, and 0.25)

The estimated frequencies of the three genotypes (A_1A_1, A_1A_2, and A_2A_2) are shown in Fig. 11.2. The large SDL on chromosome 6 was extremely strong, and it wiped out all heterozygotes and homozygotes of the other type. The allele of this locus was fixed for the A_2 allele.

Chapter 12
QTL Mapping in Other Populations

BC and F_2 populations are the most commonly used populations for QTL mapping. There are other populations which can also be used for QTL mapping. These include recombinant inbred lines (RIL), double haploids (DH), four-way crosses, diallel crosses, full-sib families, half-sib families, and random-mating populations with pedigree structures. We are going to discuss a few of them in this chapter.

12.1 Recombinant Inbred Lines

Recombinant inbred lines are derived from repeated selfings of F_2 individuals for many generations until all progeny become homozygotes. In animals (except some lower worms), selfing is impossible, and thus, RIL must be obtained by repeated brother–sister matings. For large animals with long generation intervals, RIL cannot be obtained within a reasonable amount of time. Therefore, only small laboratory animals, e.g., fruit flies and mice, are possible to have RIL. An RIL generated via selfing is called RIL1, while an RIL generated via brother–sister mating is called RIL2.

RILs are obtained by systematic inbreeding (selfing and brother–sister mating). As the number of generations increases, the inbreeding coefficient progressively increases and eventually reaches unity (pure lines) in an asymptotic fashion. Theoretically, there is no guarantee that an RIL line contains no heterozygous loci. Residual heterozygosity always exists in RIL lines. An empirical criterion of defining an inbred line is the less than 2 % residual heterozygosity rule, corresponding to an inbreeding coefficient of 98 %. The two mating designs lead to the desired level of heterozygosity in different speeds. Selfing requires about seven generations, and brother–sister mating requires about 20 generations. We first discuss systematic inbreeding via selfing. Starting from the F_1 hybrid as generation $t = 1$, the inbreeding coefficient at generation t is

S. Xu, *Principles of Statistical Genomics*, DOI 10.1007/978-0-387-70807-2_12,
© Springer Science+Business Media, LLC 2013

$$F_t = 1 - \left(\frac{1}{2}\right)^{t-1} \tag{12.1}$$

At generation $t = 7$, the inbreeding coefficient is $F_7 = 0.984375$, which is higher than the desired level of 98 %.

For brother–sister mating, the exact inbreeding coefficient at generation t must be calculated using the following recurrent equation (Falconer and Mackay 1996):

$$F_t = \frac{1}{4}(1 + 2F_{t-1} + F_{t-2}) \tag{12.2}$$

where F_t and F_{t-1} are the inbreeding coefficients at generation t and $t - 1$, respectively. Like selfing, the F_1 hybrid is designated as generation $t = 1$. Here, we must set $F_0 = F_1 = 0$, and the recurrent equation must start with $t = 2$. At $t = 20$, we get $F_{20} = 0.983109$, higher than 98 %. Although the recurrent equation is exact, one must start at $t = 2$ and calculate the inbreeding coefficient for the next generation based on the inbreeding coefficients of the previous two generations. If we only need the inbreeding coefficient of generation t, there is an alternative way of calculating F_t as a function of t. This alternative method is approximate and differs from the exact method in the first few generations. As generations progress, the difference becomes diminished. The approximate method is a general approach and applies to all mating systems. The inbreeding coefficient at generation t is

$$F_t = 1 - (1 - \Delta F)^{t-1} = 1 - \left(1 - \frac{1}{2N_e}\right)^{t-1} \tag{12.3}$$

where $\Delta F = \frac{1}{2N_e}$ is the rate of inbreeding and N_e is the effective population size. For brother–sister mating, the inbreeding rate is $\Delta F = 0.191$ (Falconer and Mackay 1996), leading to $N_e = 2.618$. Using (12.3), the approximate inbreeding coefficient at generation 20 is $F_{20} = 0.982169$, much the same as the exact value of 0.983109.

The general approach also applies to selfing when $N_e = 1$ is used as the effective population size. Interestingly, the general method is exact for selfing. Note that (12.3) is different from the usual equation commonly seen in population genetics textbooks (Falconer and Mackay 1996) where $t - 1$ in the right-hand side of (12.3) is substituted by t. This difference is due to our designation of F_1 hybrid as generation one. If we had designated F_2 as generation one, we would get the same formula as that of Falconer and Mackay (1996). Our notation has avoided some confusion because F_t represents the inbreeding coefficient of the F_t generation.

Figure 12.1 shows the increase of inbreeding coefficient as the number of generations increases for both selfing and brother–sister mating. Inbreeding coefficient represents the rate of allelic fixation. Selfing has a much faster rate of allele fixation than brother–sister mating. Figure 12.2 compares the inbreeding coefficients of the exact method and the approximate method under brother–sister mating. The difference is barely noticeable after generation four. Therefore, it is very safe to use the approximate method to calculate the inbreeding coefficient for the brother–sister mating design.

Fig. 12.1 Comparison of inbreeding coefficient of selfing with that of brother–sister mating

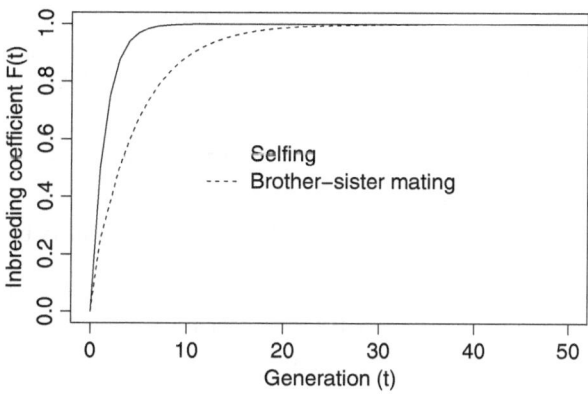

Fig. 12.2 Comparison of inbreeding coefficient of the exact method with that of the approximate method under brother–sister mating

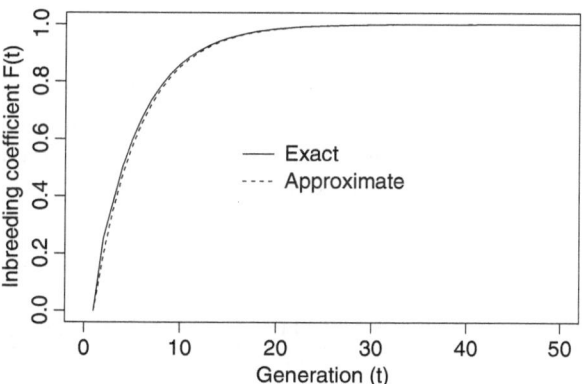

There are two possible genotypes with an equal chance in an RIL population at each locus, A_1A_1 and A_2A_2. The model is the same as that of the BC design,

$$y_j = X_j\beta + Z_j\gamma + \epsilon_j \tag{12.4}$$

except that

$$Z_j = \begin{cases} +1 & \text{for } A_1A_1 \\ -1 & \text{for } A_2A_2 \end{cases} \tag{12.5}$$

Methods for marker-trait association and interval mapping remain the same as what we have learned in the BC design. However, the conditional probability of QTL genotype given marker information is calculated using a modified recombination fraction, which reflects the cumulation of many generations of crossovers. For recombinant inbred lines generated via selfing (RIL1), the modified recombination fraction between loci A and B is

$$c_{AB} = \frac{2r_{AB}}{1 + 2r_{AB}} \tag{12.6}$$

where r_{AB} is the usual recombination fraction defined earlier. For recombinant inbred lines generated through brother–sister mating (RIL2), the modified recombination fraction is (Haldane and Waddington 1931)

$$c_{AB} = \frac{4r_{AB}}{1 + 6r_{AB}} \tag{12.7}$$

Given c_{AB}, we define a new 2×2 transition matrix between two loci by

$$
\begin{aligned}
T_{AB} &= \begin{bmatrix} \Pr(B = +1|A = +1), \Pr(B = -1|A = +1) \\ \Pr(B = +1|A = -1), \Pr(B = -1|A = -1) \end{bmatrix} \\
&= \begin{bmatrix} 1 - c_{AB} & c_{AB} \\ c_{AB} & 1 - c_{AB} \end{bmatrix}
\end{aligned} \tag{12.8}
$$

This transition matrix is used to calculate the conditional probability of QTL genotype given marker genotypes using either the three-point or multipoint analysis. For example, let AQB be three ordered loci. The conditional probability of QTL genotype given genotypes of markers A and B is

$$
\begin{aligned}
\Pr(Q = +1|A = +1, B = -1) &= \frac{\Pr(A = +1, Q = +1, B = -1)}{\Pr(A = +1, B = -1)} \\
&= \frac{\Pr(Q = +1)\Pr(A = +1|Q = +1)\Pr(B = -1|Q = +1)}{\Pr(A = +1)\Pr(B = -1|A = +1)}
\end{aligned} \tag{12.9}
$$

Substituting the marginal and conditional probabilities by the values from the transition matrix, we get

$$\Pr(Q = +1|A = +1, B = -1) = \frac{(1 - c_{AQ})c_{QB}}{c_{AB}} \tag{12.10}$$

There are two advantages of using RIL for QTL mapping compared to BC and F_2 designs. One is that RIL can map QTL in a finer scale than both BC and F_2. Note that the modified recombination fraction is always greater than the usual recombination fraction because RIL can take advantage of historically cumulative crossovers. This can be demonstrated, e.g. for selfing, by

$$c_{AB} = \frac{r_{AB}}{1/2 + r_{AB}} > r_{AB} \tag{12.11}$$

because the denominator, $1/2 + r_{AB}$, is always less than unity. This phenomenon is called genome expansion. Using an expanded genome can increase the resolution of QTL mapping. The other advantage is that the variance of variable Z_j in the RIL design is twice as large as that in the F_2 design, i.e. $\sigma_Z^2 = 1$ for the RIL whereas

$\sigma_Z^2 = \frac{1}{2}$ for the F_2. A larger σ_Z^2 means a smaller estimation error of the genetic effect. Recall that

$$\text{var}(\hat{\gamma}) = \frac{\sigma^2}{\sum(Z - \bar{Z})^2} = \frac{\sigma^2}{n\left[\frac{1}{n}\sum(Z - \bar{Z})^2\right]} \approx \frac{\sigma^2}{n\sigma_Z^2} \qquad (12.12)$$

where σ^2 is the residual error variance and, as $n \longrightarrow \infty$, $\sigma_Z^2 \approx \frac{1}{n}\sum(Z - \bar{Z})^2$. You can verify by yourself that

$$\sigma_Z^2 = E(Z^2) - E^2(Z) = 1 - 0 = 1$$

for the RIL design because Z takes $+1$ or -1 with an equal probability. One problem with the RIL design is the high cost due to the long time required to generate the individual recombinant inbred lines. The other problem is the inability to detect dominance effect, similar problem as that of the BC design.

A final comment on the two types of recombinant inbred lines (RIL1 and RIL2) is that the resolution of QTL mapping using RIL2 (brother–sister mating) is higher than that of using RIL1 (selfing). In other words, the expanded recombination of RIL2 is higher than that of RIL1 because RIL2 cumulates more historical recombinant events.

12.2 Double Haploids

Double haploids (DH) are obtained by doubling the gametes of F_1 individuals through some special cytogenetic treatment. DH can be achieved by a single generation of cytogenetic manipulation, just like a BC population. However, a DH individual is homozygous for all loci. Therefore, like RIL, a DH population contains two possible genotypes, A_1A_1 and A_2A_2. The linear model for the phenotypic value of individual j is the same as that in RIL. The Z_j variable is also the same as that defined in RIL. However, the conditional probability of QTL genotype given marker genotypes is calculated using the same formula as that used in the BC design, i.e., using the original recombination fraction (r_{AB}), not the modified recombination fraction (c_{AB}). DH mapping is more powerful than BC and F_2 due to the large σ_Z^2 compared to BC and F_2.

12.3 Four-Way Crosses

The four-way (FW) cross design of QTL mapping was first proposed by Xu (1996, 1998b) and Xie and Xu (1999). The model and method described here follow closely to that of Xu (1998b). Four-way cross design involves two families of crosses. An F_1 derived from the cross of two parents (P_1 and P_2) is further crossed with an F_1

Fig. 12.3 Sketch of four-way (FW) cross design of experiment

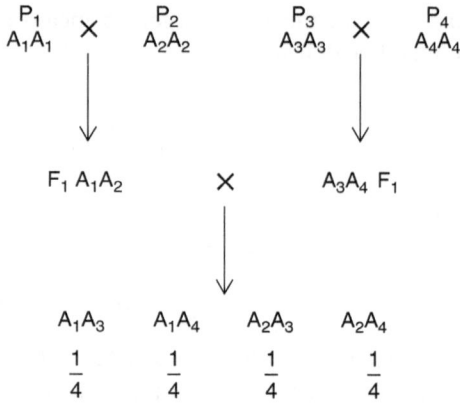

derived from the cross of two additional and independent parents (P_3 and P_4). A total of four inbred lines are involved in a four-way cross. Figure 12.3 illustrates the four-way cross design.

The genetic model of a four-way cross is different from that of an F_2 design. There are four alleles involved in a four-way cross, and only four out of $4(4+1)/2 = 10$ allelic combinations occur in the four-way cross family. Let the F_1 from the first cross be the female parent of the four-way cross progeny and the F_1 from the second cross be the male parent. We relabel the four alleles by $A_1^d = A_1$, $A_2^d = A_2$, $A_1^s = A_3$, and $A_2^s = A_4$, where the superscripts d and s stand for "dam" (mother) and "sire" (father), respectively, and the subscripts 1 and 2 stand for the first and second alleles, respectively, of each parent. The genotypes of the two F_1 parents are relabeled as $A_1^d A_2^d$ and $A_1^s A_2^s$. A progeny of the four-way cross carries one of the four possible genotypes,

$$\{A_1^d A_1^s, A_1^d A_2^s, A_2^d A_1^s, A_2^d A_2^s\}$$

Unlike the analysis of the F_2 design, in the FW cross design, we assign each allele a genetic value, called the allelic effect. No matter how many individuals in the FW cross family, they inherit a total number of four alleles from their parents. The four allelic values are denoted by a_1^s, a_2^d, a_2^s, and a_2^d, respectively, for the four alleles, A_1^s, A_2^d, A_2^s, and A_2^d. Corresponding to the four genotypes of the progeny in the FW cross family, we denote the four genotypic values by

$$G_{11} = \mu + a_1^d + a_1^s + d_{11} \text{ for } A_1^d A_1^s$$

$$G_{12} = \mu + a_1^d + a_2^s + d_{12} \text{ for } A_1^d A_2^s$$

$$G_{21} = \mu + a_2^d + a_1^s + d_{21} \text{ for } A_2^d A_1^s$$

$$G_{22} = \mu + a_2^d + a_2^s + d_{22} \text{ for } A_2^d A_2^s \tag{12.13}$$

where μ is the population mean, a_1^d and a_2^d are the effects of the two alleles of the dam, a_1^s and a_2^s are the effects of the two alleles of the sire, and d_{kl} is the interaction (dominance) effect between alleles A_k^d and A_l^s, $\forall k, l = 1, 2$. The linear model for the phenotypic value of individual j is

$$y_j = \mu + G_{kl} + \epsilon_j = \mu + a_k^d + a_l^s + d_{kl} + \epsilon_j \tag{12.14}$$

There are nine model effects, but we only have four possible genotypes, which are not sufficient to allow us to estimate the nine model effects. We must make some constraints on the model effects. Enforcing constraints on model effects is a typical treatment in analysis of variance. The following constraints are commonly used in such an analysis,

$$a_1^d + a_2^d = 0$$
$$a_1^s + a_2^s = 0$$
$$d_{11} + d_{12} + d_{21} + d_{22} = 0 \tag{12.15}$$

Additional constraints are required for the dominance effects,

$$d_{11} + d_{12} = 0$$
$$d_{11} + d_{21} = 0$$
$$d_{12} + d_{22} = 0$$
$$d_{21} + d_{22} = 0 \tag{12.16}$$

After enforcing these constraints, we have only three genetic parameters to estimate (in addition to the population mean μ), which are

$$\alpha^d = \frac{1}{2}a_1^d - \frac{1}{2}a_2^d$$
$$\alpha^s = \frac{1}{2}a_1^s - \frac{1}{2}a_2^s$$
$$\delta = \frac{1}{4}d_{11} - \frac{1}{4}d_{12} - \frac{1}{4}d_{21} + \frac{1}{4}d_{22} \tag{12.17}$$

The four estimable parameters (including μ) can be expressed as linear functions of the four genotypic values,

$$\mu = \frac{1}{4}G_{11} + \frac{1}{4}G_{12} + \frac{1}{4}G_{21} + \frac{1}{4}G_{22}$$
$$\alpha^d = \frac{1}{4}G_{11} + \frac{1}{4}G_{12} - \frac{1}{4}G_{21} - \frac{1}{4}G_{22}$$
$$\alpha^s = \frac{1}{4}G_{11} - \frac{1}{4}G_{12} + \frac{1}{4}G_{21} - \frac{1}{4}G_{22}$$
$$\delta = \frac{1}{4}G_{11} - \frac{1}{4}G_{12} - \frac{1}{4}G_{21} + \frac{1}{4}G_{22} \tag{12.18}$$

The reverse relationship of the above equations is

$$G_{11} = \mu + a^d + a^s + \delta$$

$$G_{12} = \mu + a^d - a^s - \delta$$

$$G_{21} = \mu - a^d + a^s - \delta$$

$$G_{22} = \mu - a^d - a^s + \delta \qquad (12.19)$$

Excluding μ, we have three genetic effects to estimate: a^d representing the difference between the two alleles of the dam, a^s representing the difference between the two alleles of the sire, and δ representing the interaction between the two allelic differences. The two allelic differences, a^d and a^s, may also be called allelic substitution effects (Falconer and Mackay 1996. To be consistent with the notation used early in the F_2 design of experiment, we now denote $\beta = \mu$, $\gamma = \{a^d, a^s, \delta\}$, and $X_j = 1, \forall j = 1, \ldots, n$. The linear model for y_j is now rewritten as

$$y_j = X_j \beta + Z_j \gamma + \epsilon_j \qquad (12.20)$$

where Z_j is a 1×3 vector defined as

$$Z_j = \begin{cases} H_1 & \text{for } A_1 A_3 \\ H_2 & \text{for } A_1 A_4 \\ H_3 & \text{for } A_2 A_3 \\ H_4 & \text{for } A_2 A_4 \end{cases} \qquad (12.21)$$

and $H_k, \forall k = 1, \ldots, 4$ is the kth row of matrix H, which is a 4×3 matrix defined below,

$$H = \begin{bmatrix} +1 & +1 & +1 \\ +1 & -1 & -1 \\ -1 & +1 & -1 \\ -1 & -1 & +1 \end{bmatrix} \qquad (12.22)$$

The linear model given in (12.20) is the same as that in the F_2 mating design. Assume that the residual error has a normal distribution with mean zero and variance σ^2; the maximum likelihood method can be applied to the FW cross design. The EM algorithm is also identical to that described in the F_2 design except that now the mixture distribution consists of four components instead of three. In the expectation step, we need

$$E(Z_j) = \sum_{k=1}^{4} E\left[\delta(G_j, k)\right] H_k = \sum_{k=1}^{4} p_j^*(k) H_k$$

$$E(Z_j^T Z_j) = \sum_{k=1}^{4} E\left[\delta(G_j, k)\right] H_k^T H_k = \sum_{k=1}^{4} p_j^*(k) H_k^T H_k \qquad (12.23)$$

where

$$E\left[\delta(G_j, k)\right] = p_j^*(k) = \frac{p_j(k) f_k(y_j)}{\sum_{k'}^{4} p_j(k') f_{k'}(y_j)} \qquad (12.24)$$

is the posterior probability that $G_j = k$, i.e., individual j takes the k genotype for $k = 1, \ldots, 4$. The posterior probability requires the probability density of the phenotype

$$f_k(y_j) = \frac{1}{\sqrt{2\pi\sigma^2}} \exp\left[-\frac{1}{2}(y_j - X_j\beta - H_k\gamma)^2\right] \qquad (12.25)$$

and the probability of QTL genotype conditional on marker information is

$$p_j(k) = \Pr(G_j = k | \text{marker}) \qquad (12.26)$$

This probability may be calculated using either two flanking markers (interval mapping) or all markers on the same chromosome (multipoint mapping).

Using FW cross families for QTL mapping will face a problem of noninformative markers. We have dealt with a perfect situation where four distinguished alleles are present for every marker locus. In reality, we will see some markers that may only have three or less distinguished alleles. For example, if the dam has a genotype $A_1 A_1$ and the sire has a genotype $A_2 A_3$, we only observe three alleles and two genotypes in the FW cross family. The two genotypes are $A_1 A_2$ and $A_1 A_3$. How do we use the multipoint method to calculate the four probabilities of the QTL genotypes? This situation has been discussed in Chap. 4 where we need to list all the four ordered genotypes in the progeny, $A_1 A_2$, $A_1 A_3$, $A_1 A_2$, and $A_1 A_3$. We can see that the first and the third ordered genotypes are not distinguishable, so are the second and the fourth genotypes. When we code the genotype for a progeny using the J matrix notation, an individual with observed genotype $A_1 A_3$ should be coded as $J = \{0, 1, 0, 1\}$. While the J notation can be used for coding the markers, how do we deal with the QTL if there are less than four alleles at the QTL? Our assumption for the QTL is that there are always four distinguished alleles. Therefore, the model always contains three genetic effects, α^d, α^s, and δ. If a parent, say the dam, happens to be homozygote at the QTL, α^d still occurs in the model, but the estimated α^d will be close to zero if the sample size is sufficiently large. In fact, $\alpha^d = 0$ is treated the same as no segregation for the two alleles of the dam.

The usual likelihood ratio test is adopted here to test the hypothesis $H_0 : \gamma = 0$. Once a QTL is detected, we can further test the significance of the additive effects $H_0 : \alpha^d = \alpha^s = 0$ and the significance of the dominance effect $H_0 : \delta = 0$.

The size of the QTL detected can be measured as the ratio of the genetic variance component to the total phenotypic variance. The genetic variance is defined as

$$\sigma_G^2 = \gamma^T \text{var}(Z)\gamma = \gamma^T \gamma \tag{12.27}$$

where

$$\text{var}(Z) = E(Z^T Z) - E(Z^T)E(Z)$$

$$= \left(\frac{1}{4}\sum_{k=1}^{4} H_k^T H_k\right) - \left(\frac{1}{4}\sum_{k=1}^{4} H_k^T\right)\left(\frac{1}{4}\sum_{k=1}^{4} H_k\right) = I \tag{12.28}$$

is the marginal variance matrix of Z_j across all individuals in the FW family (not the conditional variance matrix given marker information). This matrix happens to be an identity matrix because the design of the FW cross model is orthogonal. The total phenotypic variance is

$$\sigma_P^2 = \gamma^T \gamma + \sigma^2 \tag{12.29}$$

Therefore, the size of the QTL expressed as the total phenotypic variance contributed by the QTL is

$$H^2 = \frac{\sigma_G^2}{\sigma_P^2} = \frac{\gamma^T \gamma}{\gamma^T \gamma + \sigma^2} = \frac{(\alpha^d)^2 + (\alpha^s)^2 + \delta^2}{(\alpha^d)^2 + (\alpha^s)^2 + \delta^2 + \sigma^2} \tag{12.30}$$

The FW cross design is a general design of line crosses for QTL mapping. The F_2 and BC designs can all be considered as special cases. For example, the F_2 design requires selfing of the F_1 progeny of two inbred lines. From the FW cross perspective, both the sire and the dam have exactly the same genotype for all loci, A_1A_2 and A_1A_2. The four genotypes in the progeny are A_1A_1, A_1A_2, A_2A_1, and A_2A_2. We can never distinguish the two phases of the heterozygote. In addition, $\alpha^d = \alpha^s$ always holds. Therefore, to handle an F_2 population using the FW cross approach, we need a restriction in the parameters, that is, $\alpha^d = \alpha^s$. The BC design requires crossing of the F_1 back to one of the two inbred lines, say P_1. From the FW cross perspective, the dam is A_1A_1 and the sire is A_1A_2. This means that $\alpha^d = 0$ holds for all loci. Note that the dominance effect requires the interaction of the allelic difference of the dam and the allelic difference of the sire. Because the allelic difference of the dam is always absent, the dominance effect is always absent as well. Therefore, to map QTL in a BC family using the FW cross approach, we need two restrictions, $\alpha^d = 0$ and $\delta = 0$.

Finally, an advantage of the FW cross design over the F_2 design occurs in the possible high level of polymorphism. In a F_2 design, a QTL can only be detected if it is segregating in the F_2 family; that is, the two inbred lines initiating the cross must carry two different alleles for the QTL. If the two lines are identical at the QTL, the F_1 is homozygote, and thus, the QTL does not segregate in the F_2 progeny. No matter how large the sample size is, the QTL will never be detected. People may argue that the so-called QTL is not actually a QTL if it is not segregating. This argument may be right, but how do we call it if other people detected a QTL at the same location of the genome but using a different inbred lines to initiate the cross? This leads to two explanations for the failure of QTL detection. One is that a QTL (genome location) is present, but it is not segregating. The other is that the QTL is segregating, but it fails to be detected because either the effect is small or the sample size is small. If a large QTL fails to be detected in an extremely large sample, this type of error must be due the wrong choice of the inbred lines. We may loosely call it the type I error due to genetic drift. To avoid this type of error, we may either use different inbred lines to initiate the cross or use more inbred lines to initiate a complicated crossing experiment. The FW cross design is one of such cross designs involving more than two inbred lines. It combines two separate crosses. If a QTL fails to segregate in one cross but it segregates in the other cross, we still have a chance to detect such a QTL (assuming that the sample size is sufficiently large) because if any one of the three equations ($\alpha^d = \alpha^s = \delta = 0$) fails, a QTL will be claimed.

12.4 Full-Sib Family

When inbred lines are not available, e.g., in forest trees, it is impossible to initiate a line crossing experiment for QTL mapping. Instead, we can collect seeds from a single tree (female plant or dam) pollinated by another tree (male plant or sire). The seeds are planted and grow into adult trees that can be genotyped and phenotyped. The progeny from this family are full siblings. Since all trees are outbred, we can treat the female parent and the male parent as two F_1 plants from crosses of different grandparents. Therefore, the full-sib family is similar to a four-way cross family. Therefore, the statistical method described in the FW cross can be directly adopted here for the full-sib family mapping. The only difference between FW and full-sib family is that the marker linkage phases are not necessarily known for the two parents of the full-sib family while the linkage phases are known in the parents of the four-way cross progeny.

The simplest method to infer the linkage phases is the so-called two-point analysis. Let us use a three-locus genome (ABC) as an example to demonstrate the phase inference algorithm. Let $A_1 A_2 B_1 B_2 C_1 C_2$ be the genotype for the female parent and $A_3 A_4 B_3 B_4 C_3 C_4$ be the genotype for the male parent. This type of notation for genotype is phase unknown. The phase-known genotype can be written as $\frac{A_1 B_1 C_1}{A_2 B_2 C_2}$ for the female parent. However, this is only one of many different possible phases. Similarly, a possible linkage phase for the male parent is $\frac{A_3 B_3 C_3}{A_4 B_4 C_4}$. The phases of the first locus (A) are irrelevant, and thus, they can be arbitrarily assigned. Let us arbitrarily assign the phase of the female parent for locus A as $\frac{A_1}{A_2}$ and the phase of the male parent for locus A as $\frac{A_3}{A_4}$. We now consider the phases for locus B, ignoring locus C for the moment. There are two possible phases for locus

B in each parent, leading to a total number of four possible phases, which are $\left\{ \frac{A_1B_1}{A_2B_2}, \frac{A_1B_2}{A_2B_1}, \frac{A_3B_3}{A_4B_4}, \frac{A_3B_4}{A_4B_3} \right\}$. Under each possible phase, say phase k for $k = 1, \ldots, 4$, we calculate the likelihood value of all the progeny for the observed two-locus genotypes given the phase and the recombination fraction between loci A and B. The log likelihood value is $L(r_{AB}|k) = \sum_{j=1}^{n} L_j(r_{AB}|k)$ where

$$L_j(r_{AB}|k) = \ln \Pr(A_j B_j | r_{AB}) = \ln(J^T D_{A_j} T_{AB} D_{B_j} J) \qquad (12.31)$$

The joint probability of marker genotypes for a particular individual can be found in Chap. 4. The inferred linkage phase is the one that has the maximum $L(r_{AB}|k)$ for $k = 1, \ldots, 4$. Once the phase for locus B is determined, we follow the same approach to inferring the phase of locus C using the BC segment. We can infer both B and C simultaneously using the three-point analysis, but the total number of phases that need to be evaluated is $4^2 = 16$. For m markers, the two-point analysis only evaluates $4m$ different phases, while the m-point analysis must evaluate 4^m different phases.

Mapping QTL using full-sib families is the simplest design in QTL mapping for outbred populations. Remember that using the full-sib design, we can only identify the difference between the two alleles of each parent. It is impossible to identify the allelic differences between the sire and the dam. Further crosses are required to identify the difference between the two parents. Tree breeders sometimes choose two phenotypically very different trees to cross and collect the full-sib progeny from the cross for QTL mapping. Their intention for choosing the cross of two diversified trees is to identify the allelic difference between the parental trees. However, they may not realize that the full siblings are the hybrids (like F_1), although their parents are not inbred. This full-sib family can only be used to identify the allelic difference within each parent, not the allelic difference between parents. Trees of the next generation by crossing the full siblings to each other are the material for identifying the difference between the diversified parental trees. If breeders can only manage to cross for one generation, instead of crossing two diversified varieties, they should select two trees from the same variety for crossing.

12.5 F_2 Population Derived from Outbreds

In large animals, e.g., beef cattle, inbred lines are not available. We can still perform QTL mapping using F_2 derived from two different breeds. A breed in animals is just like a line in plants except that a breed consists of a group of noninbred animals who have the same genetic background. A group of female animals from breed 1 are crossed with a group of male animals from breed 2 to generate a group of F_1 individuals. These F_1 individuals are intercrossed to generate a large number of F_2

individuals, called an F_2 population. They are called an F_2 population rather than an F_2 family because these F_2 progeny may come from many different families. These F_2 individuals are the experimental material for QTL mapping.

The optimal strategy to analyze such an F_2 population is the mixed model methodology proposed by Xu and Yi (2000). Their method was developed under the general framework of pedigree analysis. The mixed model methodology partitions the total genetic variance of a QTL into a between-breed variance and a within-breed variance. Either variance being significant implies significance of the QTL. We will not discuss this complicated method here; instead, we will introduce the simplified method of Haley et al. (1994). The method is a combination of the F_2 design and the FW cross design. It adopts the F_2 model for the QTL but with the FW cross approach to inferring the QTL genotype probabilities. The F_2 model assumes that there are two alleles of the QTL, fixed alternatively in the two breeds. As a result, the classical F_2 model applies. The markers, however, are not necessarily fixed alternatively for the two breeds. There may be multiple alleles per marker locus. Recall that the F_2 population may consist of many full-sib families. The QTL model applies to all the families (regardless of the family origin of individuals), but the way to infer the QTL genotype probabilities is family dependent. Within each family, there are a maximum of four alleles per marker locus. Using the parental marker genotypes and the marker genotypes of a sibling, we can infer the QTL genotype probabilities under the full-sib family design (if the linkage phases are unknown) or the four-way cross design (if the linkage phases are known).

If the biallelic assumption (fixed alternatively in the two breeds) is violated, the simplified method still works except that the power will be reduced. If the QTL alleles are not fixed within the parental breeds, the within-breed genetic variance will contribute to the total genetic variance, in addition to the between-breed variance. The simplified analysis, however, will only capture the between-breed genetic variance (fail to capture the within-breed variance) and thus will have a lower power to detect the QTL. The justification of this simplified method is that the two breeds are usually selected based on their diversified genetic backgrounds, which indicates that the within-breed genetic variance is negligible compared to the between-breed variance.

12.6 Example

This example demonstrates the application of the interval mapping method to a double haploid population of barley. The data were originally published by Hayes et al. (1993) and retrieved from

http://www.genenetwork.org/genotypes/SxM.geno

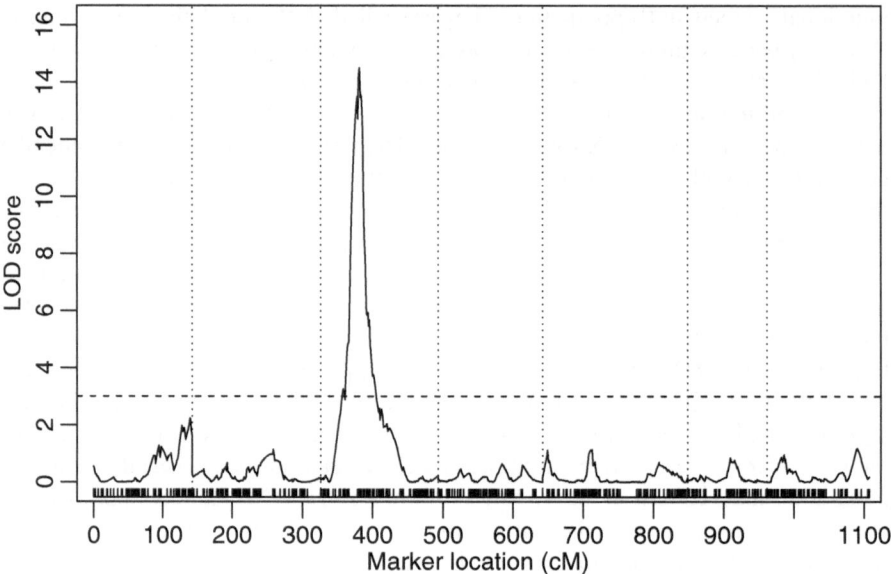

Fig. 12.4 The LOD test statistic profile for QTL mapping of the barley yield trait. The seven chromosomes are separated by the *vertical dotted lines*. The "bar-code"-like ticks on the horizontal axis indicate the marker locations

for the genotypes and

http://wheat.pw.usda.gov/ggpages/SxM/phenotypes.html

for the phenotypes. The data consist of 150 double haploid (DH) lines derived from the cross of two spring barley varieties, Steptoe and Morex. A total of eight quantitative traits, including grain yield (YIELD), heading date (HEAD), plant height (HEIGHT), lodging (LODG), grain protein (PROTEIN), alpha amylase (ALPHA), diastatic power (POWER), and malt extract (EXTRACT), were measured from multiple environments with the number of environments ranging from 6 to as many as 16. The average values of trait across the environments were considered as the original phenotypic values for the QTL mapping experiment. QTL by environment (Q × E) interaction is assumed to be absent. The total number of markers was 495 distributed along seven chromosomes of the barley genome. The genotypes of the markers were denoted by A for the Steptoe parent and B for the Morex parent. Missing values were designated by H. The mixture model maximum likelihood method was used for the interval mapping. The entire genome was scanned with a 1-cM increment. Figure 12.4 gives the LOD score profile of the barley genome for the grain yield (YIELD) trait. The highest LOD score occurs at marker ABC325 on chromosome 3 with a LOD score of 14.5. The QTL genotype was coded as $+1$ for A and -1 for B. With this coding system, the estimated QTL effect for this locus was 0.248 (see Fig. 12.5 for the QTL effect profile of the yield trait).

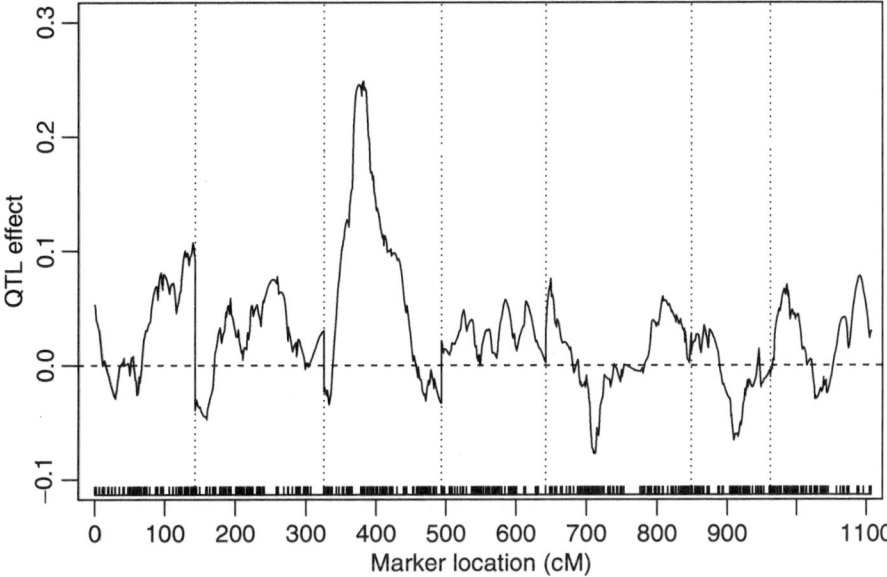

Fig. 12.5 The QTL effect profile for the barley yield trait. The seven chromosomes are separated by the *vertical dotted lines*. The "bar-code"-like ticks on the horizontal axis indicate the marker locations

The estimated residual variance was 0.169. Therefore, the proportion of the yield trait variance contributed by this single QTL is

$$h^2 = \frac{0.248^2}{0.248^2 + 0.169} = 0.2668$$

This huge QTL should be very useful for marker-assisted selection to improve the grain yield of barley.

Chapter 13
Random Model Approach to QTL Mapping

The mapping populations we have discussed so far are all initiated from crosses of two or a few lines (breeds). As a result, the number of alleles is relatively small, and thus the conclusion is drawn based on narrow genetic variation. In addition, through control of the mating design, we can control the allele frequencies. Because the number of alleles is determined by the number of inbred lines involved in a line crossing experiment and the number of lines is small, we can estimate and test the allelic effects or the average effects of allelic substitution. The linear models that allow us to estimate and test the allelic effects are called fixed effect models. Therefore, all methods we have learned so far are based on the fixed model approach.

When designed matings are impossible, we must collect data as they exist and use such data to conduct QTL mapping. Because the number of alleles involved in the mapping population is unknown and we cannot control the allelic frequencies, the fixed model approach is hard to implement. In this chapter, we introduce an alternative approach of QTL mapping that involves multiple alleles, the random model approach to QTL mapping. Under the random model framework, rather than estimating and testing the effects of allelic substitution of QTL, we estimate and test the variances of the allelic effects for the QTL.

The mapping population may consist of a few large pedigrees or many small pedigrees (e.g., nuclear families). A pedigree is a collection of genetically related individuals descending from a few ancestors. Two types of pedigrees are commonly used in QTL mapping: complicated pedigrees and simple pedigrees. A complicated pedigree is a collection of relatives that expand for multiple generations. Members in a complicated pedigree can be inbred or outbred, and their relationships can be arbitrarily complicated. A simple pedigree, however, consists of two outbred parents and their children, and thus, it is also called a nuclear family. When the phenotypic values of the parents are excluded from the analysis, the method is called full-sib analysis. In this chapter, we only discuss the random model methodology using multiple full-sib families. Extension of the method to QTL mapping for complicated pedigrees will be mentioned briefly toward the end of this chapter.

S. Xu, *Principles of Statistical Genomics*, DOI 10.1007/978-0-387-70807-2_13,
© Springer Science+Business Media, LLC 2013

A brief introduction to the milestones of the random model methodology of QTL mapping is presented in this paragraph. The random model methodology is also called variance component analysis (Searle et al. 1992). The parameters of interest are variances of the random effects rather than the effects themselves. In terms of QTL mapping, the parameter of interest is the variance of the allelic effects of the locus under investigation in the mapping population. Early works of random model QTL mapping include Goldgar (1990), Schork (1993), Amos (1994), and Xu and Atchley (1995). These studies laid the foundation for the popular QTL mapping procedures in human pedigrees (e.g., Almasy and Blangero 1998). Goldgar (1990) partitioned the entire genome into many regions (chromosome segments) and used a multipoint method to estimate the identity-by-descent (IBD) value shared by pair of relatives (siblings) for the target region. Using the maximum likelihood method, Goldgar (1990) was able to estimate the genetic variance explained by that chromosome segment. Schork (1993) extended the method and proposed to estimate variance components of multiple segments simultaneously. In addition, Schork (1993) also included a common environmental effect shared by relatives and estimated the common environmental variance. Amos' (1994) model differs from Goldgar (1990) in that fixed effects not relevant to genetics are included in the model. Therefore, Amos' (1994) method is a linear mixed model approach. Another new feature of Amos' (1994) model is that he replaced the IBD of a chromosome region by the IBD of a marker. Xu and Atchley (1995) adopted the idea of interval mapping (Lander and Botstein, 1989) to estimate the genetic variance of a particular location of the genome using flanking markers. Using genome-wide markers, Xu and Atchley (1995) were able to scan the entire genome for QTL under the random model approach.

Prior to the maximum likelihood methods of QTL variance estimation, Haseman and Elston (1972) developed a sib-pair regression method for estimating genetic variance of a polymorphic marker. They found that the squared difference between the phenotypic values of a sib pair is a linear function of the genetic variance of a marker. Therefore, they regressed the squared phenotypic difference on the IBD value of a sib pair to obtain an estimation of the genetic variance. Many people believe that the regression model of Haseman and Elston (1972) is a genius. However, the real creativity of Haseman and Elston (1972) comes from the recognition of the variation of sib-pair IBD and the method to calculate the conditional expectation of IBD values given marker information. It is the variance of locus-specific IBD that allows the separation of the QTL variance from the polygenic variance. The original sib-pair regression of Haseman and Elston (1972) is still a marker analysis. It is the sib-pair interval mapping of Fulker and Cardon (1994) that allows the entire genome to be scanned and thus puts QTL mapping of random populations in the same framework as interval mapping of line crosses (Lander and Botstein 1989).

13.1   Identity by Descent

Identity by descent is a special terminology in quantitative genetics to describe the relationship between two alleles. If two alleles are the same copy of an ancestral allele in the past, the two alleles are said to be identical by descent (IBD). In contrast, two alleles are said to be identical by state (IBS) if they have the same allelic form, regardless of their origins. Without mutation, two alleles that are IBD must also be IBS; however, the reverse is not necessarily true. We now use Fig. 13.1 (modified from Lynch and Walsh (1998)) to demonstrate the difference between IBD and IBS. This diagram shows the paths of the four alleles of the parents to the four alleles of the progeny. Such a diagram is called a descent graph. The progeny in the left (sib 1) has two A_1 alleles, but only A_1 in the left is IBD to the A_1 allele carried by the progeny in the right (sib 2). All three A_1 alleles in the progeny are IBS. The term IBD is an event describing the relationship between two alleles. In a diploid organism, however, an individual has two alleles at any given locus. To describe the relationship between two individuals, we define the IBD value as a proportion of the number of IBD alleles. Two individuals can share two IBD alleles, one IBD allele or non-IBD allele. Therefore, the IBD proportion between two individuals can be $2/2 = 1.0$, $1/2 = 0.5$, or $0/2 = 0.0$, depending on how many IBD alleles shared by the two individuals. IBD is the key of QTL mapping under the random model methodology. We now discuss the IBD value between siblings and the properties of the IBD value. Let $A_1^s A_2^s$ and $A_1^d A_2^d$ be the genotypes of the father and the mother of a nuclear family, respectively, where A_1^s and A_2^s are the two alleles of the father and A_1^d and A_2^d are the two alleles of the mother. Note that the two alleles carried by a parent are ordered. For example, A_1^s and A_2^s represent the paternal and maternal alleles of the father, respectively. Previously (Chap. 12), we used A_1, A_2, A_3, and A_4 to represent the four alleles carried by the two parents. Now, these four alleles are represented by A_1^s, A_2^s, A_1^d, and A_2^d, respectively. We use this new notation simply to represent the four different origins of the alleles. These four alleles are not necessarily different in terms of the

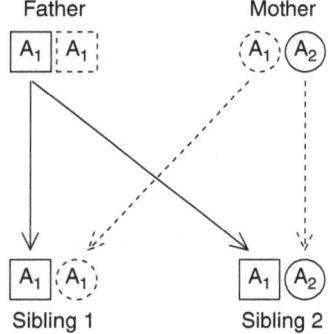

Fig. 13.1 Descent graph showing the allelic transmissions from the parents to the progeny. Sib 1 has two A_1 alleles, but only A_1 in the left is IBD to the A_1 allele carried by sib 2. All three A_1 alleles in the progeny are IBS

Table 13.1 The 16 possible pairs of siblings and the numbers of IBD alleles (in parentheses) shared by each sib pair

		Sib two		
Sib one	$A_1^s A_1^d$	$A_1^s A_2^d$	$A_2^s A_1^d$	$A_2^s A_2^d$
$A_1^s A_1^d$	$A_1^s A_1^d$-$A_1^s A_1^d$(2)	$A_1^s A_1^d$-$A_1^s A_2^d$(1)	$A_1^s A_1^d$-$A_2^s A_1^d$(1)	$A_1^s A_1^d$-$A_2^s A_2^d$(0)
$A_1^s A_2^d$	$A_1^s A_2^d$-$A_1^s A_1^d$(1)	$A_1^s A_2^d$-$A_1^s A_2^d$(2)	$A_1^s A_2^d$-$A_2^s A_1^d$(0)	$A_1^s A_2^d$-$A_2^s A_2^d$(1)
$A_2^s A_1^d$	$A_2^s A_1^d$-$A_1^s A_1^d$(1)	$A_2^s A_1^d$-$A_1^s A_2^d$(0)	$A_2^s A_1^d$-$A_2^s A_1^d$(2)	$A_2^s A_1^d$-$A_2^s A_2^d$(1)
$A_2^s A_2^d$	$A_2^s A_2^d$-$A_1^s A_1^d$(0)	$A_2^s A_2^d$-$A_1^s A_2^d$(1)	$A_2^s A_2^d$-$A_2^s A_1^d$(1)	$A_2^s A_2^d$-$A_2^s A_2^d$(2)

allelic forms (states). For example, if both A_1^s and A_2^s have the same allelic state, say A_1, then the actual genotype of the father is $A_1 A_1$, a homozygote.

Four possible genotypes in terms of the allelic origins can be generated by the mating pair, which are $A_1^s A_1^d$, $A_1^s A_2^d$, $A_2^s A_1^d$, and $A_2^s A_2^d$, each with an equal probability. If we randomly sample a pair of siblings from the family, there will be 16 possible combinations of the sib pairs. These 16 sib-pair combinations are listed in Table 13.1. We now evaluate each sib pair to see how many alleles are shared by the siblings. Take sib-pair $A_1^s A_1^d - A_1^s A_2^d$ for example. The two siblings share one common allele from their father, A_1^s, but the two alleles from their mother are different in origin. Since each individual has two alleles at any locus, the proportion of IBD alleles shared by the siblings is $\pi = 1/2 = 0.5$. However, sib-pair $A_1^s A_1^d - A_1^s A_1^d$ shares both alleles IBD, and thus, their IBD value is $\pi = 2/2 = 1.0$. Although the two individuals are siblings, they behave like identical twins at the locus of interest. Some sib pairs, e.g., $A_1^s A_1^d - A_2^s A_2^d$, do not share any IBD allele, and thus $\pi = 0/2 = 0.0$. For this locus, these two individuals act like strangers, although they are actually siblings. We can see that the IBD proportion shared by siblings, denoted by π, is a discrete variable, taking one of three possible values. Among the 16 possible sib pairs, four of them (on the major diagonals) share two IBD alleles ($\pi = 1.0$), four of them (on the minor diagonals) share non-IBD allele ($\pi = 0.0$), and the remaining eight sib pairs share one IBD allele ($\pi = 0.5$). Therefore, the expectation of the IBD value is

$$\mathrm{E}(\pi) = \frac{1}{4} \times 1.0 + \frac{1}{2} \times 0.5 + \frac{1}{4} \times 0.0 = \frac{1}{2}. \tag{13.1}$$

The variance of the IBD value is

$$\mathrm{var}(\pi) = \mathrm{E}(\pi^2) - \mathrm{E}^2(\pi) = \frac{1}{4} \times 1.0^2 + \frac{1}{2} \times 0.5^2 + \frac{1}{4} \times 0.0^2 - 0.5^2 = \frac{1}{8}. \tag{13.2}$$

If we consider the whole genome, siblings share half of their genome (genetic material) and thus have a genome-wide IBD proportion of 0.5. However, if we consider a single locus, the IBD proportion is a variable with an expectation of $\frac{1}{2}$ and a variance of $\frac{1}{8}$. This variance is the key to the random model analysis of QTL for outbred populations.

13.2 Random Effect Genetic Model

Consider a mapping population consisting of n full-sib families each with two siblings. Let y_{j1} and y_{j2} be the phenotypic values of a quantitative trait for the two siblings in family j for $j = 1, \ldots, n$. The phenotypes can be described by the following linear models:

$$y_{j1} = \mu + a_{j1} + \gamma_{j1} + \epsilon_{j1}$$
$$y_{j2} = \mu + a_{j2} + \gamma_{j2} + \epsilon_{j2} \tag{13.3}$$

where μ is the population mean of the trait, γ_{j1} and γ_{j2} are the genetic effects of a putative QTL for the two siblings, a_{j1} and a_{j2} are the polygenic effects (collective effects of all loci of the genome), and ϵ_{j1} and ϵ_{j2} are the residual errors for the two siblings. These equations can be expressed in matrix notation as

$$\begin{bmatrix} y_{j1} \\ y_{j2} \end{bmatrix} = \begin{bmatrix} \mu \\ \mu \end{bmatrix} + \begin{bmatrix} a_{j1} \\ a_{j2} \end{bmatrix} + \begin{bmatrix} \gamma_{j1} \\ \gamma_{j2} \end{bmatrix} + \begin{bmatrix} \epsilon_{i1} \\ \epsilon_{i2} \end{bmatrix} \tag{13.4}$$

The expectation of the array of phenotypic values is

$$E \begin{bmatrix} y_{j1} \\ y_{j2} \end{bmatrix} = \begin{bmatrix} \mu \\ \mu \end{bmatrix} \tag{13.5}$$

and the variance–covariance matrix of the phenotypes is

$$\text{var} \begin{bmatrix} y_{j1} \\ y_{j2} \end{bmatrix} = \text{var} \begin{bmatrix} a_{j1} \\ a_{j2} \end{bmatrix} + \text{var} \begin{bmatrix} \gamma_{j1} \\ \gamma_{j2} \end{bmatrix} + \text{var} \begin{bmatrix} \epsilon_{j1} \\ \epsilon_{j2} \end{bmatrix}, \tag{13.6}$$

where

$$\text{var} \begin{bmatrix} \gamma_{i1} \\ \gamma_{i2} \end{bmatrix} = \begin{bmatrix} 1 & \pi_j \\ \pi_j & 1 \end{bmatrix} \sigma_\gamma^2 \tag{13.7}$$

$$\text{var} \begin{bmatrix} a_{j1} \\ a_{j2} \end{bmatrix} = \begin{bmatrix} 1 & 0.5 \\ 0.5 & 1 \end{bmatrix} \sigma_a^2 \tag{13.8}$$

and

$$\text{var} \begin{bmatrix} \epsilon_{j1} \\ \epsilon_{j2} \end{bmatrix} = \begin{bmatrix} 1 & 0 \\ 0 & 1 \end{bmatrix} \sigma^2 \tag{13.9}$$

The three variance components, σ_a^2, σ_γ^2, and σ^2, are the polygenic variance, the genetic variance of the QTL, and the residual variance, respectively. Note that the covariance between the siblings at the QTL is $\text{cov}(\gamma_{j1}, \gamma_{j2}) = \pi_j \sigma_\gamma^2$ while the covariance at the polygene is $\text{cov}(a_{j1}, a_{j2}) = 0.5\sigma_a^2$. The siblings are assumed to share no common environmental effect, and thus $\text{cov}(\epsilon_{j1}, \epsilon_{j2}) = 0.0\sigma^2$. The three different coefficients, π_j, 0.5, and 0.0, in the covariance between siblings, allow us to separate the three variance components and thus to estimate and test the QTL variance σ_γ^2. Let $y_j = \{y_{j1}, y_{j2}\}$, $\gamma_j = \{\gamma_{j1}, \gamma_{j2}\}$, $a_j = \{a_{j1}, a_{j2}\}$, and $\epsilon_j = \{\epsilon_{j1}, \epsilon_{j2}\}$ be vector presentations of the corresponding contents. Let $1 = \{1, 1\}$ be a unity vector. The linear models for the two siblings can be rewritten as

$$y_j = 1\mu + a_j + \gamma_j + \epsilon_j. \tag{13.10}$$

Let

$$\Pi_j = \begin{bmatrix} 1 & \pi_j \\ \pi_j & 1 \end{bmatrix} \tag{13.11}$$

be the IBD matrix for the QTL,

$$A_j = \begin{bmatrix} 1 & 0.5 \\ 0.5 & 1 \end{bmatrix} \tag{13.12}$$

be the additive relationship matrix for the polygene, and

$$I = \begin{bmatrix} 1 & 0 \\ 0 & 1 \end{bmatrix} \tag{13.13}$$

be the identity matrix for the residual error. The compact matrix representations for the expectation and variance of y_j are

$$E(y_j) = 1\mu \tag{13.14}$$

and

$$\text{var}(y_j) = V_j = A_j\sigma_a^2 + \Pi_j\sigma_\gamma^2 + I\sigma^2, \tag{13.15}$$

respectively. Although A_j is family independent, i.e., all families have the same A_j, we still use a subscript j for notational consistency. In addition, when we deal with families with variable size, the dimension of A_j will be different from one family to another. Variable family size will be dealt with in a later section.

13.3   Sib-Pair Regression

The Haseman and Elston (1972) sib-pair regression method does not use the original phenotypic values as response variables; rather, the squared difference between the phenotypic values of a sib pair is treated as the response variable. Define $s_j = (y_{j1} - y_{j2})^2$ as the squared difference of sib-pair j. The expectation of s_j is

$$
\begin{aligned}
E(s_j) &= E\left[(y_{j1} - y_{j2})^2\right] \\
&= \sigma_a^2 + 2\sigma_\gamma^2 + 2\sigma^2 - 2\sigma_\gamma^2\pi_j
\end{aligned} \tag{13.16}
$$

Let $\beta = -2\sigma_\gamma^2$ be the regression coefficient and

$$\alpha = \sigma_a^2 + 2\sigma_\gamma^2 + 2\sigma^2 \tag{13.17}$$

be the intercept; the above model is rewritten as

$$s_j = \alpha + \beta\pi_j + \varepsilon_j \tag{13.18}$$

where ε_j is the residual for the linear model and the variance of ε_j is denoted by σ_ε^2. The residual is not normally distributed, and the residual variance here is not the environmental variance. Let $\hat{\beta}$ be the least square estimate of the regression coefficient. The estimated genetic variance for the marker is

$$\hat{\sigma}_\gamma^2 = -\frac{1}{2}\hat{\beta} \tag{13.19}$$

One may notice that the intercept given in (13.17) is different from that of Haseman and Elston (1972), which is

$$\alpha = 2\sigma_\gamma^2 + \sigma_e^2 \tag{13.20}$$

The Haseman and Elston (1972) model assumes absence of polygenic effect, and thus, σ_a^2 is excluded. The environmental error defined in Haseman and Elston (1972) is $e_j = \epsilon_{j1} - \epsilon_{j2}$, the difference between the environmental errors of the sib pair. Therefore, $\sigma_e^2 = 2\sigma^2$, twice the environmental error variance defined in this chapter.

This regression analysis looks strange because the regression coefficient $\beta = 2\sigma_\gamma^2$ also appears in the intercept (as a component). The model can be revised to the following form:

$$s_j = \alpha + \beta(\pi_j - 1) + \varepsilon_j \tag{13.21}$$

where α is redefined as $\alpha = \sigma_a^2 + 2\sigma^2$, which has excluded $2\sigma_\gamma^2$. However, the independent variable is $\pi_j - 1$, instead of π_j. The final estimation of the β remains the same.

13.4   Maximum Likelihood Estimation

We assume that π_j is known for all $j = 1, \ldots, n$. This represents a situation where we can observe the genotypes of the QTL. In QTL mapping, the genotype of a QTL is not observable, but the distribution of the genotype can be inferred from marker information. The IBD value of a putative QTL is then replaced by the IBD value estimated from marker data. The likelihood function discussed in this section is based on known IBD values. Method dealing with inferred IBD will be deferred to a later section, where a multipoint method for estimating the IBD values will also be introduced. The IBD values for all the sib pairs are treated as data. The data also include the phenotypic values of the siblings. The parameter vector is $\theta = \{\mu, \sigma_\gamma^2, \sigma_a^2, \sigma^2\}$. The log likelihood function is

$$L(\theta) = -\frac{1}{2} \sum_{j=1}^{n} \ln |V_j| - \frac{1}{2} \sum_{j=1}^{n} (y_j - 1\mu)^T V_j^{-1} (y_j - 1\mu). \tag{13.22}$$

This likelihood function is complicated, and the explicit solution for the MLE of the parameters is not available.

13.4.1 EM Algorithm

Xu and Atchley (1995) used the simplex algorithm (Nelder and Mead 1965) to find the MLE of θ. Here we introduce the EM algorithm to estimate θ. Under the random model framework, the parameters of interest are the variance components. Although QTL effects and polygenic effects both appear in the linear models, they are not the parameters of interest. Because of this, they do not appear in the log likelihood function. If these genetic effects were observed, the genetic variance components would have been estimated easily using a simple formula of variance. This reminds us the EM algorithm, in which we can take advantage of the simplicity of the variance formula by treating both the QTL effects and the polygenic effects as missing values. In this section, we introduce the EM algorithm for estimating variance components. Derivation of the method requires some complicated matrix algebra and thus will not be provided here. We will simply show the final iterative equation for each parameter. We start with all parameter values at iteration t, denoted by $\theta^{(t)} = \{\mu^{(t)}, \sigma_a^{2(t)}, \sigma_\gamma^{2(t)}, \sigma^{2(t)}\}$, and then proceed to update each parameter in sequence conditional on $\theta^{(t)}$. Let us denote V_j evaluated at $\theta^{(t)}$ by

$$V_j^{(t)} = \Pi_j \sigma_\gamma^{2(t)} + A_j \sigma_a^{2(t)} + I \sigma^{2(t)} \tag{13.23}$$

We now introduce the EM algorithm starting with $\theta = \theta^{(t)}$. If the QTL and the polygenic effects were observed for all individuals, the MLEs of the QTL and polygenic variances would be calculated using

$$\sigma_\gamma^2 = \frac{1}{2n} \sum_{j=1}^{n} \gamma_j^T \Pi_j^{-1} \gamma_j \tag{13.24}$$

and

$$\sigma_a^2 = \frac{1}{2n} \sum_{j=1}^{n} a_j^T A_j^{-1} a_j \tag{13.25}$$

respectively. The environmental error variance would be obtained using

$$\sigma^2 = \frac{1}{2n} \sum_{j=1}^{n} y_j^T (y_j - \mu - a_j - \gamma_j) \tag{13.26}$$

The EM algorithm takes advantage of these simple equations by replacing the terms involving a_j and γ_j by their expectations. However, to calculate the expectations, we need to use the phenotypic values and the current values of the parameters. Such expectations are called the posterior expectations. The EM algorithm starts with calculation of these expectations (the E-step) and then uses the simple equations to update the variance components (the M-step). In the M-step, the parameters at the $(t + 1)$th iteration are calculated using the following iteration equations:

$$\mu^{(t+1)} = \left[\sum_{i=1}^{n} 1^T (V_j^{(t)})^{-1} 1 \right]^{-1} \left[\sum_{j=1}^{n} 1^T (V_j^{(t)})^{-1} y_j \right] \tag{13.27}$$

$$\sigma_\gamma^{2(t+1)} = \frac{1}{2n} \sum_{j=1}^{n} \mathrm{E}(\gamma_j^T \Pi_j^{-1} \gamma_j)$$

$$= \frac{1}{2n} \sum_{j=1}^{n} \left\{ \mathrm{E}(\gamma_j^T) \Pi_j^{-1} \mathrm{E}(\gamma_j) + \mathrm{tr} \left[\Pi_j^{-1} \mathrm{var}(\gamma_j) \right] \right\} \tag{13.28}$$

$$\sigma_a^{2(t+1)} = \frac{1}{2n} \sum_{j=1}^{n} \mathrm{E}(a_j^T A_j^{-1} a_j)$$

$$= \frac{1}{2n} \sum_{j=1}^{n} \left\{ \mathrm{E}(a_j^T) A_j^{-1} \mathrm{E}(a_j) + \mathrm{tr} \left[A_j^{-1} \mathrm{var}(a_j) \right] \right\} \tag{13.29}$$

$$\sigma^{2(t+1)} = \frac{1}{2n} \sum_{j=1}^{n} y_j^T \left[y_j - 1\mu^{(t+1)} - \mathrm{E}(a_j) - \mathrm{E}(\gamma_j) \right] \tag{13.30}$$

These equations are the steps required in the M-step. We can see that these equations contain the expectations and variances of QTL effects and the polygenic effects. Calculating these expectations and variances is called the E-step. For the QTL effect, we have

$$\mathrm{E}(\gamma_j) = \Pi_j \sigma_\gamma^2 V_j^{-1} (y_j - \mu)$$

$$\mathrm{var}(\gamma_j) = \Pi_j \sigma_\gamma^2 \left(I - V_j^{-1} \Pi_j \sigma_\gamma^2 \right) \tag{13.31}$$

The expectation and variance for the polygenic effect are

$$\mathrm{E}(a_j) = A_j \sigma_a^2 V_j^{-1} (y_j - \mu)$$

$$\mathrm{var}(a_j) = A_j \sigma_a^2 \left(I - V_j^{-1} A_j \sigma_a^2 \right) \tag{13.32}$$

13.4.2   EM Algorithm Under Singular Value Decomposition

The EM algorithm is not always guaranteed to work because occasionally Π_j^{-1} may not exist. For example, if the two siblings share 2 IBD alleles at the QTL, $\pi_j = 1$, then Π_j will be singular. To avoid this problem, we can take a different approach. This approach requires a linear transformation of the genetic effects using singular value decomposition. Let us define Γ_j as an upper triangular matrix so that $\Gamma_j^T \Gamma_j = \Pi_j$. We call Γ_j the Cholesky decomposition. This decomposition exists even if Π_j is positive semidefinite (not necessarily positive definite). Using the Cholesky decomposition, we can avoid inverting Π_j because the inverse of Π_j does not exist if Π_j is not positive definite. Similarly, we can take the Cholesky decomposition for the additive relationship matrix, denoted by H_j, i.e., $H_j^T H_j = A_j$. For two siblings per family, both Γ_j and H_j have explicit expressions,

$$\Gamma_j = \begin{bmatrix} 1 & \pi_j \\ 0 & \sqrt{1 - \pi_j^2} \end{bmatrix} \tag{13.33}$$

and

$$H_j = \begin{bmatrix} 1 & 0.5 \\ 0 & \sqrt{1 - 0.5^2} \end{bmatrix} \tag{13.34}$$

We now rewrite the linear model as

$$y_j = 1\mu + H_j^T a_j^* + \Gamma_j^T \gamma_j^* + \epsilon_j \tag{13.35}$$

where $a_j^* \sim N(0, I\sigma_a^2)$ and $\gamma_j^* \sim N(0, I\sigma_\gamma^2)$. Note the difference between a_j and a_j^* and the difference between γ_j and γ_j^*. After the transformation, $a_j = H_j^T a_j^*$ and $\gamma_j = \Gamma_j^T \gamma_j^*$, we now deal with vectors with independent elements. The expectation and variance–covariance matrix remain the same as given before, i.e.,

$$E(y_j) = 1\mu \tag{13.36}$$

and

$$\begin{aligned} \mathrm{var}(y_j) &= H_j^T \mathrm{var}(a_j^*) H_j + \Gamma_j^T \mathrm{var}(\gamma_j^*) \Gamma_j + I\sigma^2 \\ &= H_j^T H_j \sigma_a^2 + \Gamma_j^T \Gamma_j \sigma_\gamma^2 + I\sigma^2 \\ &= A_j \sigma_a^2 + \Pi_j \sigma_\gamma^2 + I\sigma^2 \end{aligned} \tag{13.37}$$

The EM algorithm after the singular value decomposition consists of the following steps. For the maximization step, we have

$$\mu^{(t+1)} = \left[\sum_{i=1}^{n} 1^{\mathrm{T}}(V_j^{(t)})^{-1}1\right]^{-1}\left[\sum_{j=1}^{n} 1^{\mathrm{T}}(V_j^{(t)})^{-1}y_j\right] \tag{13.38}$$

$$\sigma_\gamma^{2(t+1)} = \frac{1}{2n}\sum_{j=1}^{n} \mathrm{E}(\gamma_j^{*T}\gamma_j^*)$$

$$= \frac{1}{2n}\sum_{j=1}^{n}\left\{\mathrm{E}(\gamma_j^{*T})\mathrm{E}(\gamma_j^*) + \mathrm{tr}\left[\mathrm{var}(\gamma_j^*)\right]\right\} \tag{13.39}$$

$$\sigma_a^{2(t+1)} = \frac{1}{2n}\sum_{j=1}^{n} \mathrm{E}(a_j^{*T}a_j^*)$$

$$= \frac{1}{2n}\sum_{j=1}^{n}\left\{\mathrm{E}(a_j^{*T})\mathrm{E}(a_j^*) + \mathrm{tr}\left[\mathrm{var}(a_j^*)\right]\right\} \tag{13.40}$$

$$\sigma^{2(t+1)} = \frac{1}{2n}\sum_{j=1}^{n} y_j^T\left[y_j - 1\mu^{(t+1)} - H_j^T\mathrm{E}(a_j^*) - \Gamma_j^T\mathrm{E}(\gamma_j^*)\right] \tag{13.41}$$

For the expectation step, we have

$$\mathrm{E}(a_j^*) = H_j\sigma_a^2 V_j^{-1}(y_j - 1\mu)$$

$$\mathrm{var}(a_j^*) = \sigma_a^2\left(I - H_j V_j^{-1}H_j^T\sigma_a^2\right) \tag{13.42}$$

and

$$\mathrm{E}(\gamma_j^*) = \Gamma_j\sigma_\gamma^2 V_j^{-1}(y_j - 1\mu)$$

$$\mathrm{var}(\gamma_j^*) = \sigma_\gamma^2\left(I - \Gamma_j V_j^{-1}\Gamma_j^T\sigma_\gamma^2\right) \tag{13.43}$$

Clearly, we have avoided using A_j^{-1} and Γ_j^{-1} under the singular value decomposition approach. This approach is general because it works regardless whether Π_j is singular or not.

13.4.3   Multiple Siblings

The sib-pair approach we have discussed so far can only handle two siblings per family. If a full-sib family contains more than two siblings, the extra siblings cannot be used. Additional information from the extra siblings will be wasted. Extension of the existing method to multiple siblings is quite straightforward under the general framework of random model methodology. This extension is different from the sib-pair regression approach proposed by Haseman and Elston (1972), which cannot be extended to multiple siblings. Let n_j be the number of siblings for the jth full-sib family. Assume that there are n full-sib families in the mapping population; the total sample size is $N = \sum_{j=1}^{n} n_j$. The sib-pair approach is a special case where $n_j = 2, \forall j = 1, \ldots, n$. The only difference between multiple siblings and the sib-pair situations under the random model approach is the dimensionality of the matrices. For the sib-pair method, all matrices have a dimensionality of 2×2. For multiple siblings, vectors y_j, a_j, γ_j, and ϵ_j all have a dimensionality of $n_j \times 1$, and matrices A_j and Π_j have a dimensionality of $n_j \times n_j$. For example, for three siblings per family, the A_j and Π_j matrices are

$$
A_j = \begin{bmatrix} 1 & \dfrac{1}{2} & \dfrac{1}{2} \\[2mm] \dfrac{1}{2} & 1 & \dfrac{1}{2} \\[2mm] \dfrac{1}{2} & \dfrac{1}{2} & 1 \end{bmatrix} \tag{13.44}
$$

and

$$
\Pi_j = \begin{bmatrix} 1 & \pi_{12} & \pi_{13} \\ \pi_{12} & 1 & \pi_{23} \\ \pi_{13} & \pi_{23} & 1 \end{bmatrix} \tag{13.45}
$$

where π_{23} is the IBD value between sibs 2 and 3. The corresponding changes from sib pair to multiple siblings in the EM algorithms occur in the following three places:

$$
\sigma_\gamma^{2(t+1)} = \frac{1}{N} \sum_{j=1}^{n} E(\gamma_j^{*T} \gamma_j^*)
$$

$$
= \frac{1}{N} \sum_{j=1}^{n} \left\{ E(\gamma_j^{*T}) E(\gamma_j^*) + \mathrm{tr}\left[\mathrm{var}(\gamma_j^*) \right] \right\} \tag{13.46}
$$

$$
\sigma_a^{2(t+1)} = \frac{1}{N} \sum_{j=1}^{n} E(a_j^{*T} a_j^*)
$$

$$
= \frac{1}{N} \sum_{j=1}^{n} \left\{ E(a_j^{*T}) E(a_j^*) + \mathrm{tr}\left[\mathrm{var}(a_j^*) \right] \right\} \tag{13.47}
$$

$$
\sigma^{2(t+1)} = \frac{1}{N} \sum_{j=1}^{n} y_j^T \left[y_j - 1\mu^{(t+1)} - H_j^T E(a_j^*) - \Gamma_j^T E(\gamma_j^*) \right] \tag{13.48}
$$

If the Π_j matrix is the true IBD matrix for the n_j siblings, Π_j is positive semidefinite. The EM algorithm under the singular value decomposition will work equally well as the sib-pair method. However, the Π_j matrix for a QTL is always estimated using marker information, and the method of estimation for Π_j cannot guarantee the positive semidefinite property. Therefore, the EM algorithm may not always work. As a result, the simplex method is highly recommended because the method directly evaluates the log likelihood function, which only requires a positive definite $V_j = A_j \sigma_a^2 + \Pi_j \sigma_\gamma + I \sigma^2$. The property of positive definite for V_j is guaranteed as long as σ^2 is not too small. If Π_j is positive semidefinite, V_j is always positive definite because $\sigma^2 > 0$ always holds. Another reason for using the simplex method is that the EM algorithm is sensitive to the initial values of the parameters while the simplex is robust to the initial values of the parameters. For example, using the simplex method, the initial values for $\sigma_a^2 = \sigma_\gamma^2 = 0$ usually work very well, but the EM algorithm cannot take such initial values. Unfortunately, programming the simplex algorithm is not as easy as writing the program code for the EM algorithm.

13.5   Estimating the IBD Value for a Marker

When a marker is not fully informative, the IBD values shared by siblings are ambiguous. We need a special algorithm to estimate the IBD values. We assume that every individual in a nuclear family has been genotyped for the marker, including the parents. If parents are not genotyped, their genotypes must be inferred first from the genotypes of the progeny. However, the method becomes complicated and will not be dealt with in this chapter. There are four possible allelic sharing states for a pair of siblings. The probabilities of the four states are denoted by $\Pr(s, d)$, where $s = \{1, 0\}$ indicates whether or not the siblings share their paternal alleles (alleles from the sire). If they share the paternal alleles, we let $s = 1$; otherwise, $s = 0$. Similarly, $d = \{1, 0\}$ indicates whether or not the siblings share the maternal alleles (alleles from the dam). Once we have the four probabilities, the estimated IBD value for the siblings is

$$\hat{\pi} = \frac{1}{2}\{2\Pr(1,1) + 1[\Pr(1,0) + \Pr(0,1)]\}$$

$$= \Pr(1,1) + \frac{1}{2}[\Pr(1,0) + \Pr(0,1)]. \tag{13.49}$$

Therefore, the problem of estimating the IBD value becomes a problem of calculating the four probabilities of allelic sharing states. We first convert the numbers of shared alleles for the 16 possible sib-pair combinations listed in Table 13.1 into a 4×4 matrix,

$$S = \begin{bmatrix} S_{11} & S_{12} & S_{13} & S_{14} \\ S_{21} & S_{22} & S_{23} & S_{24} \\ S_{31} & S_{32} & S_{33} & S_{34} \\ S_{41} & S_{42} & S_{43} & S_{44} \end{bmatrix} = \begin{bmatrix} 2 & 1 & 1 & 0 \\ 1 & 2 & 0 & 1 \\ 1 & 0 & 2 & 1 \\ 0 & 1 & 1 & 2 \end{bmatrix}, \tag{13.50}$$

where S_{ij} is the ith row and the jth column of matrix S. We then construct another 4×4 matrix, denoted by P, to represent the probabilities of the 16 possible sibling-pair combinations. To find matrix P, we need to find two 4×1 vectors, one for each sibling. Let $U = \{U_1, U_2, U_3, U_4\}$ be the vector for sib one and $V = \{V_1, V_2, V_3, V_4\}$ be the vector for sib two. Recall that each progeny can take one of four possible genotype configurations. Each element of vector U is the probability that the progeny takes that particular genotype. For example, if the sire and the dam of the nuclear family have genotypes of $A_1 A_2$ and $A_3 A_3$, respectively, then the four genotype configurations are $A_1 A_3$, $A_1 A_3$, $A_2 A_3$, and $A_2 A_3$. If sib one has a genotype of $A_2 A_3$, it matches the third and the fourth genotype configurations. Therefore, $U = \{0, 0, \frac{1}{2}, \frac{1}{2}\}$. If sib two has a genotype of $A_1 A_3$, then $V = \{\frac{1}{2}, \frac{1}{2}, 0, 0\}$. Both U and V are defined as column vectors. The 4×4 matrix P is defined as $P = U V^T$, i.e.,

$$P = \begin{bmatrix} P_{11} & P_{12} & P_{13} & P_{14} \\ P_{21} & P_{22} & P_{23} & P_{24} \\ P_{31} & P_{32} & P_{33} & P_{34} \\ P_{41} & P_{42} & P_{43} & P_{44} \end{bmatrix} = \begin{bmatrix} U_1 \\ U_2 \\ U_3 \\ U_4 \end{bmatrix} \begin{bmatrix} V_1 & V_2 & V_3 & V_4 \end{bmatrix}, \qquad (13.51)$$

where $P_{ij} = U_i V_j$ is the ith row and the jth column of matrix P. In the above example, $U = \{0, 0, \frac{1}{2}, \frac{1}{2}\}$ and $V = \{\frac{1}{2}, \frac{1}{2}, 0, 0\}$, and therefore,

$$P = \begin{bmatrix} 0 \\ 0 \\ \frac{1}{2} \\ \frac{1}{2} \end{bmatrix} \begin{bmatrix} \frac{1}{2} & \frac{1}{2} & 0 & 0 \end{bmatrix} = \begin{bmatrix} 0 & 0 & 0 & 0 \\ 0 & 0 & 0 & 0 \\ \frac{1}{4} & \frac{1}{4} & 0 & 0 \\ \frac{1}{4} & \frac{1}{4} & 0 & 0 \end{bmatrix}. \qquad (13.52)$$

The two matrices, S and P, contain all information for estimating the IBD value shared by the siblings. The probability that the siblings share both alleles IBD is

$$\Pr(1, 1) = P_{11} + P_{22} + P_{33} + P_{44}, \qquad (13.53)$$

which is the sum of the diagonal elements of matrix P, called the trace of matrix P. The probability that the siblings share one IBD allele from the sire is

$$\Pr(1, 0) = P_{12} + P_{21} + P_{34} + P_{43}. \qquad (13.54)$$

The probability that the siblings share one IBD allele from the dam is

$$\Pr(0, 1) = P_{13} + P_{24} + P_{31} + P_{42}. \qquad (13.55)$$

Although $\Pr(0, 0)$ is not required in estimating the IBD value, it is the probability that the siblings share no IBD allele. It is calculated using

$$\Pr(0, 0) = P_{14} + P_{23} + P_{32} + P_{41}, \qquad (13.56)$$

which is the sum of the minor diagonal elements of matrix P. In the above example, the estimated IBD value is

$$\hat{\pi} = \Pr(1, 1) + \frac{1}{2} [\Pr(1, 0) + \Pr(0, 1)] = 0 + \frac{1}{2} \left(0 + \frac{1}{2}\right) = \frac{1}{4}. \qquad (13.57)$$

In fact, we do not need to calculate $\Pr(s, d)$ for estimating the IBD value for a marker. The IBD value may be simply estimated using

$$\hat{\pi} = \frac{1}{2} \sum_{i=1}^{4} \sum_{j=1}^{4} S_{ij} P_{ij}. \qquad (13.58)$$

In matrix notation, $\hat{\pi} = \frac{1}{2}[J'(S\#P)J]$, where $J = \{1, 1, 1, 1\}$ is a column vector of unity and the symbol # represents element-wise matrix multiplication. The reason that we develop $\Pr(s, d)$ is to infer the IBD value of a putative QTL using $\Pr(s, d)$ of the markers.

13.6 Multipoint Method for Estimating the IBD Value

Multipoint method for QTL mapping is more important under the IBD-based random model framework than that under the fixed model framework using a single line cross. The reason is that fully informative markers are rare in a random-mating population, in which parents are usually randomly sampled. This is in contrast to a line crossing experiment, in which parents that initiate the cross are inbred and often selected based on maximum diversity in both marker and phenotype distributions. Consider five loci in the order of ABCDE, where locus C is a putative QTL and the other loci are markers. Let r_{AB}, r_{BC}, r_{CD}, and r_{DE} be the recombination fractions between pairs of consecutive loci. The purpose of the multipoint analysis is to estimate the IBD value shared by siblings for locus C given marker information for loci A, B, D, and E. It is not appropriate to use the estimated IBD values for markers to estimate the IBD value of the QTL. Instead, we should use the probabilities of the four possible IBD sharing states of the markers to estimate the four possible IBD sharing states of the QTL, from which the estimated IBD value of the QTL can be obtained.

Let us denote $\Pr(s, d)$ for a marker locus, say locus A, by $P_A(s, d)$. We now define $P_A = \{P_A(1, 1), P_A(1, 0), P_A(0, 1), P_A(0, 0)\}$ as a 4×1 vector for the probabilities of the four IBD sharing states of locus A. Because this locus is a marker, the four probabilities are calculated based on information of the marker genotypes. Let us define $D_A = \text{diag}\{P_A(1, 1), P_A(1, 0), P_A(0, 1), P_A(0, 0)\}$ as a diagonal matrix for locus A. Similar diagonal matrices are defined for all other makers, i.e., D_B, D_D, and D_E. These diagonal matrices represent the marker data, from which the multipoint estimate of the IBD of locus C is obtained. The transition matrix between loci A and B is

$$
T_{AB} = \begin{bmatrix}
\psi_{AB}^2 & (1-\psi_{AB})\psi_{AB} & \psi_{AB}(1-\psi_{AB}) & (1-\psi_{AB})^2 \\
(1-\psi_{AB})\psi_{AB} & \psi_{AB}^2 & (1-\psi_{AB})^2 & \psi_{AB}(1-\psi_{AB}) \\
\psi_{AB}(1-\psi_{AB}) & (1-\psi_{AB})^2 & \psi_{AB}^2 & (1-\psi_{AB})\psi_{AB} \\
(1-\psi_{AB})^2 & \psi_{AB}(1-\psi_{AB}) & (1-\psi_{AB})\psi_{AB} & \psi_{AB}^2
\end{bmatrix},
$$

$$(13.59)$$

where $\psi_{AB} = r_{AB}^2 + (1-r_{AB})^2$. The four conditional probabilities of IBD sharing states for locus C are calculated using the following equations:

$$
\Pr(1,1) = \frac{J'D_A T_{AB} D_B T_{BC} D_{(1)} T_{CD} D_D T_{DE} D_E J}{\sum_{k=1}^4 J'D_A T_{AB} D_B T_{BC} D_{(k)} T_{CD} D_D T_{DE} D_E J},
$$

$$
\Pr(1,0) = \frac{J'D_A T_{AB} D_B T_{BC} D_{(2)} T_{CD} D_D T_{DE} D_E J}{\sum_{k=1}^4 J'D_A T_{AB} D_B T_{BC} D_{(k)} T_{CD} D_D T_{DE} D_E J},
$$

$$
\Pr(0,1) = \frac{J'D_A T_{AB} D_B T_{BC} D_{(3)} T_{CD} D_D T_{DE} D_E J}{\sum_{k=1}^4 J'D_A T_{AB} D_B T_{BC} D_{(k)} T_{CD} D_D T_{DE} D_E J},
$$

$$
\Pr(0,0) = \frac{J'D_A T_{AB} D_B T_{BC} D_{(4)} T_{CD} D_D T_{DE} D_E J}{\sum_{k=1}^4 J'D_A T_{AB} D_B T_{BC} D_{(k)} T_{CD} D_D T_{DE} D_E J}. \qquad (13.60)
$$

The denominator can be simplified into $J'D_A T_{AB} D_B T_{BC} T_{CD} D_D T_{DE} D_E J$ because $\sum_{k=1}^4 D_{(k)} = I$. Let π be the IBD value shared by the siblings at locus C. The multipoint estimate of π using marker information is

$$
\hat{\pi} = \Pr(1,1) + \frac{1}{2}[\Pr(1,0) + \Pr(0,1)]. \qquad (13.61)
$$

For a family with n_j siblings, there are $\frac{1}{2}n_j(n_j-1)$ sib pairs. Therefore, for each locus, we need to calculate $\frac{1}{2}n_j(n_j-1)$ IBD values, one per sib pair. These IBD values are calculated one sib pair at a time. Because they are not calculated jointly, the IBD matrix obtained from these estimated IBD values may not be positive definite or positive semidefinite. As a result, the EM algorithm proposed early may not always work because it requires the positive semidefinite property.

Before we proceed to the next section, it is worthy to mention a possible extension of the additive variance component model to include the dominance variance. The modified model looks like

$$
y_j = 1\mu + a_j + \gamma_j + \xi_j + \epsilon_j \qquad (13.62)
$$

where $\xi_j \sim N(0, \Delta_j \sigma_\xi^2)$ is a vector of dominance effects, σ_ξ^2 is the dominance variance, and Δ_j is the dominance IBD matrix for the siblings in the jth family. The dominance IBD matrix is

$$\Delta_j = \begin{bmatrix} 1 & \delta_j \\ \delta_j & 1 \end{bmatrix} \tag{13.63}$$

where δ_j is a binary variable indicating the event of the siblings sharing IBD genotype. If the two siblings share both IBD alleles (i.e., the same IBD genotype), then $\delta_j = 1$; otherwise, $\delta = 0$. The estimated δ_j is

$$\hat{\delta}_j = \Pr(1, 1) \tag{13.64}$$

the probability that the siblings share both IBD alleles.

13.7   Genome Scanning and Hypothesis Tests

For each putative QTL position, we calculate the estimated IBD value for each sib pair, denoted by $\hat{\pi}_j$ for the jth family. We then construct the IBD matrix

$$\hat{\Pi}_j = \begin{bmatrix} 1 & \hat{\pi}_j \\ \hat{\pi}_j & 1 \end{bmatrix} \tag{13.65}$$

for the jth family. These estimated IBD matrices are used in place of the true IBD matrices for QTL mapping.

The null hypothesis is $H_0 : \sigma_\gamma^2 = 0$. Again, a likelihood ratio test statistic is adopted here for the hypothesis test. The likelihood ratio test statistic is

$$\lambda = -2 \left[L_0(\hat{\hat{\mu}}, \hat{\hat{\sigma}}_a^2, \hat{\hat{\sigma}}^2) - L_1(\hat{\mu}, \hat{\sigma}_\gamma^2, \hat{\sigma}_a^2, \hat{\sigma}^2) \right]. \tag{13.66}$$

Note that the MLE of parameters under the null model differs from the MLE of parameters under the full model by wearing double hats.

The genome is scanned for every putative position with a one- or two-centiMorgan increment. The test statistic will form a profile along the genome. The estimated QTL position takes the location of the genome where the peak of the test statistic profile occurs, provided that the peak is higher than a predetermined critical level. The critical value is usually obtained with permutation test (Churchill and Doerge 1994) that has been described in Chap. 7.

The random model approach to interval mapping utilizes a single QTL model. When multiple QTL are present, the QTL variance will be overestimated. This bias can be eliminated through fitting a multiple QTL model, which will be presented in the next section. If the multiple QTL are not tightly linked, the interval mapping approach still provides reasonable estimates for the QTL variances. Multiple peaks of the test statistic profile indicate the presence of multiple QTL. The positions of the genome where the multiple peaks occur represent estimated positions of the multiple QTL.

One difference between the random model approach and the fixed model approach to interval mapping is that the variances of QTL in other chromosomes will be absorbed by the polygenic variance in the random model analysis rather than by the residual variance in the fixed model analysis. Therefore, the random model approach of interval mapping handles multiple QTL better than the fixed model approach.

13.8 Multiple QTL Model

Assume that there are p QTL in the model, the multiple QTL model is

$$y_j = \mu + \sum_{k=1}^{p} \gamma_{jk} + \epsilon_j \tag{13.67}$$

where $\gamma_{jk} \sim N(0, \Pi_{jk}\sigma_k^2)$ is an $n_j \times 1$ vector for the kth QTL effects, Π_{jk} is an $n_j \times n_j$ IBD matrix for the kth QTL, and σ_k^2 is the genetic variance for the kth QTL. Previously, we had a polygenic effect in the single QTL model to absorb effects of all other QTL not included in the model. With the multiple QTL model, the polygenic effect has disappeared (not needed). The expectation of the model remains the same, i.e., $E(y_j) = \mu$. The variance matrix is

$$\text{var}(y_j) = V_j = \sum_{k=1}^{p} \Pi_{jk}\sigma_k^2 + I\sigma^2 \tag{13.68}$$

The likelihood function and the maximum likelihood estimation of the parameters can follow what has been described previously in the single QTL model.

The complications with the multiple QTL model come from the genome locations of the QTL. Under the single QTL model, we can scan the genome to find the peaks of the test statistic profile. Under the multiple QTL model, the multiple dimensional scanning is hard to implement. Therefore, an entirely different method, called the Bayesian method, is required to simultaneously search for multiple QTL. The Bayesian method will be introduced in Chap. 15. An ad hoc method is discussed here for parameter estimation when the QTL positions are assumed to be known. We can put a finite number of QTL that evenly cover the entire genome and hope that some proposed QTL will sit nearby a true QTL. We may put one QTL in every d cM, say $d = 20$, of the entire genome. This may end up with too many proposed QTL (more than the actual number of QTL). Some of the proposed QTL may be nearby one or more true QTL, and thus, they can absorb the QTL effects. These proposed QTL are useful and must be included in the model. Majority of the proposed QTL may not be able to pick up any information at all because they may not be close to any QTL. These proposed QTL are fake ones and, theoretically, should be excluded from the model. Therefore, model selection appears to be necessary to delete those false QTL. However, the random model methodology introduced in this chapter is special in the sense that it allows a model to include many proposed QTL without encountering much technical problem. The number of proposed QTL can even be larger than the sample size. This is clearly different from the fixed effect linear model where the number of model effects must be substantially smaller than the sample size to be able to produce any meaningful result.

Under the multiple QTL model, the parameter vector is $\theta = \{\mu, \sigma_1^2, \ldots, \sigma_p^2, \sigma^2\}$ with a dimensionality $(p + 2) \times 1$. The high dimensionality of θ is a serious problem with regard to choosing the appropriate algorithm for parameter estimation. The majority of the proposed QTL are fake, and thus, their estimated variance components should be near zero. Since the EM algorithm does not allow the use of zero as the initial value for a variance component, it is not an option. The simplex algorithm, although allows the use of zero as initial values, can only handle a few variance components, say $p = 20$ or less. The easiest and simplest algorithm is the sequential search with one component at a time (Han and Xu 2010). We search for the optimal value of one component, conditional on the values of all other components. When every variance component is sequentially optimized, another round of search begins. The iteration continues until a certain criterion of convergence is satisfied. The sequential search algorithm is summarized as follows Han and Xu (2010):

1. Set $t = 0$, and initialize $\theta = \theta^{(t)} = \{\mu^{(t)}, \sigma_1^{2(t)}, \ldots, \sigma_p^{2(t)}, \sigma^{2(t)}\}$, where

$$\mu^{(0)} = \bar{y}$$

$$\sigma^{2(0)} = \frac{1}{N} \sum_{j=1}^{n} (y_j - 1\bar{y})^T (y_j - 1\bar{y})$$

$$\sigma_k^{2(0)} = 0, \forall k = 1, \cdots, p \tag{13.69}$$

2. Define $V_j^{(t)} = \sum_{j=1}^{n} \Pi_{jk} \sigma_k^{2(t)} + I\sigma^{2(t)}$, and update μ by maximizing

$$L(\mu) = -\frac{1}{2} \sum_{j=1}^{n} \ln \left| V_j^{(t)} \right| - \frac{1}{2} \sum_{j=1}^{n} (y_j - 1\mu)^T \left(V_j^{(t)} \right)^{-1} (y_j - 1\mu) \tag{13.70}$$

3. Define $V_j(\sigma_k^2) = V_j^{(t)} - \Pi_{jk}(\sigma_k^{2(t)} - \sigma_k^2)$, and update σ_k^2 by maximizing

$$L(\sigma_k^2) = -\frac{1}{2}\sum_{j=1}^{n}\ln\left|V_j(\sigma_k^2)\right|$$

$$-\frac{1}{2}\sum_{j=1}^{n}\left(y_j - 1\mu^{(t)}\right)^T V_j^{-1}(\sigma_k^2)\left(y_j - 1\mu^{(t)}\right) \qquad (13.71)$$

for all $k = 1,\ldots,p$

4. Define $V_j(\sigma^2) = V_j^{(t)} - I(\sigma^{2(t)} - \sigma^2)$, and update σ^2 by maximizing

$$L(\sigma^2) = -\frac{1}{2}\sum_{j=1}^{n}\ln\left|V_j(\sigma^2)\right|$$

$$-\frac{1}{2}\sum_{j=1}^{n}\left(y_j - 1\mu^{(t)}\right)^T V_j^{-1}(\sigma^2)\left(y_j - 1\mu^{(t)}\right) \qquad (13.72)$$

5. Set $t = t+1$, and repeat from Steps 2 to 4 until a certain criterion of convergence is satisfied.

The solution for μ in Step 2 is explicit, as shown below:

$$\mu = \left[\sum_{j=1}^{n}1^T V_j^{(t)}1\right]^{-1}\left[\sum_{j=1}^{n}1^T V_j^{(t)}y_j\right] \qquad (13.73)$$

However, the solutions for σ_k^2 within Step 3 and σ^2 within Step 4 must be obtained via some numerical algorithm. Since the dimensionality is low (a single variable), any algorithm will work well. The simplex algorithm, although designed for multiple variable, works very well for a single variable.

Although the multiple QTL model implemented via the sequential search algorithm can handle an extremely large number of QTL, most of the estimated variance components will be close to zero. A QTL with a zero variance component is equivalent to being excluded from the model. It is still a model selection strategy but only conducted implicitly rather than explicitly. The caveat of the ad hoc sequential search is that the estimated residual variance, σ^2, approaches to zero as p grows. This, however, does not affect the relative contribution of each identified QTL because the relative contribution of the kth QTL is expressed as

$$h_k^2 = \frac{\sigma_k^2}{\sum_{k'}^{p}\sigma_{k'}^2 + \sigma^2} \qquad (13.74)$$

13.9 Complex Pedigree Analysis

The random model approach to QTL mapping can also be extended to handle large families with complicated relationships. Such families are called pedigrees. A pedigree may consist of relatives with arbitrary relationships, and the members may expand for several generations. The model and algorithm remain the same as the nuclear family analysis except that methods for calculating the IBD matrix for a putative QTL are much more involved. The multiple regression method developed by Fulker and Cardon (1994) for sib-pair analysis and later extended to pedigree analysis by Almasy and Blangero (1998) can be adopted. The regression method for calculating the IBD matrix, however, is not optimal. The Markov chain Monte Carlo method for calculating the IBD matrix implemented in the software package Lokie (Heath 1997) and the program named in SimWalk2 (Sobel et al., 2001) are recommended.

6.9 Complex Pedigree Analysis

Chapter 14
Mapping QTL for Multiple Traits

Multiple traits are measured virtually in all line crossing experiments of QTL mapping. Yet, almost all data collected for multiple traits are analyzed separately for different traits. Joint analysis for multiple traits will shed new light in QTL mapping by improving the statistical power of QTL detection and increasing the accuracy of QTL localization when different traits segregating in the mapping population are genetically related. Joint analysis for multiple traits is defined as a method that includes all traits simultaneously in a single model, rather than analyzing one trait at a time and reporting the results in a format that appears to be multiple-trait analysis. In addition to the increased power and resolution of QTL detection, joint mapping can provide insights into fundamental genetic mechanisms underlying trait relationships such as pleiotropy versus close linkage and genotype by environment ($G \times E$) interaction, which would otherwise be difficult to address if traits are analyzed separately. Substantial work has been done in joint mapping for multiple quantitative traits (Jiang and Zeng 1995; Mangin et al. 1998; Almasy and Blangero 1998; Henshall and Goddard 1999; Williams et al. 1999; Knott and Haley 2000; Hackett et al. 2001; Korol et al. 1995, 2001). In general, there are two ways to handle joint mapping. One way is the true multivariate analysis in which a multivariate normal distribution is assumed for the multiple traits, and thus, a Gaussian model is applied to construct the likelihood function. Parameter estimation is conducted via either the expectation-maximization (EM) algorithm (Dempster et al. 1977) or the multiple-trait least-squares method (Knott and Haley 2000). One problem with the multivariate analysis is that if the number of traits is large, there will be too many hypotheses to test and interpretation of the results will become cumbersome. The other way to handle multiple traits is to utilize a dimension reduction technique, e.g., the principal component analysis, to transform the data into fewer variables, i.e., "super traits," that explain majority of the total variation of the entire set of traits. Analyzing the "super traits" requires little additional work (Mangin et al. 1998; Korol et al. 1995, 2001) compared to the single-trait genetic mapping statistics. However, as pointed by Hackett et al. (2001), inferences based on the "super traits" might result in detection of spurious QTL. Furthermore, parameters of the super traits are often difficult to interpret

S. Xu, *Principles of Statistical Genomics*, DOI 10.1007/978-0-387-70807-2_14,
© Springer Science+Business Media, LLC 2013

biologically. Nevertheless, joint mapping provides a good opportunity to answer more questions about the genetic architecture of complex trait sets and deserves continued efforts from investigators in the QTL mapping community.

The previously reviewed methods are all based on the maximum likelihood method except of Knott and Haley (2000) who took a least-squares approach. Recently, Banerjee et al. (2008) and Xu et al. (2009) developed Bayesian methods implemented via the Markov chain Monte Carlo algorithm for mapping multiple quantitative traits. Xu et al. (2009) also included multiple binary trait mapping. The Bayesian methods will not be described here. In the following sections, we will only introduce the maximum likelihood method developed by Xu et al. (2005). This method was originally proposed for mapping multiple binary traits. For multiple quantitative traits, an additional step of estimating the residual variance–covariance matrix is needed.

14.1   Multivariate Model

Let $y_j = y_1, \ldots, y_m$ be an $1 \times m$ vector for the phenotypic values of m traits measured from the jth individual of a mapping population. Note that y_j is now a row vector, not a column vector as presented earlier. Only F_2 populations are considered in the text. Methods for other mapping populations are simple extension of that for the F_2 population. The linear model for the m traits is

$$y_j = X_j \beta + Z_j \gamma + \epsilon_j \tag{14.1}$$

where X_j is a $1 \times p$ design matrix for some systematic effects (not related to QTL), β is a $p \times m$ matrix for the systematic effects, $Z_j = \{Z_{j1}, Z_{j2}\}$ is a 1×2 vector determined by the genotypes of a putative QTL, γ is a $2 \times m$ matrix for the QTL effects, and ϵ_j is a $1 \times m$ vector for the residual errors. The Z variables are defined as

$$Z_{j1} = \begin{cases} +1 & \text{for } A_1 A_1 \\ 0 & \text{for } A_1 A_2 \\ -1 & \text{for } A_2 A_2 \end{cases}, Z_{j2} = \begin{cases} 0 & \text{for } A_1 A_1 \\ 1 & \text{for } A_1 A_2 \\ 0 & \text{for } A_2 A_2 \end{cases} \tag{14.2}$$

When Z_{j1} and Z_{j2} are written together as a vector, we have

$$Z_j = \begin{cases} H_1 & \text{for } A_1 A_1 \\ H_2 & \text{for } A_1 A_2 \\ H_3 & \text{for } A_2 A_2 \end{cases} \tag{14.3}$$

where H_i is the ith row of matrix

$$H = \begin{bmatrix} +1 & 0 \\ 0 & 1 \\ -1 & 0 \end{bmatrix} \tag{14.4}$$

The QTL effect matrix is

$$\gamma = \begin{bmatrix} a_1 & \cdots & a_m \\ d_1 & \cdots & d_m \end{bmatrix} \tag{14.5}$$

where a_i and d_i are the additive and dominance effects, respectively, for the ith trait. The vector of residual errors is assumed to be $\epsilon_j \sim N(0, \Sigma)$ where

$$\Sigma = \begin{bmatrix} \sigma_1^2 & \cdots & \sigma_{1m} \\ \vdots & \ddots & \vdots \\ \sigma_{1m} & \cdots & \sigma_m^2 \end{bmatrix} \tag{14.6}$$

is an $m \times m$ variance matrix. Let

$$\mu_{jk} = X_j \beta + H_k \gamma \tag{14.7}$$

The log likelihood function for the jth individual is

$$L_j(\theta) = -\frac{1}{2} \ln |\Sigma| + \ln \sum_{k=1}^{3} p_j(2-k) \exp\left[-\frac{1}{2}(y_j - \mu_{jk})\Sigma^{-1}(y_j - \mu_{jk})^T \right] \tag{14.8}$$

where $p_j(2-k) = \Pr(G_j = k|\text{marker})$ is the conditional probability that individual j takes the kth genotype for the QTL given the marker information for $k = 1, 2, 3$. The log likelihood function for the entire population is

$$L(\theta) = \sum_{j=1}^{n} L_j(\theta) \tag{14.9}$$

which is the observed log likelihood function to be used in significance test.

14.2 EM Algorithm for Parameter Estimation

The missing value is Z_j, which is redundantly expressed by $\delta(G_j, k)$. In other words, if $Z_j = H_k$, then $\delta(G_j, k) = 1$; otherwise, $\delta(G_j, k) = 0$. Let

$$p_j(2-k) = \Pr\left[\delta(G_j, k) = 1|\text{marker}\right] = \mathrm{E}\left[\delta(G_j, k)|\text{marker}\right] \tag{14.10}$$

be the prior probability before y_j is observed. The posterior probability that $\delta(G_j, k) = 1$ after y_j is observed becomes

$$
\begin{aligned}
p_j^*(2-k) &= \mathrm{E}\left[\delta(G_j, k)\right] \\
&= \frac{p_j(2-k)\exp\left[-\frac{1}{2}\left(y_j - \mu_{jk}\right)\Sigma^{-1}\left(y_j - \mu_{jk}\right)^T\right]}{\sum_{k'=1}^{3} p_j(2-k')\exp\left[-\frac{1}{2}\left(y_j - \mu_{jk'}\right)\Sigma^{-1}\left(y_j - \mu_{jk'}\right)^T\right]}
\end{aligned}
\tag{14.11}
$$

This posterior probability will be used in calculating all expectation terms that involve the missing value Z_j. We introduce the formulas in the maximization step first because the expectation step then follows naturally. The maximization step of the EM algorithm consists of the following equations:

$$
\beta = \left[\sum_{j=1}^{n} X_j^T X_j\right]^{-1}\left[\sum_{j=1}^{n} X_j^T\left(y_j - E(Z_j)\gamma\right)\right]
$$

$$
\gamma = \left[\sum_{j=1}^{n} E\left(Z_j^T Z_j\right)\right]^{-1}\left[\sum_{j=1}^{n} E(Z_j^T)\left(y_j - X_j\beta\right)\right]
$$

$$
\Sigma = \frac{1}{n}\sum_{j=1}^{n} E\left[\left(y_j - X_j\beta - Z_j\gamma\right)^T\left(y_j - X_j\beta - Z_j\gamma\right)\right]
\tag{14.12}
$$

where all the expectations in the above equations are taken with respect to the missing value Z_j using the posterior probability of QTL genotype. In the expectation step, we calculate the following quantities:

$$
E(Z_j) = \sum_{k=1}^{3} p_j^*(2-k) H_k
$$

$$
E(Z_j^T Z_j) = \sum_{k=1}^{3} p_j^*(2-k) H_k^T H_k
\tag{14.13}
$$

and

$$
E\left[\left(y_j - X_j\beta - Z_j\gamma\right)^T\left(y_j - X_j\beta - Z_j\gamma\right)\right] = \sum_{k=1}^{3} p_j^*(2-k)\left(y_j - \mu_{jk}\right)^T\left(y_j - \mu_{jk}\right)
\tag{14.14}
$$

Derivation of the formulas in the maximization step is deferred to the last section of this chapter.

14.3 Hypothesis Tests

The null hypothesis is $H_0 : \gamma = 0$. The log likelihood values under the full model and the reduced model are

$$L_1(\hat{\beta}, \hat{\gamma}, \hat{\Sigma}) = \sum_{j=1}^{n} L_j(\hat{\beta}, \hat{\gamma}, \hat{\Sigma}) \tag{14.15}$$

and

$$L_0(\hat{\beta}, \hat{\Sigma}) = \sum_{j=1}^{n} L_j(\hat{\beta}, \hat{\Sigma}) \tag{14.16}$$

respectively. The MLE of β and Σ under the reduced model is

$$\hat{\beta} = \left[\sum_{j=1}^{n} X_j^T X_j \right]^{-1} \left[\sum_{j=1}^{n} X_j^T y_j \right]$$

$$\hat{\Sigma} = \frac{1}{n} \sum_{j=1}^{n} \left(y_j - X_j \hat{\beta} \right)^T \left(y_j - X_j \hat{\beta} \right) \tag{14.17}$$

There are many other interesting hypotheses that can be tested under the multivariate QTL mapping framework. The most important one of them is the null hypothesis that the genetic correlation of traits is caused by linkage (not by pleiotropy). Recall that

$$\gamma = \begin{bmatrix} a_1 \ a_2 \ \cdots \ a_m \\ d_1 \ d_2 \ \cdots \ d_m \end{bmatrix} \tag{14.18}$$

which can be written as

$$\gamma = \begin{bmatrix} \gamma_1 \ \gamma_2 \ \cdots \ \gamma_m \end{bmatrix} \tag{14.19}$$

where

$$\gamma_i = \begin{bmatrix} a_i \\ d_i \end{bmatrix} \tag{14.20}$$

is the vector of genetic effects for the ith trait. The null hypothesis is complicated because there are many different situations for linkage. We may only want to test the hypothesis for one pair of traits at a time. Let us take $m = 2$ as an example to show various hypotheses. The pleiotropic model says $\gamma_1 \neq 0$ and $\gamma_2 \neq 0$. The linkage model is stated as either $\gamma_1 = 0$ and $\gamma_2 \neq 0$ or $\gamma_0 \neq 0$ and $\gamma_2 = 0$. For each null hypothesis, we need to estimate one set of parameters with another set of parameters constrained. The likelihood ratio test statistic can be constructed for each test. For example, if the null hypothesis is $\gamma_1 = 0$ and $\gamma_2 \neq 0$, the new QTL effects matrix under the constraint is

$$
\gamma^* = \begin{bmatrix} 0 & a_2 \\ 0 & d_2 \end{bmatrix}
\tag{14.21}
$$

The two log likelihood functions used for construction of the likelihood ratio test are

$$
L_1(\hat{\beta}, \hat{\gamma}, \hat{\Sigma}) = \sum_{j=1}^{n} L_j(\hat{\beta}, \hat{\gamma}, \hat{\Sigma})
\tag{14.22}
$$

and

$$
L_0(\hat{\beta}, \hat{\gamma}^*, \hat{\Sigma}) = \sum_{j=1}^{n} L_j(\hat{\beta}, \hat{\gamma}^*, \hat{\Sigma})
\tag{14.23}
$$

respectively.

For more than two traits, the number of different likelihood functions to be evaluated can be very large, tremendously increasing the computational burden. Therefore, it is better to shift to the Wald tests (1943), where only the estimated parameters from the full likelihood function are required. The Wald test, however, requires calculation of the variance–covariance matrix of the estimated parameters.

14.4   Variance Matrix of Estimated Parameters

The general formula for the variance–covariance matrix is difficult to derive. Fortunately, we can use some nonparametric method to calculate this matrix. The bootstrap method (Efron 1979) is the ideal method for the nonparametric estimation of the variance matrix. This method is a sampling-based method and thus is time consuming. However, we do not need to calculate this matrix for every point of the genome. Only points that show some evidence of QTL are needed for estimating the variance–covariance matrix. The bootstrap method requires repeated sampling of the original data. For n individuals, we sample n individuals randomly with replacement from the original data. Sampling with replacement means that some individuals may be selected several times while others may not be selected at all in a particular sample. Once a bootstrap sample is collected, the same procedure is used to estimate the parameter vector θ. Suppose that we sample R times to generate R bootstrap samples. For the rth sample, we estimate the parameter vector, denoted by $\theta^{(r)}$, for $r = 1, \ldots, R$. The sequence of θ, denoted by $\{\theta^{(1)}, \theta^{(2)}, \ldots, \theta^{(R)}\}$, forms an empirical sample for the parameters. From this parameter sample (with size R), a variance–covariance matrix is calculated. This estimated variance matrix, denoted by

$$\mathrm{var}(\hat{\theta}) = \frac{1}{R-1} \sum_{r=1}^{R} (\theta^{(r)} - \bar{\theta})(\theta^{(r)} - \bar{\theta})^{T}, \qquad (14.24)$$

is the bootstrap-estimated variance matrix for the MLE of θ, where $\bar{\theta}$ is the algebraic mean of the sampled θ. Given the variance matrix of $\hat{\theta}$, we can choose a subset of this matrix, e.g., $\mathrm{var}(\gamma_i)$, the variance matrix for QTL of the ith trait. The Wald test for $H_0 : \gamma_i = 0$ is then established using

$$W = \hat{\gamma}_i^{T} \left[\mathrm{var}(\hat{\gamma}_i) \right]^{-1} \hat{\gamma}_i \qquad (14.25)$$

The Wald tests can be converted into LOD scores for graphical presentation of the mapping result. To calculate the Wald test, we only need the maximum likelihood analysis of the full model. This can eliminate much computational load compared to the likelihood ratio test, in which various reduced models must be evaluated. On the other hand, the Wald test also requires the variance matrix of estimated parameters. If the variance matrix is calculated for every point scanned using the bootstrap method, the overall computational burden may even be higher than that of the likelihood ratio test. For the first moment parameters, e.g., the QTL effects, the estimated parameters are usually independent of the variances of the estimated parameters. Therefore, the variance matrix for the estimated QTL effects of a genome location may be roughly a constant across the genome provided that the marker density is high and there are very few missing genotypes. In this case, we may only need to calculate the variance matrices for a few points of the genome and use the average as the variance matrix of the estimated QTL effects. This average variance matrix can be used to construct the Wald-test statistic for whole genome scanning of quantitative trait loci. The high marker density assumption is to make sure that the information content of markers is approximately constant across the genome because the marker information content affects the variance matrix of the estimated QTL effects.

14.5   Derivation of the EM Algorithm

The EM algorithm does not directly maximize the observed log likelihood function; rather it maximizes the expectation of the complete-data log likelihood function. The latter is denoted by $Q = E[L(\theta, \delta)]$, as shown below:

$$Q = -\frac{n}{2} \ln |\Sigma| - \frac{1}{2} \sum_{j=1}^{n} E\left[(y_j - X_j\beta - Z_j\gamma) \Sigma^{-1} (y_j - X_j\beta - Z_j\gamma)^{T} \right]$$

$$(14.26)$$

The equations involved in the maximization step are obtained by setting the partial derivatives of the above function with respect to the parameters to zero and solving for the parameters. The partial derivative of Q with respect to β is

$$\frac{\partial Q}{\partial \beta} = \sum_{j=1}^{n} E\left[X_j^T(y_j - X_j\beta - Z_j\gamma)\Sigma^{-1}\right]$$

$$= \sum_{j=1}^{n} E\left[X_j^T(y_j - X_j\beta - Z_j\gamma)\right]\Sigma^{-1} \qquad (14.27)$$

Let

$$\frac{\partial Q}{\partial \beta} = \sum_{j=1}^{n} E\left[X_j^T(y_j - X_j\beta - Z_j\gamma)\right]\Sigma^{-1} = 0 \qquad (14.28)$$

Multiplying both sides of the above equation by Σ leads

$$\sum_{j=1}^{n} E\left[X_j^T(y_j - Z_j\gamma)\right] - \sum_{j=1}^{n} X_j^T X_j\beta = 0 \qquad (14.29)$$

Solving for β results in

$$\beta = \left[\sum_{j=1}^{n} X_j^T X_j\right]^{-1} \left[\sum_{j=1}^{n} X_j^T\left(y_j - E(Z_j)\gamma\right)\right] \qquad (14.30)$$

The partial derivative of Q with respect to γ is

$$\frac{\partial Q}{\partial \gamma} = \sum_{j=1}^{n} E\left[Z_j^T(y_j - X_j\beta - Z_j\gamma)\right]\Sigma^{-1} \qquad (14.31)$$

Let $\frac{\partial Q}{\partial \beta} = 0$; we get

$$\sum_{j=1}^{n} E(Z_j^T)(y_j - X_j\beta) - \sum_{j=1}^{n} E\left(Z_j^T Z_j\right)\gamma = 0 \qquad (14.32)$$

Therefore,

$$\gamma = \left[\sum_{j=1}^{n} E(Z_j^T Z_j)\right]^{-1} \left[\sum_{j=1}^{n} E(Z_j^T)(y_j - X_j\beta)\right] \qquad (14.33)$$

Let us rewrite the expected complete-data log likelihood function by

$$Q = -\frac{n}{2} \ln |\Sigma| - \frac{1}{2} \text{tr} \left(\Sigma^{-1} V \right) \tag{14.34}$$

where

$$V = \sum_{j=1}^{n} E \left[(y_j - X_j \beta - Z_j \gamma)(y_j - X_j \beta - Z_j \gamma)^T \right] \tag{14.35}$$

The partial derivative of Q with respect to Σ is

$$\frac{\partial Q}{\partial \Sigma} = -\frac{n}{2} \frac{\partial}{\partial \Sigma} \ln |\Sigma| - \frac{1}{2} \frac{\partial}{\partial \Sigma} \text{tr}(\Sigma^{-1} V) \tag{14.36}$$

Based on the rules of matrix calculus (Steeb and Hardy 2011), we have

$$\frac{\partial}{\partial \Sigma} \ln |\Sigma| = \Sigma^{-1} \tag{14.37}$$

and

$$\frac{\partial}{\partial \Sigma} \text{tr}(\Sigma^{-1} V) = -\Sigma^{-1} V \Sigma^{-1} \tag{14.38}$$

Therefore,

$$\frac{\partial Q}{\partial \Sigma} = -\frac{n}{2} \Sigma^{-1} + \frac{1}{2} \Sigma^{-1} V \Sigma^{-1} \tag{14.39}$$

Setting $\frac{\partial Q}{\partial \Sigma} = 0$ yields

$$-\frac{n}{2} \Sigma^{-1} + \frac{1}{2} \Sigma^{-1} V \Sigma^{-1} = 0 \tag{14.40}$$

Multiplying both sides of the equation by $\frac{2}{n} \Sigma$ and making appropriate rearrangement, we get

$$\frac{1}{n} V \Sigma^{-1} = I \tag{14.41}$$

Therefore,

$$\Sigma = \frac{1}{n} V = \frac{1}{n} \sum_{j=1}^{n} E \left[(y_j - X_j \beta - Z_j \gamma)(y_j - X_j \beta - Z_j \gamma)^T \right] \tag{14.42}$$

This concludes the derivation of the EM algorithm.

14.6   Example

The barley data analyzed in Chap. 12 are used here again for demonstration of the multivariate analysis. In addition to the yield trait, the authors (Hayes et al. 1993) recorded seven additional quantitative traits. The eight traits recorded are Yield (y_1), Lodging (y_2), Height (y_3), Heading (y_4), Protein (y_5), Extract (y_6), Amylase (y_7), and Power (y_8). Details about the traits and the experiment can be found from the original paper (Hayes et al. 1993). The genome was scanned using the maximum likelihood method with a 1-cM increment. The LOD score profile for the overall test (all eight traits simultaneously) is shown in Fig. 14.1. It appears that there are many QTL controlling the variation of the eight traits. The highest peak occurs on chromosome 2 at location 36.3 cM (overlapping with marker Tef4). Figure 14.2 gives the LOD score profiles for individual traits, also from the joint analysis. Figure 14.2 shows that this marker (Tef4) has a QTL controlling three traits (Heading, Height, and Protein).

We now focus on this particular marker (Tef4) and present the estimated QTL effects and other parameters. Since it is difficult to calculate the covariance matrix of the estimated QTL effects, we used the bootstrap method to calculate the

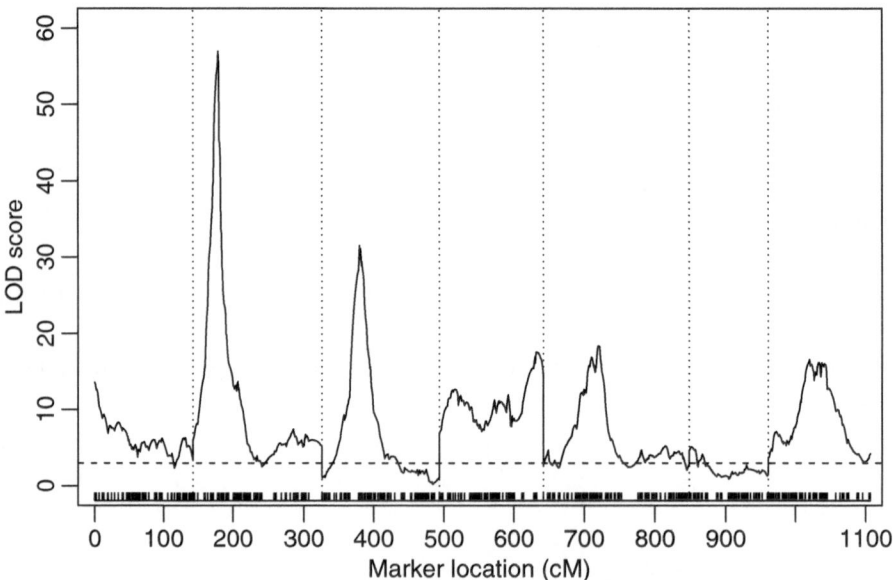

Fig. 14.1 The overall LOD score profile for the barley multivariate analysis for eight quantitative traits. The chromosomes are separated by the *vertical reference lines*

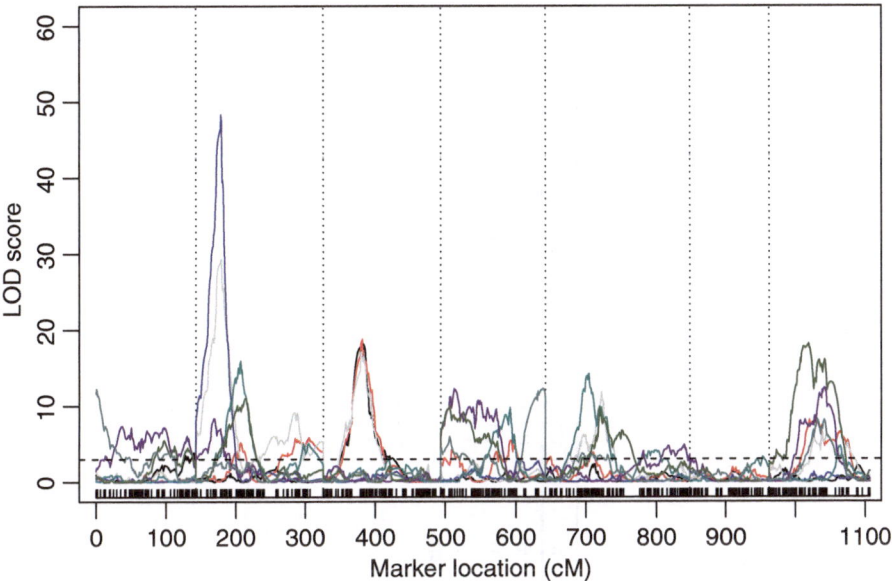

Fig. 14.2 The LOD score profiles for individual traits in the multivariate analysis of the barley data. The chromosomes are separated by the *vertical reference lines*

Table 14.1 Estimated QTL effects (locus Tef4) and population means for the eight quantitative traits of the barley experiment along with their standard errors generated from 1,000 bootstrap samples

Trait	LOD score	QTL effect	StdErr(QTL)	Mean	StdErr(Mean)
Yield	0.2887	0.0281	0.0348	5.2884	0.0353
Lodging	0.1518	0.7332	1.1890	37.3285	1.1957
Height	29.2368	−5.4817	0.4838	95.2229	0.4932
Heading	48.3004	−2.9150	0.1686	181.3121	0.1665
Protein	3.9084	−0.1663	0.0613	12.9244	0.0614
Extract	6.3618	−0.2956	0.0917	74.7246	0.0910
Amylase	4.4276	−0.7740	0.2571	29.0392	0.2611
Power	2.5375	−3.5541	1.3229	87.2563	1.3470

empirical covariance matrix. The standard error of any estimated parameter is the square root of the bootstrap-generated variance for that parameter. Table 14.1 gives the LOD scores, the estimated QTL effects, and the estimated population means along with the bootstrap-generated standard errors of the corresponding estimates.

Table 14.2 Residual covariance matrix for the eight traits of the barley experiment

Trait	Yield	Lodging	Height	Heading	Protein	Extract	Amylase	Power
Yield	0.1690	−2.4937	−0.7415	0.0738	−0.0969	0.1022	−0.2519	−0.8839
Lodging	−2.4937	212.6377	42.6552	4.2482	−3.2903	−2.0391	−6.1413	−21.9567
Height	−0.7415	42.6552	35.0471	3.2483	0.3731	−1.9680	−1.8318	7.2616
Heading	0.0738	4.2482	3.2483	3.9869	−0.5405	−0.5220	−2.1274	−3.4172
Protein	−0.0969	−3.2903	0.3731	−0.5405	0.5615	−0.2195	0.7016	3.7534
Extract	0.1022	−2.0391	−1.9680	−0.5220	−0.2195	1.2031	1.9120	7.105
Amylase	−0.2519	−6.1413	−1.8318	−2.1274	0.7016	1.9120	9.7640	28.2087
Power	−0.8839	−21.9567	7.2616	−3.4172	3.7534	7.1054	28.2087	270.7794

Table 14.3 Bootstrap-generated covariance matrix for the estimated QTL effects (locus Tef4) for the eight traits of the barley experiment

Trait	Yield	Lodging	Height	Heading	Protein	Extract	Amylase	Power
Yield	0.0012	−0.0186	−0.0052	0.0003	−0.0007	0.0009	−0.0014	−0.0034
Lodging	−0.0186	1.4138	0.2793	0.0313	−0.0199	−0.0240	−0.0573	−0.2646
Height	−0.0052	0.2793	0.2341	0.0245	0.0025	−0.0165	−0.0188	−0.0096
Heading	0.0003	0.0313	0.0245	0.0284	−0.0038	−0.0045	−0.0162	−0.0343
Protein	−0.0007	−0.0199	0.0025	−0.0038	0.0038	−0.0015	0.0046	0.0242
Extract	0.0009	−0.0240	−0.0165	−0.0045	−0.0015	0.0084	0.0134	0.0483
Amylase	−0.0014	−0.0573	−0.0188	−0.0162	0.0046	0.0134	0.0661	0.1862
Power	−0.0034	−0.2646	−0.0096	−0.0343	0.0242	0.0483	0.1862	1.7501

The estimated residual variance–covariance matrix (Σ) is given in Table 14.2. Finally, the bootstrap-generated covariance matrix for the estimated QTL effects is given in Table 14.3. Note that the standard errors of the estimated QTL effects presented in Table 14.1 are the square roots of the diagonals of Table 14.3.

Chapter 15
Bayesian Multiple QTL Mapping

So far we have learned the least-squares method, the weighted least squares method, and the maximum likelihood method for QTL mapping. These methods share a common problem in handling multiple QTL, that is, the problem of multicollinearity. Therefore, a model can include only a few QTL. Recently, Bayesian method has been developed for mapping multiple QTL (Satagopan et al. 1996; Heath 1997; Sillanpää and Arjas 1998; Sillanpää and Arjas 1999; Xu 2003; Yi 2004; Wang et al. 2005b; Yi and Shriner 2008). Under the Bayesian framework, the model can tolerate a much higher level of multicollinearity than the maximum likelihood method. As a result, the Bayesian method can handle highly saturated model. This chapter is focused on the Bayesian method via the Markov chain Monte Carlo (MCMC) algorithm. Before introducing the methods of Bayesian mapping, it is necessary to review briefly the background knowledge of Bayesian statistics.

15.1 Bayesian Regression Analysis

We will learn the basic principle and method of Bayesian analysis using a simple regression model as an example. The simple regression model has the following form:

$$y_j = X_j\beta + \epsilon_j, \forall j = 1, \ldots, n \tag{15.1}$$

where y_j is the response (dependent) variable, X_j is the regressor (independent variable), β is the regression coefficient, and ϵ_j is the residual error with an assumed $N(0, \sigma^2)$ distribution. This model is a special case of

$$y_j = \alpha + X_j\beta + \epsilon_j, \forall j = 1, \ldots, n \tag{15.2}$$

with $\alpha = 0$, i.e., regression through the origin. We use this special model to derive the Bayesian estimates of parameters. In subsequent sections, we will extend

S. Xu, *Principles of Statistical Genomics*, DOI 10.1007/978-0-387-70807-2_15,
© Springer Science+Business Media, LLC 2013

the model to the usual regression with nonzero intercept and also regression with multiple explanatory variables (multiple regression). The log likelihood function is

$$L(\theta) = -\frac{n}{2}\log(\sigma^2) - \frac{1}{2\sigma^2}\sum_{j=1}^{n}(y_j - X_j\beta)^2 \qquad (15.3)$$

where $\theta = \{\beta, \sigma^2\}$. The MLEs of θ are

$$\hat{\beta} = \left(\sum_{j=1}^{n}X_j^2\right)^{-1}\left(\sum_{j=1}^{n}X_jy_j\right) \qquad (15.4)$$

and

$$\hat{\sigma}^2 = \frac{1}{n}\sum_{j=1}^{n}(y_j - X_j\hat{\beta})^2 \qquad (15.5)$$

In the maximum likelihood analysis, parameters are estimated from the data. Sometimes investigators have prior knowledge of the parameters. This prior knowledge can be incorporated into the analysis to improve the estimation of parameters. This is the primary purpose of Bayesian analysis. The prior knowledge is formulated as a prior distribution of the parameters. Let $p(\beta, \sigma^2)$ be the joint prior density of θ. Usually, we assume that β and σ^2 are independent so that

$$p(\beta, \sigma^2) = p(\beta)p(\sigma^2) \qquad (15.6)$$

The choice of $p(\beta)$ and $p(\sigma^2)$ depends on investigator's knowledge on the problem and mathematical attractiveness. In the simple regression analysis, the following priors are both legitimate and attractive, which are

$$p(\beta) = N(\beta|\mu_\beta, \sigma_\beta^2) \qquad (15.7)$$

and

$$p(\sigma^2) = \text{Inv} - \chi^2(\sigma^2|\tau, \omega) \qquad (15.8)$$

where $N(\beta|\mu_\beta, \sigma_\beta^2)$ is the notation for the normal density of variable β with mean μ_β and variance σ_β^2, and $\text{Inv} - \chi^2(\sigma^2|\tau, \omega)$ is the probability density for the scaled inverse chi-square distribution of variable σ^2 with degree of freedom τ and scale parameter ω. The notation for a distribution and the notation for the probability density of the distribution are now consistent. For example, $x \sim N(\mu, \sigma^2)$ means that x is normally distributed with mean μ and variance σ^2, which is equivalently described as $p(x) = N(x|\mu, \sigma^2)$. The exact forms of these distributions are

$$p(\beta) = N(\beta|\mu_\beta, \sigma_\beta^2) = \frac{1}{\sqrt{2\pi\sigma_\beta^2}}\exp\left[-\frac{1}{2\sigma_\beta^2}(\beta - \mu_\beta)^2\right] \qquad (15.9)$$

and

$$p(\sigma^2) = \text{Inv} - \chi^2(\sigma^2|\tau,\omega) = \frac{(\tau\omega/2)^{\tau/2}}{\Gamma(\tau/2)}(\sigma^2)^{-(\tau/2+1)}\exp\left(-\frac{\tau\omega}{2\sigma^2}\right) \quad (15.10)$$

where $\Gamma(\tau/2)$ is the gamma function with argument $\tau/2$. Conditional on the parameter θ, the data vector y has a normal distribution with probability density

$$p(y|\theta) = \prod_{j=1}^{n} N(y_j|\mu,\sigma^2) \propto \frac{1}{(\sigma^2)^{n/2}}\exp\left[-\frac{1}{2\sigma^2}\sum_{j=1}^{n}(y_j - X_j\beta)^2\right] \quad (15.11)$$

We now have the probability density of the data and the density of the prior distribution of the parameters. We treat both the data and the parameters as random variables and formulate the joint distribution of the data and the parameters,

$$p(y,\theta) = p(y|\theta)p(\theta) \quad (15.12)$$

where $p(\theta) = p(\beta)p(\sigma^2)$. The purpose of Bayesian analysis is to infer the conditional distribution of the parameters given the data and draw conclusion about the parameters from the conditional distribution. The conditional distribution of the parameters has the form of

$$p(\theta|y) = \frac{p(y,\theta)}{p(y)} \propto p(y,\theta) \quad (15.13)$$

which is also called the posterior distribution of the parameters. The denominator, $p(y)$, is the marginal density of the data, which is irrelevant to the parameters and can be ignored because we are only interested in the estimation of parameters. Note that the above conditional density is rewritten as

$$p(\beta,\sigma^2|y) = \frac{p(y,\beta,\sigma^2)}{p(y)} \propto p(y,\beta,\sigma^2) \quad (15.14)$$

which is still a joint posterior density with regard to the two components of the parameter vector. The ultimate purpose of the Bayesian analysis is to infer the marginal posterior distribution for each component of the parameter vector. The marginal posterior density for β is obtained by integrating the joint posterior distribution over σ^2,

$$p(\beta|y) = \int_0^\infty p(\beta,\sigma^2|y)d\sigma^2 \quad (15.15)$$

The integration has an explicit form, which turns out to be the kernel of a t-distribution with $n + \tau - 1$ degrees of freedom (Sorensen and Gianola 2002). The β itself is not a t-distributed variable. It is $(\beta - \tilde{\beta})/\sigma_{\tilde{\beta}}$ that has a t-distribution, where

$$E(\beta|y) = \tilde{\beta} = \left(\frac{1}{\sigma_{\hat{\beta}}^2} + \frac{1}{\sigma_\beta^2} \right)^{-1} \left(\frac{\hat{\beta}}{\sigma_{\hat{\beta}}^2} + \frac{\mu_\beta}{\sigma_\beta^2} \right) \tag{15.16}$$

is the marginal posterior mean of β and

$$\text{var}(\beta|y) = \sigma_{\tilde{\beta}}^2 = \left(\frac{1}{\sigma_{\hat{\beta}}^2} + \frac{1}{\sigma_\beta^2} \right)^{-1} \tag{15.17}$$

is the marginal posterior variance of β. Both the mean and the variance contain $\hat{\beta}$ and $\hat{\sigma}^2$, the MLEs of β and σ^2, respectively. The role that $\hat{\sigma}^2$ plays in the above equations is through

$$\sigma_{\hat{\beta}}^2 = \left(\sum_{j=1}^n X_j^2 \right)^{-1} \hat{\sigma}^2 \tag{15.18}$$

The density of the t-distributed variable with mean $\tilde{\beta}$ and variance $\sigma_{\tilde{\beta}}^2$ is denoted by

$$p(\beta|y) = t_{n+\tau-1}(\beta|\tilde{\beta}, \sigma_{\tilde{\beta}}^2) \tag{15.19}$$

The marginal posterior density for σ^2 is obtained by integrating the joint posterior over β,

$$p(\sigma^2|y) = \int_{-\infty}^{\infty} p(\beta, \sigma^2|y)d\beta \tag{15.20}$$

which happens to be a scaled inverse chi-square distribution with

$$\tau^* = n + \tau - 1 \tag{15.21}$$

degrees of freedom and a scale parameter (Sorensen and Gianola 2002)

$$\omega^* = \frac{\tau\omega + \sum_{j=1}^n (y_j - X_j\tilde{\beta})^2}{\tau + n - 1} \tag{15.22}$$

The density of the new scaled inverse chi-square variable is denoted by

$$p(\sigma^2|y) = \text{Inv} - \chi^2(\sigma^2|\tau^*, \omega^*) \tag{15.23}$$

The mean and variance of the above distribution are

$$E(\sigma^2|y) = \tilde{\sigma}^2 = \frac{\tau\omega + \sum_{j=1}^n (y_j - X_j\tilde{\beta})^2}{\tau + n - 3} \tag{15.24}$$

and

$$\text{var}(\sigma^2|y) = \frac{2[\tau\omega + \sum_{j=1}^{n}(y_j - X_j\tilde{\beta})^2]^2}{(\tau + n - 3)^2(\tau + n - 5)} \tag{15.25}$$

respectively (Sorensen and Gianola 2002).

The marginal posterior distribution of each parameter contains all the information we have gathered for that parameter. The Bayesian estimate of that parameter can be either the posterior mean, the posterior mode, or the posterior median, depending on the preference of the investigator. The marginal posterior distribution of a parameter itself can also be treated as an estimate of the parameter. Assume that the marginal posterior mean of a parameter is considered as the Bayesian estimate of that parameter. The Bayesian estimates of β and σ^2 are $\tilde{\beta}$ and $\tilde{\sigma}^2$, respectively.

The simple regression analysis (regression through origin) discussed above is the simplest case of Bayesian analysis where the marginal posterior distribution of each parameter is known. In most situations, especially when the dimensionality of the parameter θ is high, the marginal posterior distribution of a single parameter involves high-dimensional multiple integration, and often the integration does not have an explicit expression. Therefore, the posterior distribution of a parameter often has an unknown form, which makes the Bayesian inference difficult. Thanks to the ever-growing computing power, we can perform multiple numerical integrations very efficiently. We can even utilize Monte Carlo integration by repeatedly simulating multivariate random variables. For extremely high-dimensional problems, Monte Carlo integration is perhaps the only way to implement the Bayesian method.

Let us now discuss the relationship between the joint distribution and the marginal distribution. Let $\theta = \{\theta_1, \theta_2, \ldots, \theta_m\}$ be an m dimensional multiple variables. Let $p(\theta) = p(\theta_1, \ldots, \theta_m|y)$ be the joint posterior distribution. The marginal posterior distribution for the kth component is

$$p(\theta_k|y) = \int \ldots \int p(\theta_1, \ldots, \theta_m|y)\mathrm{d}\theta_1 \ldots \mathrm{d}\theta_{k-1}\mathrm{d}\theta_{k+1} \ldots \mathrm{d}\theta_m \tag{15.26}$$

If the multiple integration has an explicit form and we can recognize the marginal distribution of θ_k, i.e., $p(\theta_k|y)$ is the density of a well-known distribution, then the expectation (or mode) of this distribution is what we want to know in the Bayesian analysis. Suppose that we know neither the joint posterior distribution nor the marginal posterior distribution, but somehow we have a joint posterior sample of multivariate θ with size N. In other words, we are only given N joint observations of θ. The sample is denoted by $\{\theta^{(1)}, \theta^{(2)}, \ldots, \theta^{(N)}\}$. We can imagine that the data in the sample are arranged in a $N \times m$ matrix. Each row represents an observation, while each column represents a variable. What is the estimated marginal expectation of θ_k drawn from this sample? Remember that this sample is supposed to be generated from the joint posterior distribution. The answer is simple; we only need to calculate the algebraic mean of variable θ_k from this sample, i.e.,

$$\bar{\theta}_k = \frac{1}{N}\sum_{j=1}^{N}\theta_k^{(j)} \tag{15.27}$$

This average value of θ_k is an empirical marginal posterior mean of θ_k, i.e., a Bayesian estimate of θ_k. We can see that as long as we have a joint sample of θ, we can infer the marginal mean of a single component of θ simply by calculating the mean of that component from the sample. While calculating the mean only requires knowledge learned from elementary school, generating the joint sample of θ becomes the main focus of the Bayesian analysis.

15.2   Markov Chain Monte Carlo

There are many different ways to generate a sample of θ from the joint distribution. The classical method is to use the following sequential approach to generate the first observation, denoted by $\theta^{(1)}$:

- Simulate $\theta_1^{(1)}$ from $p(\theta_1|y)$
- Simulate $\theta_2^{(1)}$ from $p(\theta_2|\theta_1^{(1)}, y)$
- Simulate $\theta_3^{(1)}$ from $p(\theta_3|\theta_1^{(1)}, \theta_2^{(1)}, y)$
- ....
- Simulate $\theta_m^{(1)}$ from $p(\theta_m|\theta_1^{(1)}, \ldots, \theta_{m-1}^{(1)}, y)$

The process is simply repeated N times to simulate an entire sample of θ. Observations generated this way are independent. We can see that we still need the marginal distribution for θ_1 and various levels of marginality of other components. Only θ_m is generated from a fully conditional posterior, which does not involve any integration. Therefore, this sequential approach of generating random sample is not what we want.

The MCMC approach draws all variables from their fully conditional posterior distributions. To draw a variable from a conditional distribution, we must have some values of the variables that are conditioned on. For example, to draw y from $p(y|x)$, the value of x must be known. Let $\theta^{(0)}$ be the initial value of multivariate θ. The first observation of θ is drawn using the following process:

- Simulate $\theta_1^{(1)}$ from $p(\theta_1|\theta_{-1}^{(0)}, y)$
- Simulate $\theta_2^{(1)}$ from $p(\theta_2|\theta_{-2}^{(0)}, y)$
- Simulate $\theta_3^{(1)}$ from $p(\theta_3|\theta_{-3}^{(0)}, y)$
- ....
- Simulate $\theta_m^{(1)}$ from $p(\theta_m|\theta_{-m}^{(0)}, y)$

where $\theta_{-k}^{(0)}$ is a subset of vector $\theta^{(0)}$ that excludes the kth element, i.e.,

$$\theta_{-k}^{(0)} = \{\theta_1^{(0)}, \ldots, \theta_{k-1}^{(0)}, \theta_{k+1}^{(0)}, \ldots, \theta_m^{(0)}\}$$

This special notation (negative subscript) has tremendously simplified the expressions of the MCMC sampling algorithm. The above process concludes the simulation for the first observation. The process is repeated N times to generate a sample of θ with size N. The sampled $\theta^{(t)}$ depends on $\theta^{(t-1)}$, i.e., the sampled θ in the current cycle only depends on the θ in the previous cycle. Therefore, the sequence

$$\{\theta^{(0)} \rightarrow \theta^{(1)} \rightarrow \cdots \rightarrow \theta^{(N)}\}$$

forms a Markov chain, which explains why the method is called Markov chain Monte Carlo. Because of the Markov chain property, the observations are not independent, and the first few hundred (or even thousand) observations highly depend on the initial value $\theta^{(0)}$ used to start the chain. Once the chain is stabilized, i.e., the sampled θ does not depend on the initial value, we say that the chain has reached its stationary distribution. The period from the beginning to the time when the stationary distribution is reached is called the burn-in period. Observations in the burn-in period should be deleted. After the burn-in period, the observations are presumably sampled from the joint distribution. The observations may still be correlated; such a correlation is called serial correlation or autocorrelation. We can save one observation in every sth cycle to remove the serial correlation, where $s = 20$ or $s = 50$ or any other integers, depending on the particular problem. This process is called trimming or thinning the Markov chain. After burn-in deleting and chain trimming, we collect N^* observations from the total of N observations simulated. The sample of θ with N^* observations is the posterior sample (sampled from the $p(\theta|y)$ distribution). Any Bayesian statistics can be inferred empirically from this posterior sample.

Recall that the marginal posterior for β is a t-distribution and the marginal posterior for σ^2 is a scaled inverse chi-square distribution. Both distributions have complicated forms of expression. The MCMC sampling process only requires the conditional posterior distribution, not the marginal posterior. Let us now look at the conditional posterior distribution of each parameter of the simple regression analysis.

As previously shown, the MLE of β is

$$\hat{\beta} = \left(\sum_{j=1}^{n} X_j^2\right)^{-1} \left(\sum_{j=1}^{n} X_j y_j\right) \tag{15.28}$$

and the variance of the estimate is

$$\sigma_{\hat{\beta}}^2 = \left(\sum_{j=1}^{n} X_j^2\right)^{-1} \sigma^2 \tag{15.29}$$

Note that $\sigma_{\hat{\beta}}^2$ differs from that defined in (15.18) in that σ^2 is used here in place of $\hat{\sigma}^2$. So, just from the data without any prior information, we can infer β. The estimated β itself is a variable, which follows a normal distribution denoted by

$$\beta \sim N_1(\hat{\beta}, \sigma_{\hat{\beta}}^2) \tag{15.30}$$

The subscript 1 means that this is an estimate drawn from the first source of information. Before we observed the data, the prior information about β is considered the second source of information, which is denoted by

$$\beta \sim N_2(\mu_\beta, \sigma_\beta^2) \tag{15.31}$$

The posterior distribution of β is obtained by combining the two sources of information (Box and Tiao 1973), which remains normal and is denoted by

$$\beta \sim N(\bar{\beta}, \sigma_{\bar{\beta}}^2) \tag{15.32}$$

where

$$\bar{\beta} = \left(\frac{1}{\sigma_{\hat{\beta}}^2} + \frac{1}{\sigma_\beta^2} \right)^{-1} \left(\frac{\hat{\beta}}{\sigma_{\hat{\beta}}^2} + \frac{\mu_\beta}{\sigma_\beta^2} \right) \tag{15.33}$$

and

$$\sigma_{\bar{\beta}}^2 = \left(\frac{1}{\sigma_{\hat{\beta}}^2} + \frac{1}{\sigma_\beta^2} \right)^{-1} \tag{15.34}$$

We now have the conditional posterior distribution for β denoted by

$$p(\beta|\sigma^2, y) = N(\beta|\bar{\beta}, \sigma_{\bar{\beta}}^2) \tag{15.35}$$

from which a random β is sampled.

Given β, we now evaluate the conditional posterior distribution of σ^2. The prior for σ^2 is a scaled inverse chi-square distribution with τ degrees of freedom and a scale parameter ω, denoted by

$$p(\sigma^2) = \text{Inv} - \chi^2(\sigma^2|\tau, \omega) \tag{15.36}$$

The posterior distribution remains a scaled inverse chi-square with a modified degree of freedom and a modified scale parameter, denoted by

$$p(\sigma^2|\beta, y) = \text{Inv} - \chi^2(\sigma^2|\tau^*, \omega^*) \tag{15.37}$$

where

$$\tau^* = \tau + n \tag{15.38}$$

and

$$\omega^* = \frac{\tau\omega + \sum_{j=1}^{n} (y_j - X_j\beta)^2}{\tau + n} \tag{15.39}$$

Note that ω^* defined here differs from that defined in (15.22) in that β is used here while $\bar{\beta}$ is used in (15.22). The conditional posterior of β is normal, which belongs to the same distribution family as the prior distribution. Similarly, the conditional posterior of σ^2 remains a scaled inverse chi-square, also the same type of distribution as the prior. These priors are called conjugate priors because they lead to the conditional posterior distributions of the same type.

The MCMC sampling process is summarized as:

1. Initialize $\beta = \beta^{(0)}$ and $\sigma^2 = \sigma^{2(0)}$
2. Simulate $\beta^{(1)}$ from $N(\beta | \bar{\beta}, \sigma_{\bar{\beta}}^2)$
3. Simulate $\sigma^{2(1)}$ from $\text{Inv} - \chi^2(\sigma^2 | \tau^*, \omega^*)$
4. Repeat Steps (2) and (3) until N observations of the posterior sample are collected.

It can be seen that the MCMC sampling-based regression analysis only involves two distributions, a normal distribution and a scaled inverse chi-square distribution. Most software packages have built-in functions to generate random variables from some simple distributions, e.g., $N(0, 1)$ and $\chi^2(\tau)$. Let $Z \sim N(0, 1)$ be a realized value drawn from the standardized normal distribution and $X \sim \chi^2(\tau^*)$ be a realized value drawn from a chi-square distribution with τ^* degrees of freedom. To sample β from $N(\bar{\beta}, \sigma_{\bar{\beta}}^2)$, we sample Z first and then take

$$\beta = \sigma_{\bar{\beta}} Z + \bar{\beta} \tag{15.40}$$

To sample σ^2 from $\text{Inv} - \chi^2(\tau^*, \omega^*)$, we first sample X and then take

$$\sigma^2 = \frac{\tau^* \omega^*}{X} \tag{15.41}$$

In summary, the MCMC process requires sampling a parameter only from the fully conditional posterior distribution, which usually has a simple form, e.g., normal or chi-square, and it draws one variable at a time. This type of MCMC sampling is also called Gibbs sampling (Geman and Geman 1984). With the MCMC procedure, we turn ourselves into experimentalists. Like plant breeders who plant seeds, let the seeds grow into plants, and measure the average plant yield, we plant the seeds of parameters in silico, let the parameters "grow," and measure the average of each parameter. The Bayesian posterior mean of a parameter simply takes the algebraic mean of a parameter in the posterior sample collected from the in silico experiment. Once the Bayesian method is implemented via the MCMC algorithm, it is no longer owned by a few "Bayesians"; rather, it has become a popular tool that can be used by people in all areas, including engineers, biologists, plant and animal breeders, social scientists, and so on.

Before we move on to the next section, let us demonstrate the MCMC sampling process using the simple regression as an example. The values of x and y for 20 observations are given in Table 15.1.

The model is

$$y_j = X_j \beta + \epsilon_j, \quad \forall j = 1, \ldots, 20$$

Table 15.1 Data used in the
text to demonstrate the
MCMC sampling process

x	y	x	y
1	2.95	−1	−1.23
1	0.61	1	1.06
1	4.61	1	0.41
1	3.46	−1	−3.09
1	1.12	−1	−2.08
1	4.15	−1	−1.55
−1	−2.46	1	1.07
1	4.49	−1	−5.39
1	3.34	−1	−1.26
−1	−1.44	−1	−4.46

The sample size is $n = 20$. Before introducing the prior distributions, we provide
the MLEs of the parameters, which are

$$\hat{\beta} = \left(\sum_{j=1}^{n} X_j^2\right)^{-1} \sum_{j=1}^{n} X_j y_j = 2.5115$$

$$\hat{\sigma}^2 = \frac{1}{n}\sum_{j=1}^{n}(y_j - X_j\hat{\beta})^2 = 2.3590$$

The variance of $\hat{\beta}$ is

$$\sigma_{\hat{\beta}}^2 = \left(\sum_{j=1}^{n} X_j^2\right)^{-1}\hat{\sigma}^2 = 0.1180$$

Let us choose the following prior distributions:

$$p(\beta) = N(\beta|\mu_\beta, \sigma_\beta^2) = N(\beta|0.1, 1.0)$$

and

$$p(\sigma^2) = \text{Inv} - \chi^2(\sigma^2|\tau, \omega) = \text{Inv} - \chi^2(\sigma^2|3, 3.5)$$

The marginal posterior mean and posterior variance of β are

$$E(\beta|y) = \tilde{\beta} = \left(\frac{1}{\sigma_{\hat{\beta}}^2} + \frac{1}{\sigma_\beta^2}\right)^{-1}\left(\frac{\hat{\beta}}{\sigma_{\hat{\beta}}^2} + \frac{\mu_\beta}{\sigma_\beta^2}\right) = 2.2571$$

and

$$\text{var}(\beta|y) = \sigma_{\tilde{\beta}}^2 = \left(\frac{1}{\sigma_{\hat{\beta}}^2} + \frac{1}{\sigma_\beta^2}\right)^{-1} = 0.1055$$

respectively. The marginal poster mean and posterior variance of σ^2 are

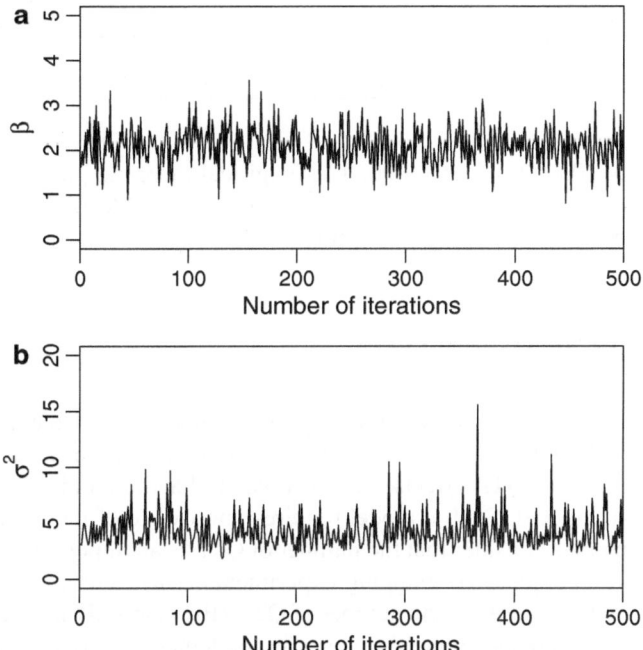

Fig. 15.1 Changes of the sampled parameters over the number of iterations since the MCMC starts. The *top panel* is the change for β, and the *bottom panel* is that for σ^2

$$E(\sigma^2|y) = \tilde{\sigma}^2 = \frac{\tau\omega + \sum_{j=1}^n (y_j - X_j\tilde{\beta})^2}{\tau + n - 3} = 2.8308$$

and

$$\text{var}(\sigma^2|y) = \frac{2[\tau\omega + \sum_{j=1}^n (y_j - X_j\tilde{\beta})^2]^2}{(\tau + n - 3)^2(\tau + n - 5)} = 0.8904$$

respectively.

We now use the MCMC sampling approach to generating the joint posterior sample for β and σ^2 and calculate the empirical marginal posterior means and posterior variances for the two parameters. For a problem as simple as this, the burn-in period can be very short or even without burn-in. Figure 15.1 shows the first 500 cycles of MCMC sampler (including the burn-in period) for the two parameters, β and σ^2. The chains converge immediately to the stationary distribution. To be absolutely sure that we actually collect samples from the stationary distribution, we set the burn-in period to 1,000 iterations (very safe), and the chain was subsequently trimmed to save one observation in every 50 iterations after the burn-in. The posterior sample size was 10,000. The total number of MCMC cycles was

Table 15.2 Empirical
marginal posterior means and
posterior variances for the
two parameters, β and σ^2

Parameter	Posterior mean	Posterior variance
β	2.2171	0.1320
σ^2	2.8489	0.9497

$1,000 + 50 \times 10,000 = 5,01,000$. The empirical marginal posterior means and marginal posterior variances for β and σ^2 are given in Table 15.2, which are very close to the theoretical values given before.

15.3   Mapping Multiple QTL

Although interval mapping (under the single QTL model) can detect multiple QTL by evaluating the number of peaks in the test statistic profile, it cannot provide accurate estimates of QTL effects. The best way to handle multiple QTL is to use a multiple QTL model. Such a model requires knowledge of the number of QTL. Most QTL mappers consider that the number of QTL is an important parameter and should be estimated in QTL mapping experiments. Therefore, model selection is often conducted to determine the number of QTL (Broman and Speed 2002). Under the Bayesian framework, model selection is implemented through the reversible jump MCMC algorithm (Sillanpää and Arjas 1998). Xu (2003) and Wang et al. (2005b) had a quite different opinion, in which the number of QTL is not considered as an important parameter. According to Wang et al. (2005b), we can propose a model that includes as many QTL as the model can handle. Such a model is called an oversaturated model. Some of the proposed QTL may be real, but most of them are spurious. As long as we can force the spurious QTL to have zero or close to zero estimated effects, the oversaturated model is considered satisfactory. The selective shrinkage Bayesian method can generate the result of QTL mapping exactly the same as we expect, that is, spurious QTL effects are shrunken to zero while true QTL have effects subject to no shrinkage.

15.3.1   Multiple QTL Model

The multiple QTL model can be described as

$$y_j = \sum_{i=1}^{q} X_{ji}\beta_i + \sum_{k=1}^{p} Z_{jk}\gamma_k + \epsilon_j \tag{15.42}$$

where y_j is the phenotypic value of a trait for individual j for $j = 1,\ldots,n$ and n is the sample size. The non-QTL effects are included in vector $\beta = \{\beta_1,\ldots,\beta_q\}$ with matrix $X_j = \{X_{j1},\ldots,X_{jq}\}$ being the design matrix to connect β and y_j. The effect of the kth QTL is denoted by γ_k for $k = 1,\ldots,p$ where p is the proposed number of QTL in the model. Vector $Z_j = \{Z_{j1}, \ldots, Z_{jp}\}$ is determined by the genotypes of the proposed QTL in the model. The residual error ϵ_j is assumed to be i.i.d. $N(0, \sigma^2)$. Let us use a BC population as an example. For the kth QTL, $Z_{jk} = 1$ for one genotype and $Z_{jk} = -1$ for the alternative genotype. Extension to F_2 population and adding the dominance effects are straightforward (only requires adding more QTL effects and increasing the model dimension). The proposed number of QTL is p, which must be larger than the true number of QTL to make sure that large QTL will not be missed. The optimal strategy is to put one QTL in every d cM of the genome, where d can be any value between 5 and 50. If $d < 5$, the model will be ill conditioned due to multicollinearity. If $d > 50$, some genome regions may not be visited by the proposed QTL even if there are true QTL located in those regions. Of course, a larger sample size is required to handle a larger model (more QTL).

15.3.2 Prior, Likelihood, and Posterior

The data involved in QTL mapping include the phenotypic values of the trait and marker genotypes for all individuals in the mapping population. Unlike Wang et al. (2005b) who expressed marker genotypes explicitly as data in the likelihood, here we suppress the marker genotypes from the data to simplify the notation. The linkage map of markers and the marker genotypes only affect the way to calculate QTL genotypes. We first use the multipoint method to calculate the genotype probabilities for all putative loci of the genome. These probabilities are then treated as the prior probabilities of QTL genotypes, from which the posterior probabilities are calculated by incorporating the phenotype and the current parameter values. Therefore, the data used to construct the likelihood are represented by $y = \{y_j, \ldots, y_n\}$. The vector of parameters is denoted by θ, which consists of the positions of the proposed QTL denoted by $\lambda = \{\lambda_1, \ldots, \lambda_p\}$, the effects of the QTL denoted by $\gamma = \{\gamma_1, \ldots, \gamma_p\}$, the non-QTL effects denoted by $\beta = \{\beta_1, \ldots, \beta_q\}$, and the residual error variance σ^2. Therefore, $\theta = \{\lambda, \beta, \gamma, \psi, \sigma^2\}$, where $\psi = \{\sigma_1^2, \ldots, \sigma_p^2\}$, will be defined later. The QTL genotypes $Z_j = \{Z_{j1}, \ldots, Z_{jp}\}$ are not parameters but missing values. The missing genotypes can be redundantly expressed as $\delta_j = \{\delta_{j1}, \ldots, \delta_{jp}\}$ where

$$\delta_{jk} = \delta(G_{jk}, \kappa)$$

is the δ function. If $G_{jk} = \kappa$, then $\delta(G_{jk}, \kappa) = 1$, else $\delta(G_{jk}, \kappa) = 0$, where G_{jk} is the genotype of the kth QTL for individual j and $\kappa = 1, 2, 3$ for an F_2 population (three genotypes per locus). The probability density of δ is

$$p(\delta_j | \lambda) = \prod_{k=1}^{p} p(\delta_{jk} | \lambda_k) \tag{15.43}$$

The independence of the QTL genotype across loci is due to the fact that they are the conditional probabilities given marker information. So, the marker information has entered here to infer the QTL genotypes. The prior for the β is

$$p(\beta) = \prod_{i=1}^{q} p(\beta_i) = \text{constant} \tag{15.44}$$

This is a uniform prior or, more appropriately, uninformative prior. The reason for choosing uninformative prior for β is that the dimensionality of β is usually very low so that β can be precisely estimated from the data alone without resorting to any prior knowledge. The prior for the QTL effects is

$$p(\gamma|\psi) = \prod_{k=1}^{p} p(\gamma_k|\sigma_k^2) = \prod_{k=1}^{p} N(\gamma_k|0, \sigma_k^2) \tag{15.45}$$

where σ_k^2 is the variance of the prior distribution for the kth QTL effect. Collectively, these variances are denoted by $\psi = \{\sigma_1^2, \ldots, \sigma_p^2\}$. This is a highly informative prior because of the zero expectation of the prior distribution. The variance of the prior distribution determines the relative weights of the prior information and the data. If σ_k^2 is very small, the prior will dominate the data, and thus, the estimated γ_k will be shrunken toward the prior expectation, that is, zero. If σ_k^2 is large, the data will dominate the prior so that the estimated γ_k will be largely unaltered (subject to no shrinkage). The key difference between this prior and the prior commonly used in Bayesian regression analysis is that different regression coefficient has a different prior variance and thus different level of shrinkage. Therefore, this method is also called the selective shrinkage method (Wang et al. 2005b). The classical Bayesian regression method, however, often uses a common prior for all regression coefficients, i.e., $\sigma_1^2 = \sigma_2^2 = \cdots = \sigma_p^2 = \sigma_\gamma^2$, which is also called ridge regression (Hoerl and Kennard 1970). The problem with this selective shrinkage method is that there are too many prior variances and it is hard to choose the appropriate values for the variances. There are two approaches to choosing the prior variances, empirical Bayesian (Xu 2007) and hierarchical modeling (Gelman 2006). The empirical Bayesian approach attempts to estimate the prior variances under the mixed model methodology by treating each regression coefficient as a random effect. The hierarchical modeling treats the prior variances as parameters and assigns a higher level prior to each variance component. By treating the variances as parameters, rather than as hyperparameters, we can estimate the variances along with the regression coefficients. Here, we take the hierarchical model approach and assign each σ_k^2 a prior distribution. The empirical Bayesian method will be discussed in the next chapter. The scaled inverse chi-square distribution is chosen for each variance component,

$$p(\sigma_k^2) = \text{Inv} - \chi^2(\sigma_k^2|\tau, \omega), \quad \forall k = 1, \ldots, p \tag{15.46}$$

The degree of freedom τ and the scale parameter ω are hyperparameters, and their influence on the estimated regression coefficients is much weaker because the influence is through the σ_k^2's. It is now easy to choose τ and ω. The degree of freedom τ is also called the prior belief. Although the proper prior should have $\tau > 0$ and $\omega > 0$, our past experience showed that an improper prior works better than the proper prior. Therefore, we choose $\tau = \omega = 0$, which leads to

$$p(\sigma_k^2) \propto \frac{1}{\sigma_k^2}, \quad \forall k = 1, \ldots, p \tag{15.47}$$

The joint prior for all the σ_k^2 is

$$p(\psi) = \prod_{k=1}^{p} p(\sigma_k^2) \tag{15.48}$$

The residual error variance is also assigned to the improper prior,

$$p(\sigma^2) \propto \frac{1}{\sigma^2} \tag{15.49}$$

The positions of the QTL depend on the number of QTL proposed, the number of chromosomes, and the size of each chromosome. Based on the average coverage per QTL (e.g., 30 cM per QTL), the number of QTL allocated to each chromosome can be calculated. Let p_c be the number of QTL proposed for the cth chromosome. These p_c QTL should be placed evenly along the chromosome. We can let the positions fixed throughout all the MCMC process so that the positions are simply constants (not parameters of interest). In this case, more QTL should be proposed to make sure that the genome is well covered by the proposed QTL. The alternative and also more efficient approach is to allow QTL position to move along the genome during the MCMC process. There is a restriction for the moving range of each QTL. The positions are disjoint along the chromosome. The first QTL must move between the first marker and the second QTL. The last QTL must move between the last marker and the second last QTL. All other QTL must move between the QTL in the left and the QTL in the right of the current QTL, i.e., the QTL that flank the current QTL. Based on this search strategy, the joint prior probability is

$$p(\lambda) = p(\lambda_1)p(\lambda_2|\lambda_1) \ldots p(\lambda_{p_c}|\lambda_{p_c-1}) \tag{15.50}$$

Given the positions of all other QTL, the conditional probability of the position of QTL k is

$$p(\lambda_k) = \frac{1}{\lambda_{k+1} - \lambda_{k-1}} \tag{15.51}$$

If QTL k is located at either end of a chromosome, the above prior needs to be modified by replacing either λ_{k-1} or λ_{k+1} by the position of the nearest end marker.

We now have a situation where the prior probability of one variable depends on values of other variables. This type of prior is called adaptive prior.

Since marker information has been used to calculate the prior probabilities of QTL genotypes, they are no longer expressed as data. The only data appearing explicitly in the model are the phenotypic values of the trait. Conditional on all parameters and the missing values, the probability density of y_j is normal. Therefore, the joint probability density of all the y_j's (called the likelihood) is

$$
p(y|\theta,\delta) = \prod_{j=1}^{n} p(y_j|\theta,\delta_j)
$$

$$
= \prod_{j=1}^{n} N\left(y_j \left| \sum_{i=1}^{q} X_{ji}\beta_i + \sum_{k=1}^{p} Z_{jk}\gamma_k, \sigma^2 \right.\right)
\tag{15.52}
$$

The fully conditional posterior of each variable is defined as

$$
p(\theta_i|\theta_{-i},\delta,y) \propto p(\theta_i,\theta_{-i},\delta,y)
\tag{15.53}
$$

where θ_i is a single element of the parameter vector θ and θ_{-i} is the collection of the remaining elements. The symbol $\propto$ means that a constant factor (not a function of parameter θ_i) has been ignored. The joint probability density $p(\theta_i,\theta_{-i},\delta,y) = p(\theta,\delta,y)$ is expressed as

$$
\begin{aligned}
p(\theta,\delta,y) &\propto p(y|\theta,\delta)p(\delta|\theta)p(\theta)\\
&= p(y|\theta,\delta)p(\beta|\psi)p(\psi)p(\delta|\lambda)p(\lambda)p(\sigma^2)
\end{aligned}
\tag{15.54}
$$

The fully conditional posterior probability density for each variable is simply derived by treating all other variables as constants and comparing the kernel of the density with a standard distribution. After some algebraic manipulation, we obtain the fully conditional distribution for most of the unknown variables (including parameters and missing values).

The fully conditional posterior for the non-QTL effect is

$$
p(\beta_i|\ldots) = N(\beta_i|\hat{\beta}_i,\sigma^2_{\hat{\beta}_i})
\tag{15.55}
$$

The special notation $p(\beta_i|\ldots)$ is used to express the fully conditional probability density. The three dots $(\ldots)$ after the vertical bar mean everything else except the variable of interest. The posterior mean and posterior variance are calculated using (15.58) and (15.59) given below:

$$
\hat{\beta}_i = \left(\sum_{j=1}^{n} X_{ji}^2\right)^{-1} \sum_{j=1}^{n} X_{ji}\left(y_j - \sum_{i'\neq i}^{q} X_{ji'}\beta_{i'} - \sum_{k=1}^{p} Z_{jk}\gamma_k\right)
\tag{15.56}
$$

and

$$\sigma_{\beta_i}^2 = \left(\sum_{j=1}^{n} X_{ji}^2 \right)^{-1} \sigma^2 \tag{15.57}$$

The fully conditional posterior for the kth QTL effect is

$$p(\gamma_k | \dots) = N(\gamma_k | \hat{\gamma}_k, \sigma_{\hat{\gamma}_k}^2) \tag{15.58}$$

where

$$\hat{\gamma}_k = \left(\sum_{j=1}^{n} Z_{jk}^2 + \frac{\sigma^2}{\sigma_k^2} \right)^{-1} \sum_{j=1}^{n} Z_{ji} \left(y_j - \sum_{i=1}^{q} X_{ji}\beta_i - \sum_{k' \neq k}^{p} Z_{jk'}\gamma_{k'} \right) \tag{15.59}$$

and

$$\sigma_{\hat{\gamma}_k}^2 = \left(\sum_{j=1}^{n} Z_{jk}^2 + \frac{\sigma^2}{\sigma_k^2} \right)^{-1} \sigma^2 \tag{15.60}$$

Comparing the conditional posterior distributions of β_i and γ_k, we notice the difference between a normal prior and a uniform prior with respect to the effects on the posterior distributions. When a normal prior is used, a shrinkage factor, $\frac{\sigma^2}{\sigma_k^2}$, is added to $\sum_{j=1}^{n} Z_{jk}^2$. If σ_k^2 is very large, the shrinkage factor disappears, meaning no shrinkage. On the other hand, if σ_k^2 is small, the shrinkage factor will dominate over $\sum_{j=1}^{n} Z_{jk}^2$, and in the end, the denominator will become infinitely large, leading to zero expectation and zero variance for the conditional posterior distribution γ_k. As such, the estimated γ_k is completely shrunken to zero. The conditional posterior distribution for each of the variance component σ_k^2 is a scaled inverse chi-square variable with probability density

$$p(\sigma_k^2 | \dots) = \text{Inv} - \chi^2 \left(\sigma_k^2 \middle| \tau + 1, \frac{\tau \omega + \gamma_k^2}{\tau + 1} \right) \tag{15.61}$$

where $\tau = \omega = 0$. The conditional posterior density for the residual error variance is

$$p(\sigma^2 | \dots) = \text{Inv} - \chi^2 \left(\sigma^2 \middle| \tau + n, \frac{\tau \omega + n S_e^2}{\tau + n} \right) \tag{15.62}$$

where

$$S_e^2 = \frac{1}{n} \sum_{j=1}^{n} \left(y_j - \sum_{i=1}^{q} X_{ji}\beta_i + \sum_{k=1}^{p} Z_{jk}\gamma_k \right)^2 \tag{15.63}$$

The next step is to sample QTL genotypes, which determine the values of the Z_j variables. Let us again use a BC population as an example and consider sampling the kth QTL genotype given that every other variable is known. There are two sources of information available to infer the probability for each of the two genotypes of the QTL. One information comes from the markers denoted by $p_j(+1)$ and $p_j(-1)$, respectively, for the two genotypes, where $p_j(+1) + p_j(-1) = 1$. These two probabilities are calculated from the multipoint method (Jiang and Zeng 1997). The other source of information comes from the phenotypic value. The connection between the phenotypic value and the QTL genotype is through the probability density of y_j given the QTL genotype. For the two alternative genotypes of the QTL , i.e., $Z_{jk} = 1$ and $Z_{jk} = -1$, the two probability densities are

$$p(y_j|Z_{jk} = +1) = N\left(y_j \left| \sum_{i=1}^{q} X_{ji}\beta_i + \sum_{k' \neq k}^{p} Z_{jk'}\gamma_{k'} + \gamma_k, \sigma^2 \right. \right)$$

$$p(y_j|Z_{jk} = -1) = N\left(y_j \left| \sum_{i=1}^{q} X_{ji}\beta_i + \sum_{k' \neq k}^{p} Z_{jk'}\gamma_{k'} - \gamma_k, \sigma^2 \right. \right) \quad (15.64)$$

Therefore, the conditional posterior probabilities for the two genotypes of the QTL are

$$p_j^*(+1) = \frac{p_j(+1)p(y_j|Z_{jk} = +1)}{p_j(+1)p(y_j|Z_{jk} = +1) + p_j(-1)p(y_j|Z_{jk} = -1)}$$

$$p_j^*(-1) = \frac{p_j(-1)p(y_j|Z_{jk} = -1)}{p_j(+1)p(y_j|Z_{jk} = +1) + p_j(-1)p(y_j|Z_{jk} = -1)} \quad (15.65)$$

where $p_j^*(+1) = p(Z_{jk} = +1|\ldots)$ and $p_j^*(-1) = p(Z_{jk} = -1|\ldots)$ are the posterior probabilities of the two genotypes. The genotype of the QTL is $Z_{jk} = 2u - 1$, where u is sampled from a Bernoulli distribution with probability $p_j^*(+1)$. So far we have completed the sampling process for all variables except the QTL positions. If we place a large number of QTL evenly distributed along the genome, say one QTL in every 10 cM, we can let the positions fixed (not moving) across the entire MCMC process. Although this fixed-position approach does not generate accurate result, it does provide some general information about the ranges where the QTL are located. Suppose that the trait of interest is controlled by only 5 QTL and we place 100 QTL evenly distributed on the genome, then majority of the assumed QTL are spurious. The Bayesian shrinkage method allows the spurious QTL to be shrunken to zero. This is why the Bayesian shrinkage method does not need variable selection. A QTL with close to zero estimated effect is equivalent to being excluded from the model. When the assumed QTL positions are fixed, investigators actually prefer to put the QTL at marker positions because marker positions contain the maximum information. This multiple-marker analysis is recommended before conducting detailed fully Bayesian analysis with QTL positions moving. Result of the detailed analysis is more or less the same as that of the multiple-marker analysis. Further detailed analysis is only conducted after the investigators get a general picture of the result.

We now discuss several different ways to allow QTL positions to move across the genome. If our purpose of QTL mapping is to find the regions of the genome that most likely carry QTL, the number of QTL is irrelevant and so are the QTL identities. If we allow QTL positions to move, the most important information we want to capture is how many times a particular segment (position) of the genome is hit or visited by nonspurious QTL. A position can be visited many times by different QTL, but all these QTL have negligible effects; such a position is not of interest. We are interested in positions that are visited repeatedly by QTL with large effects. Keeping this in mind, we propose the first strategy of QTL moving, the random walking strategy. We start with a "sufficient" number of QTL evenly placed on the genome. How sufficient is sufficient enough? This perhaps depends on the marker density and sample size of the mapping population. Putting one QTL in every 10 cM seems to work well. Each QTL is allowed to travel freely between the left and the right QTL, i.e., the QTL are distributed along the genome in a disjoint manner. The positions of the QTL are moving but the order of the QTL is preserved. This is the simplest method of QTL traveling. Let us take the kth QTL for example; the current position of the QTL is denoted by λ_k. The new position can be sampled from the following distribution:

$$\lambda_k^* = \lambda \pm \Delta\lambda \qquad (15.66)$$

where $\Delta\lambda \sim U(0, \delta)$ and δ is the maximum distance (in cM) that the QTL is allowed to move away from the current position. The following restriction $\lambda_{k-1} < \lambda_k^* < \lambda_{k+1}$ is enforced to preserve the current order of the QTL. Empirically, $\delta = 2\,\text{cM}$ seems to work well. The new position is always accepted, regardless whether it is more likely or less likely to carry a true QTL relative to the current position. The Markov chain should be sufficiently long to make sure that all putative positions are visited a number of times. Theoretically, there is no need to enforce the disjoint distribution for the QTL positions. The only reason for such a restriction is the convenience of programming if the order is preserved. With the random walk strategy of QTL moving, the frequency of hits by QTL at a position is not of interest; instead, the average effect of all the QTL hitting that position is the important information. The random walk approach does not distinguish "hot regions" (regions containing QTL) and "cold regions" (regions without QTL) of the genome. All regions are visited with equal frequency. The hot regions, however, are supposed to be visited more often than the cold regions to get a more accurate estimate of the average QTL effects for those regions. The random walk approach does not discriminate against the cold regions and thus needs a very long Markov chain to ensure that the hot regions are sufficiently visited for accurate estimation of the QTL effects.

The optimal strategy for QTL moving is to allow QTL to visit the hot regions more often than the cold regions. This sampling strategy cannot be accomplished using the Gibbs sampler because the conditional posterior of the position of a QTL does not have a well-known form of the distribution. Therefore, the Metropolis–Hastings algorithm (Metropolis et al. 1953; Hastings 1970) is adopted here to sample the QTL positions. Again, the new position is randomly generated in the neighborhood of the old position using the same approach as used in the random walk approach, but the new position λ_k^* is only accepted with a certain probability.

The acceptance probability is determined based on the Metropolis–Hastings rule, denoted by $\min\left[1, \alpha(\lambda_k^*, \lambda_k)\right]$. The new position λ_k^* has an $1 - \min\left[1, \alpha(\lambda_k^*, \lambda_k)\right]$ chance to be rejected, where

$$\alpha(\lambda_k^*, \lambda_k) = \frac{\prod_{j=1}^{n} \left[\sum_{l=-1,+1} \Pr(Z_{jk} = l|\lambda_k^*)p(y_j|Z_{jk} = l)\right] q(\lambda_k|\lambda_k^*)}{\prod_{j=1}^{n} \left[\sum_{l=-1,+1} \Pr(Z_{jk} = l|\lambda_k)p(y_j|Z_{jk} = l)\right] q(\lambda_k^*|\lambda_k)} \tag{15.67}$$

If it is rejected, the QTL remains at the current position, i.e., $\lambda_k^* = \lambda_k$. If the new position is accepted, the old position is replaced by the new position, i.e., $\lambda_k^* = \lambda \pm \Delta\lambda$. Whether the new position is accepted or not, all other variables are updated based on the information from position λ_k^*, where $\Pr(Z_{jk} = -1|\lambda_k)$ and $\Pr(Z_{jk} = +1|\lambda_k)$ are the conditional probabilities that $Z_{jk} = -1$ and $Z_{jk} = +1$, respectively, calculated from the multipoint method. These probabilities depend on position λ_k. Previously, these probabilities were denoted by $p_j(-1) = \Pr(Z_{jk} = -1|\lambda_k)$ and $p_j(+1) = \Pr(Z_{jk} = +1|\lambda_k)$, respectively. For the new position λ_k^*, these probabilities are $\Pr(Z_{jk} = -1|\lambda_k^*)$ and $\Pr(Z_{jk} = +1|\lambda_k^*)$, respectively. The proposal probabilities $q(\lambda_k^*|\lambda_k)$ and $q(\lambda_k|\lambda_k^*)$ are usually equal to $\frac{1}{2\delta}$ and thus are canceled out each other. However, once λ_k and λ_k^* are near the boundaries, these two probabilities may not be the same. Since the new position is always restricted to the interval where the old position occurs, the proposal density $q(\lambda_k^*|\lambda_k)$ and its reverse partner $q(\lambda_k|\lambda_k^*)$ may be different. Let us denote the positions of the left and right QTL by λ_{k-1} and λ_{k+1}, respectively. If λ_k is close to the left QTL so that $\lambda_k - \lambda_{k-1} < \delta$, then the new position must be sampled from $\lambda_k^* \sim U(\lambda_k - \lambda_{k-1}, \lambda_k + \delta)$ to make sure that the new position is within the required sample space. Similarly, if λ_k is close to the right QTL so that $\lambda_{k+1} - \lambda_k < \delta$, then the new position must be sampled from $\lambda_k^* \sim U(\lambda_k - \delta, \lambda_{k+1})$. In either case, the proposal density should be modified. The general formula of the proposal density after incorporating the modification is

$$q(\lambda_k|\lambda_k^*) = \begin{cases} \frac{1}{\delta+(\lambda_k-\lambda_{k-1})} & \text{if } \lambda_k - \lambda_{k-1} < \delta \\ \frac{1}{\delta+(\lambda_{k+1}-\lambda_k)} & \text{if } \lambda_{k+1} - \lambda_k < \delta \\ \frac{1}{2\delta} & \text{otherwise} \end{cases} \tag{15.68}$$

The assumption of using the above proposal density is that the distance between any two QTL must be larger than δ. The reverse partner of this proposal density is

$$q(\lambda_k^*|\lambda_k) = \begin{cases} \frac{1}{\delta+(\lambda_k^*-\lambda_{k-1})} & \text{if } \lambda_k^* - \lambda_{k-1} < \delta \\ \frac{1}{\delta+(\lambda_{k+1}-\lambda_k^*)} & \text{if } \lambda_{k+1} - \lambda_k^* < \delta \\ \frac{1}{2\delta} & \text{otherwise} \end{cases} \tag{15.69}$$

The differences between sampling λ_k and sampling other variables are the following: (1) The proposed new position may or may not be accepted, while the new values of all other variables are always accepted, and (2) when calculating the acceptance probability for a new position, the likelihood does not depend on the QTL genotype, while the conditional posterior probabilities of all other variables depend on sampled QTL genotypes.

15.3.3 Summary of the MCMC Process

The MCMC process is summarized as follows:

1. Choose the number of QTL to be placed in the model, p.
2. Initialize parameters and missing values, $\theta = \theta^{(0)}$ and $Z_j = Z_j^{(0)}$.
3. Sample β_i from $N(\beta_i | \hat{\beta}_i, \sigma_{\hat{\beta}_i}^2)$.
4. Sample γ_k from $N(\gamma_k | \hat{\gamma}_k, \sigma_{\hat{\gamma}_k}^2)$.
5. Sample σ_k^2 from $\text{Inv} - \chi^2(\sigma_k^2 | 1, \gamma_k^2)$.
6. Sample σ^2 from $\text{Inv} - \chi^2(\sigma^2 | n, S_e^2)$.
7. Sample Z_{jk} from its conditional posterior distribution.
8. Sample λ_k using the Metropolis–Hastings algorithm.
9. Repeat Step (3) to Step (8) until the chain reaches a desired length.

The length of the chain should be sufficiently long to make sure that, after burn-in deleting and chain trimming, the posterior sample size is large enough to allow accurate estimation of the posterior means (modes or medians) of all QTL parameters. Methods and computer programs are available to check whether the chain has converged to the stationary distribution (Gelfand et al. 1990; Gilks et al. 1996). Our past experience showed that the burn-in period may only contain a few thousand observations. The trimming frequency of saving one in every 20 observations is sufficient. The posterior sample size of 1,000 usually works well. However, if the model is not very large, it is always a good practice to delete more observations for the burn-in and trim more observations to make the chain thinner.

15.3.4 Post-MCMC Analysis

The MCMC process is much like doing an experiment. It only generates data for further analysis. The Bayesian estimates will only be available after summarizing the data (posterior sample). The parameter vector θ is very long, but not all parameters are of interest. Unlike other methods in which the number of QTL is an important parameter, the Bayesian shrinkage method uses a fixed number of QTL, and thus, p is not a parameter of interest. Although the variance component for the kth QTL, σ_k^2, is a parameter, it is also not a parameter of interest. It only serves as a factor to shrink the estimated QTL effect. Since the marginal posterior of σ_k^2 does not exist, the empirical posterior mean or mode of σ_k^2 does not have any biological meaning. In some observations, the sampled σ_k^2 can be very large, and in others, it may be very small. The residual error variance σ^2 is meaningful only if the number of QTL placed in the model is small to moderate. When p is very large, the residual error variance will be absorbed by the very large number of spurious QTL. The only parameters that are of interest are the QTL effects and QTL positions. However, the QTL identity, k, is also not something of interest. Since the kth QTL may jump all of places over the chromosome where it is originally placed, the average effect γ_k does not have any meaningful biological interpretation. The only things left are the positions of the genome that are hit frequently by QTL with large effects. Let us consider a fixed position of a genome. A position of a genome is only a point or a locus. Since the QTL position is a continuous variable, a particular point of the genome that is hit by a QTL has a probability of zero. Therefore, we define a genome position by a bin with a width of d cM, where d can be 1 or 2 or any other suitable value. The middle point value of the bin represents the genome location. For example, if $d = 2$ cM, the genome location 15 cM actually represents the bin covering a region of the genome from 14 cM to 16 cM, where $14 = 15 - \frac{1}{2}d$ and $16 = 15 + \frac{1}{2}d$. Once we define the bin width of a genome location, we can count the number of QTL that hit the bin. For each hit, we record the effect of that hit. The same location may be hit many times by QTL with the same or different identities. The average effect of the QTL hitting the bin is the most important parameter in the Bayesian shrinkage analysis. Each and every bin of the genome has an average QTL effect. We can then plot the effect against the genome location to form a QTL (effect) profile. This profile represents the overall result of the Bayesian mapping. In the BC example of Bayesian analysis, the kth QTL effect is denoted by γ_k. Since the QTL identity k is irrelevant, it is now replaced by the average QTL effect at position λ, which is a continuous variable. The λ without a subscript indicates a genome location. The average QTL effect at position λ can be expressed as $\gamma(\lambda)$ to indicate that the effect is a function of the genome location. The QTL effect profile is now represented by $\gamma(\lambda)$. If we use $\gamma(\lambda)$ to denote the posterior mean of QTL effect at position λ, we may use $\sigma^2(\lambda)$ to denote the posterior variance of QTL effect at position λ. If QTL moving is not random but guided by the Metropolis–Hastings rule, the posterior sample size at position λ should be a useful piece of information to indicate how often position λ is hit by a QTL. Let $n(\lambda)$ be the posterior sample size at λ; the standard error of the QTL effect at λ should be $\sigma(\lambda)/\sqrt{n(\lambda)}$. Therefore, another useful profile is the so-called t-test statistic profile expressed as

$$t(\lambda) = \sqrt{n(\lambda)} \frac{\gamma(\lambda)}{\sigma(\lambda)} \tag{15.70}$$

The corresponding F-test statistic profile is

$$F(\lambda) = n(\lambda) \frac{\gamma^2(\lambda)}{\sigma^2(\lambda)} \tag{15.71}$$

The t-test statistic profile is more informative than the F-test statistic profile because it also indicates the direction of the QTL effect (positive or negative) while the F-test statistic profile is always positive. On the other hand, the F-test statistic can be extended to multiple effects per locus, e.g., additive and dominance in an F_2 design. Both the t-test and F-test statistic profiles can be interpreted as kinds of weighted QTL effect profiles because they incorporated the posterior frequency of the genome location.

Before moving on to the next section, let us use a simulated example to demonstrate the behavior of the Bayesian shrinkage mapping and its difference from the maximum likelihood interval mapping. The mapping population was a simulated BC family with 500 individuals. A single chromosome of 2,400 cM in length was evenly covered by 121 markers (20 cM per marker interval). The positions and effects of 20 simulated QTL are demonstrated in Fig. 15.2 (top panel). In the Bayesian model, we placed one QTL in every 25 cM to start the search. The QTL positions constantly moved according to the Metropolis–Hastings rule. The burn-in period was set at 2,000, and one observation was saved in every 50 iterations after the burn-in. The posterior sample size was 1,000. We also analyzed the same data set using the maximum likelihood interval mapping procedure. The QTL effect profiles for both the Bayesian and ML methods are demonstrated in Fig. 15.2 also (see the panels in the middle and at the bottom). The Bayesian shrinkage estimates of the QTL effects are indeed smaller than the true values, but the resolution of the signal is much clearer that the maximum likelihood estimates. The Bayesian method has separated closely linked QTL in several places of the genome very well, which is clearly in contrast to the maximum likelihood method. The ML interval mapping provides exaggerated estimates of the QTL effects across the entire genome.

15.4 Alternative Methods of Bayesian Mapping

15.4.1 Reversible Jump MCMC

Reversible jump Markov chain Monte Carlo (RJMCMC) was originally developed by Green (1995) for model selection. It allows the model dimension to change during the MCMC sampling process. Most people believe that QTL mapping is a model selection problem because the number of QTL is not known a priori. Sillanpää and Arjas (1998, 1999) are the first people to apply the RJMCMC algorithm to QTL mapping. They treated the number of QTL, denoted by p, as an unknown parameter and infer the posterior distribution of p. The assumption is that p is a small number for a quantitative trait and thus can be assigned a Poisson prior distribution with mean ρ. Sillanpää and Arjas (1998) used the Metropolis–Hastings algorithm to sample all parameters, even though most QTL parameters have known forms of the fully conditional posterior distributions. The justification for use of M–H sampling strategy is that it is a general sampling approach while the Gibbs sampling is only a special case of the M–H sampling. The M–H sampler does not require derivation of the conditional posterior distribution for a parameter. However,

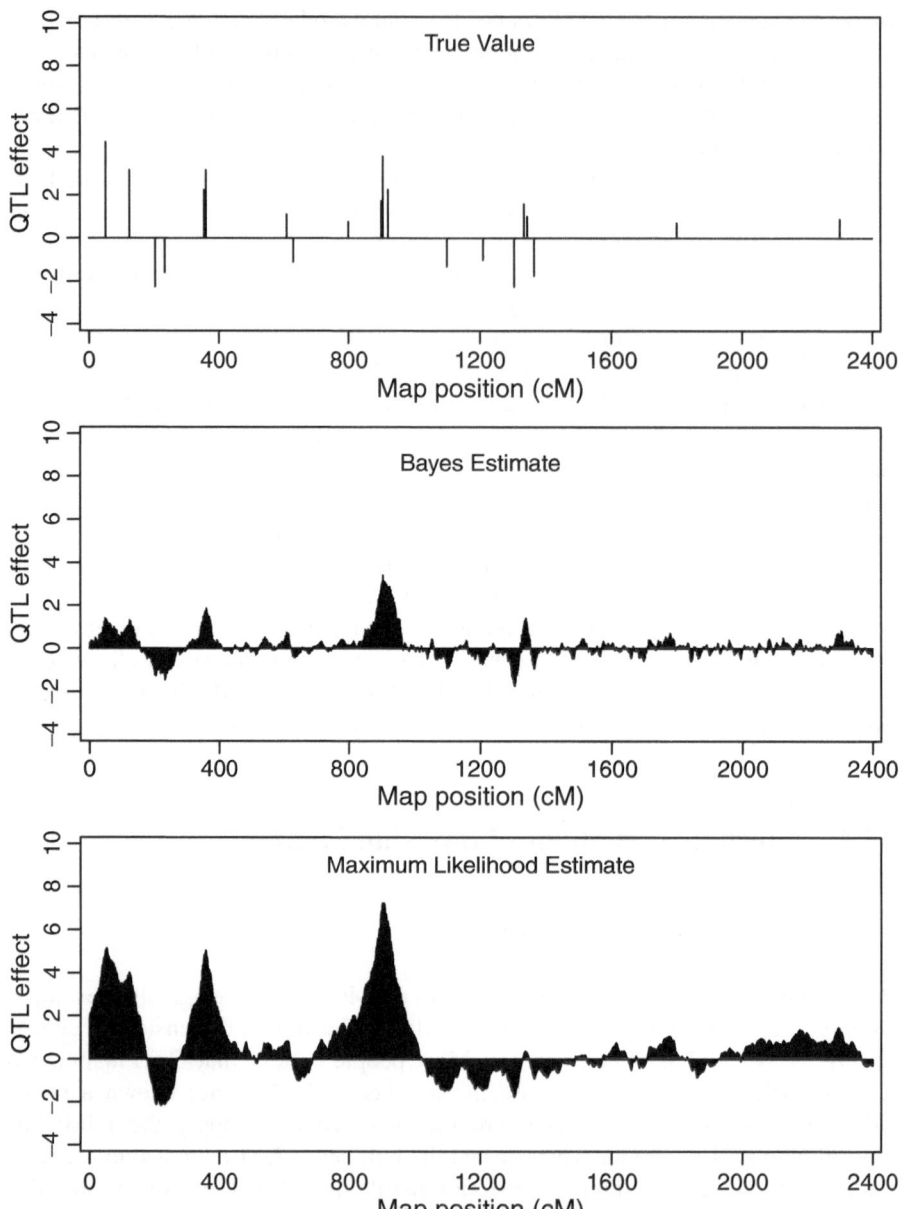

Fig. 15.2 Plots of QTL effect against genome location (QTL effect profiles) for the simulated BC population. The *top panel* shows the true locations and effects of the simulated QTL. The panel in the *middle* shows the Bayesian shrinkage estimates of the QTL effects. The panel at the *bottom* gives the maximum likelihood estimates of the QTL effects

the acceptance probability for a proposed new value of a parameter is usually less than unity because the proposal distribution from which the new value is sampled is a uniform distribution in the neighborhood of the old value and not from the conditional posterior distribution. Therefore, the M–H sampler is computationally less efficient. Yi and Xu (1999, 2000, 2001) extended RJMCMC to QTL mapping for binary traits in line crosses and random mating populations using Gibbs sampler for all parameters except the number of QTL and the location of QTL. In this section, we only introduce the RJMCMC for sampling the number of QTL. All other variables are sampled using the same method as described in the Bayesian shrinkage analysis. Another difference between the RVJMCMC and the Bayesian shrinkage method is that γ_k is assigned a uniform prior distribution for the RVJMCMC method while a $N(0, \sigma_k^2)$ prior is chosen for the shrinkage method. The conditional posterior distribution of γ_k remains normal but with mean and variance defined as

$$\hat{\gamma}_k = \left(\sum_{j=1}^{n} Z_{jk}^2 \right)^{-1} \sum_{j=1}^{n} Z_{jk} \left(y_j - \sum_{i=1}^{q} X_{ji} \beta_i - \sum_{k' \neq k}^{p} Z_{jk'} \gamma_{k'} \right) \qquad (15.72)$$

and

$$\sigma_{\hat{\gamma}_k}^2 = \left(\sum_{j=1}^{n} Z_{jk}^2 \right)^{-1} \sigma^2 \qquad (15.73)$$

respectively.

We now introduce the reversible jump MCMC. The prior distribution for p is assumed to be a truncated Poisson with mean ϕ and maximum P. The probability distribution function of p is

$$\Pr(p) = \left(\frac{\Gamma(P+1, \phi)}{P!} \right)^{-1} \left(\frac{\phi^p e^{-\phi}}{p!} \right) \propto \frac{\phi^p e^{-\phi}}{p!} \qquad (15.74)$$

where $\Gamma(P+1, \phi)$ is an incomplete Gamma function and

$$\frac{\Gamma(P+1, \phi)}{P!} = \sum_{p=0}^{P} \Pr(p) \qquad (15.75)$$

is the cumulative Poisson distribution up to P, which is irrelevant to p and thus a constant. We make a random choice among three move types of the dimensionality change: (1) Do not change the dimension, but update all other parameters except p with probability p_0; (2) add a QTL to the model with probability p_a; and (3) delete a QTL from the model with probability p_d. The three probabilities of move types sum to one, i.e., $p_0 + p_a + p_d = 1$. The following values of the probabilities may be chosen, $p_0 = p_a = p_d = \frac{1}{3}$. If no change is proposed, all other parameters are sampled from their conditional posterior distributions. If adding a QTL is proposed, we choose a chromosome to place the QTL. The probability of each chromosome being chosen is proportional to the size of the chromosome. Once a chromosome is chosen, we place the proposed new QTL randomly on the chromosome. All parameters associated with this new QTL are sampled from their prior distributions. The new QTL is then accepted with a probability determined by $\min[1, \alpha(p+1, p)]$, where

$$\alpha(p+1, p) = \frac{\prod_{j=1}^{n} p(y_j | p + 1)}{\prod_{j=1}^{n} p(y_j | p)} \times \frac{\phi}{p + 1} \times \frac{p_d}{(p + 1)p_a} \tag{15.76}$$

There are three ratios occurring in the above equation. The first ratio is the likelihood ratio, the second one is the prior ratio of the number of QTL, and the third ratio is the proposal ratio. The likelihood is defined as

$$p(y_j | p + 1) = N\left(y_j \left| \sum_{i=1}^{q} X_{ji}\beta_i + \sum_{k=1}^{p} Z_{jk}\gamma_k + Z_{j(p+1)}\gamma_{(p+1)}, \sigma^2\right.\right) \tag{15.77}$$

and

$$p(y_j | p) = N\left(y_j \left| \sum_{i=1}^{q} X_{ji}\beta_i + \sum_{k=1}^{p} Z_{jk}\gamma_k, \sigma^2\right.\right) \tag{15.78}$$

The prior probability for p is

$$\Pr(p) = \frac{\phi^p e^{-\phi}}{p!} \tag{15.79}$$

and the prior probability for $p + 1$ is

$$\Pr(p + 1) = \frac{\phi^{p+1} e^{-\phi}}{(p + 1)!} \tag{15.80}$$

Therefore, the prior ratio is

$$\frac{\Pr(p + 1)}{\Pr(p)} = \frac{\phi^{p+1} e^{-\phi}}{(p + 1)!} \frac{p!}{\phi^p e^{-\phi}} = \frac{\phi}{p + 1} \tag{15.81}$$

The proposal probability for adding a QTL is $\xi(p+1, p) = p_a$. The reverse partner is $\xi(p, p+1) = \frac{p_d}{p+1}$. It is easy to understand $\xi(p+1, p) = p_a$ because we already defined that p_a is the probability of adding a QTL. However, the reverse partner is not p_d but $p_d/(p + 1)$, which is hard to understand if we do not understand the Hastings' adjustment for the proposal probability. This probability says that if a deletion has occurred (with probability p_d) given that we have $p + 1$ QTL in the model, the probability that the newly added QTL (not any other QTL) is deleted is $1/(p+1)$ due to the fact that each QTL has an equal chance to be deleted. Therefore, the probability that the newly added QTL (not others) is deleted is $p_d/(p + 1)$. As a result, the proposal ratio is

$$\frac{\xi(p, p + 1)}{\xi(p + 1, p)} = \frac{p_d/(p + 1)}{p_a} = \frac{p_d}{(p + 1)p_a} \tag{15.82}$$

Note that the proposal ratio is the probability of deleting a QTL to the probability of adding a QTL, not the other way around. This Hastings' adjustment is important to prevent the Markov chain from being trapped at a particular QTL number. This is the very reason for the name "reversible jump." The dimension of the model can jump in either direction without being stuck at a local value of p.

If deleting a QTL is proposed, we randomly select one of the p QTL to be deleted. Suppose that the kth QTL happens to be the unlucky one. The number of QTL would change from p to $p - 1$. The reduced model with $p - 1$ QTL is accepted with probability $\min[1, \alpha(p - 1, p)]$, where

$$\alpha(p - 1, p) = \frac{\prod_{j=1}^{n} p(y_j|p)}{\prod_{j=1}^{n} p(y_j|p - 1)} \times \frac{p}{\phi} \times \frac{p_a p}{p_d} \tag{15.83}$$

where

$$p(y_j|p - 1) = N\left(y_j \left| \sum_{i=1}^{q} X_{ji}\beta_i + \sum_{k'\neq k}^{p} Z_{jk'}\gamma_{k'}, \sigma^2\right.\right) \tag{15.84}$$

The prior ratio is

$$\frac{\Pr(p - 1)}{\Pr(p)} = \frac{\phi^{p-1}e^{-\phi}}{(p - 1)!} \frac{p!}{\phi^p e^{-\phi}} = \frac{p}{\phi} \tag{15.85}$$

The proposal ratio is

$$\frac{\xi(p, p - 1)}{\xi(p - 1, p)} = \frac{p_a}{p_d/p} = \frac{p_a p}{p_d} \tag{15.86}$$

The reversible jump MCMC requires more cycles of simulations because of the frequent change of model dimension. When a QTL is deleted, all parameters associated with this QTL are gone. The chain does not memorize this QTL. In the future, if a new QTL is added to the neighborhood of this deleted QTL, the parameter associated to this added QTL must be sampled anew from the prior distribution. Even if the newly added QTL occupies exactly the same location as a previously deleted QTL, the information of the previously deleted QTL is gone permanently and cannot be reused. An improved RJMCMC may be developed to memorize the information associated with deleted QTL. If the position of a deleted QTL is sampled again later in the MCMC process (a new QTL is added to a previously deleted QTL), the parameters associated with that deleted QTL can be used again to facilitate the sampling for the newly added QTL. The improved method can substantially improve the mixing of the Markov chain and speed up the MCMC process. The tradeoff is the increased computer memory requirement for the improved method.

With the RJMCMC, the QTL number is a very important parameter. Its posterior distribution is always reported. Each QTL occurring in the model is deemed to be important and counted. In addition, the positions of QTL are usually determined by the so-called QTL intensity profile, which is simply the plot of a scaled posterior sample at a particular location $n(\lambda)$ against the genome location λ.

15.4.2 Stochastic Search Variable Selection

Stochastic search variable selection (SSVS) is a variable selection strategy for large models. The method was originally developed by George and McCulloch (1993, 1997) and applied to QTL mapping for the first time by Yi et al. (2003). The difference between this method and many other methods of model selection is that the model dimension is fixed at a predetermined value, just like the Bayesian shrinkage analysis. Model selection is actually conducted by introducing a series of binary variables, one for each model effect, i.e., the QTL effect. For p QTL effects, p indicator variables are required. Let η_k be the indicator variable for the kth QTL. If $\eta_k = 1$, the QTL is equivalent to being included in the model, and the effect will not be shrunken. If $\eta_k = 0$, the effect will be forced to take a value closed to, but not exactly equal to, zero. Essentially, the prior distribution of the kth QTL takes one of two normals. The switching button is variable η_k, as given below:

$$p(\gamma_k) = \eta_k N(\gamma_k | 0, \Delta) + (1 - \eta_k) N(\gamma_k | 0, \delta) \qquad (15.87)$$

where δ is a small positive number closed to zero, say 0.0001, and Δ is a large positive value, say 1,000. The two variances (δ and Δ) are constant hyperparameters. The indicator variable is not known, and thus, the above distribution is a mixture of two normal distributions. Let $p(\eta_k = 1) = \rho$ be the probability that γ_k comes from the first distribution; the mixture distribution is

$$p(\gamma_k) = \rho N(\gamma_k | 0, \Delta) + (1 - \rho) N(\gamma_k | 0, \delta) \qquad (15.88)$$

The mixture proportion ρ is unknown and is treated as a parameter. When the indicator variable (η_k) is known, the posterior distribution of γ_k is $p(\gamma_k | \cdots) = N(\gamma_k | \hat{\gamma}_k, \sigma^2_{\hat{\gamma}_k})$. The mean and variance of this normal are

$$\hat{\gamma}_k = \left(\sum_{j=1}^{n} Z_{jk}^2 + \frac{\sigma^2}{\upsilon_k} \right)^{-1} \sum_{j=1}^{n} Z_{jk} \left(y_j - \sum_{i=1}^{q} X_{ji} \beta_i - \sum_{k' \neq k}^{p} Z_{jk'} \gamma_{k'} \right) \qquad (15.89)$$

and

$$\sigma_{\hat{\gamma}_k}^2 = \left(\sum_{j=1}^{n} Z_{jk}^2 + \frac{\sigma^2}{\upsilon_k} \right)^{-1} \sigma^2 \tag{15.90}$$

respectively, where

$$\upsilon_k = \eta_k \Delta + (1 - \eta_k)\delta \tag{15.91}$$

is the actual variance of the posterior distribution. Let the prior distribution for η_k be

$$p(\eta_k) = \text{Bernoulli}(\eta_k | \rho) \tag{15.92}$$

The conditional posterior distribution of $\eta_k = 1$ is

$$p(\eta_k = 1 | \cdots) = \frac{\rho N(\gamma_k | 0, \Delta)}{\rho N(\gamma_k | 0, \Delta) + (1 - \rho) N(\gamma_k | 0, \delta)} \tag{15.93}$$

There is another parameter ρ involved in the conditional posterior distribution. Yi et al. (2003) treated ρ as a hyperparameter and set $\rho = \frac{1}{2}$. This prior works well for small models but fails most often for large models. The optimal strategy is to assign another prior to ρ so that ρ can be estimated from the data. Xu (2007) took a beta prior for ρ, i.e.,

$$p(\rho) = \text{Beta}(\rho | \zeta_0, \zeta_1) = \frac{\Gamma(\zeta_0 + \zeta_1)}{\Gamma(\zeta_0) \Gamma(\zeta_1)} \rho^{\zeta_1 - 1} (1 - \rho)^{\zeta_0 - 1} \tag{15.94}$$

Under this prior, the conditional posterior distribution for ρ remains beta,

$$p(\rho | \cdots) = \text{Beta} \left(\rho \left| \zeta_0 + p - \sum_{k=1}^{p} \eta_k, \zeta_1 + \sum_{k=1}^{p} \eta_k \right. \right) \tag{15.95}$$

The values of the hyperparameters were chosen by Xu (2007) as $\zeta_0 = 1$ and $\zeta_1 = 1$, leading to an uninformative prior for ρ, i.e.,

$$p(\rho) = \text{Beta}(\rho | 1, 1) = \text{constant} \tag{15.96}$$

The Gibbs sampler for σ_k^2 in the Bayesian shrinkage analysis is replaced by sampling η_k from

$$p(\eta_k | \cdots) = \text{Bernoulli} \left(\eta_k \left| \frac{\rho N(\gamma_k | 0, \Delta)}{\rho N(\gamma_k | 0, \Delta) + (1 - \rho) N(\gamma_k | 0, \delta)} \right. \right) \tag{15.97}$$

and sampling ρ from

$$p(\rho | \cdots) = \text{Beta} \left(\rho \left| 1 + p - \sum_{k=1}^{p} \eta_k, 1 + \sum_{k=1}^{p} \eta_k \right. \right) \tag{15.98}$$

in the SSVS analysis.

The additional information extracted from SSVS is the probabilistic statement about a QTL. If the marginal posterior mean of η_k is large, say $p(\eta_k|\text{data})>95\,\%$, the evidence of locus k being a QTL is strong. If the QTL position is allowed to move, η_k does not have any particular meaning. Instead, the number of hit of a particular location of the genome by QTL with $\eta(\lambda) = 1$ is more informative.

15.4.3   Lasso and Bayesian Lasso

Lasso

Lasso refers to a method called least absolute shrinkage and selection operator (Tibshirani 1996). The method can handle extremely large models by minimizing the residual sum of squares subject to a predetermined constraint, the constraint that the sum of absolute values of all regression coefficients is smaller than a predetermined shrinkage factor. Mathematically, the solution of regression coefficients is obtained by

$$\min_{\gamma} \sum_{j=1}^{n} \left(y_j - \sum_{k=1}^{p} Z_{jk}\gamma_k \right)^2 \tag{15.99}$$

subject to constraint

$$\sum_{k=1}^{p} |\gamma_k| \leq t \tag{15.100}$$

where $t > 0$. When $t = 0$, all regression coefficients must be zero. As t increases, the number of nonzero regression coefficients progressively increases. As $t \to \infty$, the Lasso estimates of the regression coefficients are equivalent to the ordinary least-squares estimates. Another expression of the problem is

$$\min_{\gamma} \left[\sum_{j=1}^{n} \left(y_j - \sum_{k=1}^{p} Z_{jk}\gamma_k \right)^2 + \lambda \sum_{k=1}^{p} |\gamma_k| \right] \tag{15.101}$$

where $\lambda \geq 0$ is a Lagrange multiplier (unknown) which relates implicitly to the bound t and controls the degree of shrinkage. The effect of λ on the level of shrinkage is just the opposite of t, with $\lambda = 0$ being no shrinkage and $\lambda \to \infty$ being the strongest shrinkage where all γ_k are shrunken down to zero. Note that the Lasso model does not involve $X_j\beta$, the non-QTL effect described earlier in the chapter. The non-QTL effect in the original Lasso refers to the population mean. For simplicity, Tibshirani (1996) centered y_j and all the independent variables. The centered y_j is simply the original y_j subtracted by $\bar{y}$, the population mean. The corresponding centered independent variables are also obtained by subtraction of $\bar{Z}_k$ from Z_{jk}. The Lasso estimates of regression coefficients can be efficiently computed via quadratic programming with linear constraints. An efficient algorithm called LARS (least angle regression) was developed by Efron et al. (2004) to implement the Lasso method. The Lagrange multiplier λ or the original t is called the Lasso parameter. The original Lasso estimates λ using the fivefold cross validation approach. One can also use any other fold cross validations, for example, the n-fold (leave-one-out) cross validation. Under each λ value, the fivefold cross validation is used to calculate the prediction error (PE),

$$
\mathrm{PE} = \frac{1}{n} \sum_{j=1}^{n} \left(y_j - \sum_{k=1}^{p} Z_{jk} \hat{\gamma}_k \right)^2 \tag{15.102}
$$

This formula appears to be the same as the estimated residual error variance. However, the prediction error differs from the residual error in that the individuals predicted do not contribute to parameter estimation. With the fivefold cross validation, we use $\frac{4}{5}$ of the sample to estimate γ_k and then use the estimated γ_k to predict the errors for the remaining $\frac{1}{5}$ sample. In other words, when we calculate $\left(y_j - \sum_{k=1}^{p} Z_{jk} \hat{\gamma}_k \right)^2$, the γ_k is estimated from $\frac{4}{5}$ of the sample that excludes y_j. Under each λ, the PE is calculated, denoted by $\mathrm{PE}(\lambda)$. We vary λ from 0 to large value. The λ value that minimizes $\mathrm{PE}(\lambda)$ is the optimal value of λ.

Bayesian Lasso

Lasso can be interpreted as Bayesian posterior mode estimation of regression coefficients when each regression coefficient is assigned an independent double-exponential prior (Tibshirani 1996; Yuan and Lin 2005; Park and Casella 2008). However, Lasso provides neither the estimate for the residual error variance nor the interval estimate for a regression coefficient. These deficiencies of Lasso can be overcome by the Bayesian Lasso (Park and Casella 2008). The double-exponential prior for γ_k is

$$
p(\gamma_k | \lambda) = \frac{\lambda}{2} \exp(-\lambda |\gamma_k|) \tag{15.103}
$$

where λ is the Lagrange multiplier in the classical Lasso method (see (15.101)). This prior can be derived from a two-level hierarchical model. The first level is

$$
p(\gamma_k | \sigma_k^2) = N(\gamma_k | 0, \sigma_k^2) \tag{15.104}
$$

and the second level is

$$
p(\sigma_k^2 | \lambda) = \frac{\lambda^2}{2} \exp\left(-\sigma_k^2 \frac{\lambda^2}{2} \right) \tag{15.105}
$$

Therefore,

$$p(\gamma_k|\lambda) = \int_0^\infty p(\gamma_k|\sigma_k^2)p(\sigma_k^2|\lambda)\mathrm{d}\sigma_k^2 = \frac{\lambda}{2}\exp(-\lambda|\gamma_k|) \qquad (15.106)$$

The Bayesian Lasso method uses the same model as the Lasso method. However, centralization of independent variables is not required, although it is still recommended. The model is described as follows:

$$y_j = \sum_{i=1}^q X_{ji}\beta_i + \sum_{k=1}^p Z_{jk}\gamma_k + \epsilon_j \qquad (15.107)$$

where β_i remains in the model and can be estimated along with the residual variance σ^2 and all QTL effects. Bayesian Lasso provides the posterior distributions for all parameters. The marginal posterior mean of each parameter is the Bayesian Lasso estimate, which is different from the posterior mode estimate obtained from the Lasso analysis. The Bayesian Lasso differs from the Bayesian shrinkage analysis only in the prior distribution for σ_k^2. Under the Bayesian Lasso, the prior for σ_k^2 is

$$p(\sigma_k^2|\lambda) = \frac{\lambda^2}{2}\exp\left(-\sigma_k^2\frac{\lambda^2}{2}\right) \qquad (15.108)$$

The Lasso parameter λ needs a prior distribution so that we can estimate λ from the data rather than choosing an arbitrary value a priori. Park and Casella (2008) choose the following gamma prior for λ^2 (not λ):

$$p(\lambda^2|a,b) = \mathrm{Gamma}(\lambda^2|a,b) = \frac{b^a}{\Gamma(a)}(\lambda^2)^{a-1}\exp\left(-b\lambda^2\right) \qquad (15.109)$$

The reason for choosing such a prior is to enjoy the conjugate property. The hyperparameters, a and b, are sufficiently remote from σ_k^2 and γ_k, and thus, their values can be chosen in an arbitrary fashion. Yi and Xu (2008) used several different sets of values for a and b and found no significant differences among those values. For convenience, we may simply set $a = b = 1$, which is sufficiently different from 0. Note that $a = b = 0$ produces an improper prior for λ^2. Once a and b values are chosen, everything else can be estimated from the data.

The fully conditional posterior distributions for most variables remain the same as the Bayesian shrinkage analysis except that the following variables must be sampled using the posterior distribution derived under the Bayesian Lasso prior distribution. For the kth QTL variance, it is better to deal with $\alpha_k = \frac{1}{\sigma_k^2}$. The conditional posterior for α_k is an inverse Gaussian distribution,

$$p(\alpha_k|\cdots) = \mathrm{Inv} - \mathrm{Gassian}\left(\alpha_k\left|\sqrt{\frac{\lambda^2\sigma^2}{\gamma_k^2}},\lambda^2\right.\right) \qquad (15.110)$$

Algorithm for sampling a random variable from an inverse Gaussian is available. Once α_k is sampled, σ_k^2 simply takes the inverse of α_k. The fully conditional posterior distribution for λ^2 remains gamma because of the conjugate property of the gamma prior,

$$p(\lambda^2|\cdots) = \text{Gamma}\left(\lambda^2 \,\middle|\, p + a, \frac{1}{2}\sum_{k=1}^{p} \sigma_k^2 + b\right) \qquad (15.111)$$

The Bayesian Lasso can potentially improve the estimation of regression coefficients for the following reasons: (1) It assigns an exponential prior, rather than a scaled inverse chi-square prior, distribution to σ_k^2, and (2) it increases the hierarchy of the prior to another level so that the hyperparameters do not have strong influence on the Bayesian estimates of the regression coefficients.

15.5   Example: Arabidopsis Data

The first example is the recombinant inbred line data of *Arabidopsis* data (Loudet et al. 2002), where the two parents initiating the line cross were Bay-0 and Shahdara with Bay-0 as the female parent. The recombinant inbred lines were actually F_7 progeny of single-seed descendants of the F_2 plants. Flowering time was recorded for each line in two environments: long day (16-h photoperiod) and short day (8-h photoperiod). We used the short-day flowering time as the quantitative trait for QTL mapping. The two parents had very little difference in short-day flowering time. The sample size (number of recombinant inbred lines) was 420. A couple of lines did not have the phenotypic records, and their phenotypic values were replaced by the population mean for convenience of data analysis. A total of 38 microsatellite markers were used for the QTL mapping. These markers are more or less evenly distributed along five chromosomes with an average 10.8 cM per marker interval. The marker names and positions are given in the original article (Loudet et al. 2002). We inserted a pseudomarker in every 5 cM of the genome. Including the inserted pseudomarkers, the total number of loci subject to analysis was 74 (38 true markers plus 36 pseudomarkers). All the 74 putative loci were evaluated simultaneously in a single model. Therefore, the model for the short-day flowering time trait is

$$y = X\beta + \sum_{k=1}^{74} Z_k \gamma_k + \epsilon$$

where X is a 420×1 vector of unity, Z_k coded as 1 for one genotype and 0 for the other genotype for locus k. If locus k is a pseudomarker, $Z_k = \text{Pr(genotype} = 1)$, which is the conditional probabilities of marker k being of genotype 1. Finally, γ_k is the QTL effect of locus k. For the original data analysis, the burn-in period was 1,000. The thinning rate was 10. The posterior sample size was 10,000, and thus, the total number of iterations was $1,000 + 10,000 \times 10 = 101,000$.

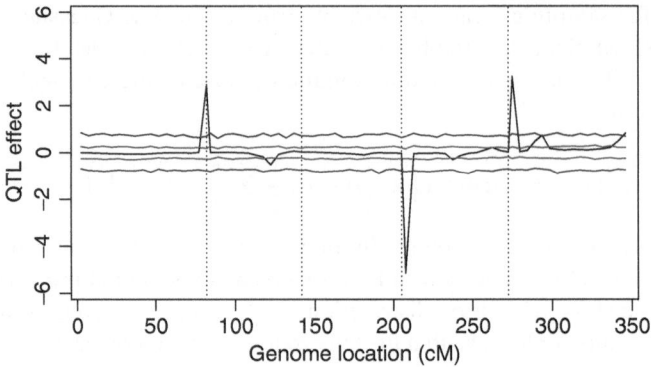

Fig. 15.3 The estimated QTL effects (*black*) and the permutation generated 1 % (*blue*) and 5 % (*red*) confidence intervals for the Arabidopsis short-time flowering time trait. The *dotted reference lines* separate the five chromosomes

We also performed a permutation analysis (Che and Xu 2010) to generate empirical quantiles of the QTL effects under the null model. The posterior sample size in permutation analysis was 80,000. The total number of iterations was $1,000 + 80,000 \times 10 = 801,000$. The estimated QTL effects and the permutation generated 0.5 % and 99.5 % (corresponding to a type I error of 0.01) and 2.5 % and 97.5 % (corresponding to a type I error of 0.05) are shown in Fig. 15.3. Based on the 0.01 criterion, a total of five QTL were detected on four chromosomes (1, 3, 4, and 5).

Chapter 16
Empirical Bayesian QTL Mapping

Empirical Bayesian is still a Bayesian method, but the hyperparameters (the parameters of the prior distribution) are not preselected by the investigators; instead, they are estimated from the same dataset as that used in the Bayesian analysis. Once the hyperparameters are estimated, they are used in the usual Bayesian analysis as if they were the true hyperparameters of the prior distributions. The data are actually used twice, once for estimating the hyperparameters and once for estimating the Bayesian posterior means of the parameters of interest. In QTL mapping, the parameters of interest are the QTL effects. A normal prior distribution is assigned to each QTL effect. The variance in the normal prior is a hyperparameter. In the Bayesian shrinkage analysis described earlier, the variance is assigned a higher level of prior distribution so that a posterior distribution of the variance parameter can be derived and the variance is then sampled via the MCMC sampling algorithm. The posterior distribution of the variance depends on the QTL effect. In the empirical Bayesian analysis, we estimate the prior variance before the Bayesian analysis. The estimated prior variance does not depend on the QTL effect. This is the key difference between the Bayesian analysis and the empirical Bayesian analysis. Elimination of the dependency of the variance on the QTL effect can increase the probability of global convergence of QTL effects during the iteration process and reduce the chance for parameters of being trapped in the locality of the initial values.

16.1 Classical Mixed Model

In the QTL mapping problem, the flat prior for β does not have any hyperparameters. If uniform prior is used for the residual variance σ^2, there is also no hyperparameter for the uniform prior. Here, we only need to concern the prior for each QTL effect. Let us assume independent normal prior,

$$p(\gamma_k|\sigma_k^2) = N(\gamma_k|0, \sigma_k^2), \forall k = 1, \ldots, p \qquad (16.1)$$

S. Xu, *Principles of Statistical Genomics*, DOI 10.1007/978-0-387-70807-2_16,
© Springer Science+Business Media, LLC 2013

Because it is a prior for γ_k, the parameter σ_k^2 is called the hyperparameter. If the number of QTL is p, we need to choose p hyperparameters. In the fully Bayesian method under the hierarchical model, σ_k^2 is estimated simultaneously along with γ_k. In the empirical Bayes method, we estimate σ_k^2 first, independent of γ_k, from the same dataset. Recall that the linear model for y_j is

$$y_j = \sum_{i=1}^{q} X_{ji}\beta_i + \sum_{k=1}^{p} Z_{jk}\gamma_k + \epsilon_j \qquad (16.2)$$

where $\epsilon_j \sim N(0, \sigma^2)$ is assumed. The compact matrix notation of this model is

$$y = X\beta + \sum_{k=1}^{p} Z_k \gamma_k + \epsilon \qquad (16.3)$$

Since γ_k is treated as a random effect (due to the normal prior assigned to it), the expectation of γ_k is zero. Therefore,

$$E(y) = X\beta \qquad (16.4)$$

The variance–covariance matrix of y is

$$\mathrm{var}(y) = V = \sum_{k=1}^{p} Z_k Z_k^T \sigma_k^2 + I\sigma^2 \qquad (16.5)$$

Let us define $G = \mathrm{diag}(\sigma_1^2, \ldots, \sigma_p^2)$ and $\gamma = [\gamma_1, \gamma_2, \ldots, \gamma_p]^T$ so that the joint normal prior for γ is written as

$$p(\gamma|G) = N(\gamma|0, G) \qquad (16.6)$$

The variance–covariance matrix is rewritten as

$$\mathrm{var}(y) = V = ZGZ^T + I\sigma^2 \qquad (16.7)$$

Define $\theta = \{\beta, \sigma^2, G\}$ as the parameter vector. The log likelihood function for the parameters is

$$L(\theta) = -\frac{1}{2}\ln|V| - \frac{1}{2}(y - X\beta)^T V^{-1}(y - X\beta) \qquad (16.8)$$

Both β and σ^2 are assumed to have a uniformly distributed prior distribution. Therefore, the maximum likelihood estimates of the parameters (θ) can be obtained by maximizing $L(\theta)$ with respect to θ. The standard mixed model approach (Searle et al. 1992) can be used to estimate the parameters. The mixed procedure in SAS

(SAS Institute 2008b) is a typical program to estimate variance components. Here, we introduce three special algorithms for the maximum likelihood estimation of the parameters.

16.1.1 Simultaneous Updating for Matrix G

Here, we partition the parameters into three sets, β, σ^2, and G. Parameters within each set are estimated simultaneously but conditional on parameter values of the other sets. To update β conditional on values of σ^2 and G at the tth iteration, we need to define

$$V^{(t)} = ZG^{(t)}Z^T + I\sigma^{2(t)} \tag{16.9}$$

which is not a function of β. The log likelihood function expressed as a function of β is

$$L(\beta) = -\frac{1}{2}\ln|V^{(t)}| - \frac{1}{2}(y - X\beta)^T (V^{(t)})^{-1}(y - X\beta) \tag{16.10}$$

The partial derivative of $L(\beta)$ with respect to β is

$$\frac{\partial L(\beta)}{\partial \beta} = X^T(V^{(t)})^{-1}(y - X\beta) \tag{16.11}$$

Setting $\frac{\partial L(\beta)}{\partial \beta} = 0$ and solving for β, we obtain the updated β,

$$\beta^{(t)} = [X^T(V^{(t)})^{-1}X]^{-1}[X^T(V^{(t)})^{-1}y] \tag{16.12}$$

There are many different ways to update σ^2, but all of which require current values of γ, the QTL effect vector. Any one of the existing methods can be adopted here. The purpose of the empirical Bayesian analysis is to estimate the variance components before γ is estimated. Therefore, all existing methods of updating σ^2 appear to be counterintuitive. Here we introduce a different method to update σ^2 without relying on γ. Let us rewrite V as

$$V = \left(\frac{1}{\sigma^2}ZGZ^T + I\right)\sigma^2 \tag{16.13}$$

When G and σ^2 in the parentheses are substituted by $G^{(t)}$ and $\sigma^{2(t)}$, we get

$$V = (ZG^{(t)}Z^T + I\sigma^{2(t)})\frac{\sigma^2}{\sigma^{2(t)}} = V^{(t)}\frac{\sigma^2}{\sigma^{2(t)}} \tag{16.14}$$

The variance matrix is now expressed a function of σ^2 and $V^{(t)}$. The inverse and determinant of V are

$$V^{-1} = (V^{(t)})^{-1} \frac{\sigma^{2(t)}}{\sigma^2} \tag{16.15}$$

and

$$|V| = |V^{(t)}| \left(\frac{\sigma^{2(t)}}{\sigma^2} \right)^n \tag{16.16}$$

respectively. Ignoring terms that are irrelevant to σ^2, we get the following log likelihood function:

$$L(\sigma^2) = -\frac{n}{2} \log(\sigma^2) - \frac{1}{2\sigma^2} \sigma^{2(t)} (y - X\beta^{(t)})^T (V^{(t)})^{-1} (y - X\beta^{(t)}) \tag{16.17}$$

The partial derivative of this log likelihood function with respect to σ^2 is

$$\frac{\partial L(\sigma^2)}{\partial \sigma^2} = -\frac{n}{2\sigma^2} + \frac{1}{2\sigma^4} \sigma^{2(t)} (y - X\beta^{(t)})^T (V^{(t)})^{-1} (y - X\beta^{(t)}) \tag{16.18}$$

Setting $\frac{\partial L(\sigma^2)}{\partial \sigma^2} = 0$ and solving for σ^2 yield

$$\sigma^2 = \frac{\sigma^{2(t)}}{n} (y - X\beta^{(t)})^T (V^{(t)})^{-1} (y - X\beta^{(t)}) \tag{16.19}$$

This updating method was proposed by Xu (2007) who first introduced the empirical Bayesian mapping procedure.

Once β and σ^2 are updated, their values are denoted by $\beta^{(t)}$ and $\sigma^{2(t)}$ and treated as known quantities for updating G. Using the Sherman–Morrison–Woodbury matrix identities (Golub and Van Loan 1996), we reformulated V^{-1} and $|V|$ by

$$V^{-1} = \frac{1}{\sigma^2} I - \frac{1}{\sigma^2} Z \left(Z^T Z \frac{1}{\sigma^2} + G^{-1} \right)^{-1} Z^T \frac{1}{\sigma^2} \tag{16.20}$$

and

$$|V| = (\sigma^2)^n \left| Z^T Z \frac{1}{\sigma^2} + G^{-1} \right| |G| \tag{16.21}$$

respectively. The log likelihood function for G conditional on σ^2 and β is

$$L(G) = -\frac{1}{2} \ln \left| Z^T Z \frac{1}{\sigma^2} + G^{-1} \right| - \frac{1}{2} \ln |G|$$

$$+ \frac{1}{2} (y - X\beta)^T \frac{1}{\sigma^2} Z \left(Z^T Z \frac{1}{\sigma^2} + G^{-1} \right)^{-1} Z^T \frac{1}{\sigma^2} (y - X\beta) \tag{16.22}$$

where terms that are irrelevant to G have been ignored. Let

$$h = Z^T \frac{1}{\sigma^2}(y - X\beta) \tag{16.23}$$

and

$$s = Z^T Z \frac{1}{\sigma^2} \tag{16.24}$$

and further define

$$\psi = Z^T Z \frac{1}{\sigma^2} + G^{-1} = s + G^{-1} \tag{16.25}$$

The log likelihood function can be simplified into

$$
\begin{aligned}
L(G) &= -\frac{1}{2}\ln|\psi| - \frac{1}{2}\ln|G| + \frac{1}{2}h^T \psi^{-1} h \\
&= -\frac{1}{2}\ln|\psi| - \frac{1}{2}\ln|G| + \frac{1}{2}\text{tr}(hh^T \psi^{-1})
\end{aligned}
\tag{16.26}
$$

We now need matrix calculus to derive the solution for G. Note that

$$\frac{\partial \psi}{\partial G} = \frac{\partial G^{-1}}{\partial G} = -G^{-1}G^{-1} \tag{16.27}$$

The partial derivative of the log likelihood with respect to G is

$$
\begin{aligned}
\frac{\partial L(G)}{\partial G} &= \frac{1}{2}G^{-1}\psi^{-1}G^{-1} + \frac{1}{2}G^{-1}\psi^{-1}hh^T\psi^{-1}G^{-1} - \frac{1}{2}G^{-1} \\
&= \frac{1}{2}G^{-1}(\psi^{-1} + \psi^{-1}hh^T\psi^{-1})G^{-1} - \frac{1}{2}G^{-1}
\end{aligned}
\tag{16.28}
$$

Setting $\frac{\partial L(G)}{\partial G} = 0$ leads to

$$G^{-1}(\psi^{-1} + \psi^{-1}hh^T\psi^{-1})G^{-1} - G^{-1} = 0 \tag{16.29}$$

Rearranging the above equation yields

$$G^{-1} = G^{-1}(\psi^{-1} + \psi^{-1}hh^T\psi^{-1})G^{-1} \tag{16.30}$$

Pre- and post-multiplying the above equation by G give

$$GG^{-1}G = GG^{-1}(\psi^{-1} + \psi^{-1}hh^T\psi^{-1})G^{-1}G \tag{16.31}$$

and thus

$$G = \psi^{-1} + \psi^{-1}hh^T\psi^{-1} \tag{16.32}$$

This equation is not explicit in terms of G because ψ^{-1} is a function of G. It may be used as an iterative equation to achieve a solution for G, as demonstrated later when we deal with the hierarchical linear mixed model. We now try to find an explicit solution for G. Pre- and post-multiplying both sides of the above equation by ψ, we have

$$\psi G \psi = \psi + hh^T \tag{16.33}$$

Substituting ψ by $\psi = s + G^{-1}$ and simplifying the final equation yield the following explicit solution:

$$G = s^{-1} hh^T s^{-1} - s^{-1} = s^{-1}(hh^T - s)s^{-1} \tag{16.34}$$

The solution is explicit because s and h are not functions of G. The fact that G is a diagonal matrix has not been taken into consideration. The solution given in (16.34) does not guarantee the diagonality of G. More rigorous derivation should have G restricted as a diagonal matrix. An ad hoc approach is simply to diagonalize G as

$$G = \text{diag}[s^{-1}(hh^T - s)s^{-1}] \tag{16.35}$$

The iteration process is summarized as follows:

Step (1)   Initialize $\theta = \theta^{(t)}$ for $t = 0$.
Step (2)   Update β using (16.12).
Step (3)   Update σ^2 using (16.19).
Step (4)   Update G using (16.35).
Step (5)   Repeat Steps (2)–(4) until a certain criterion of convergence is reached.

There are a couple of limitations of the algorithm that need attention: (1) The algorithm only works when the model dimension is small, say $p < n$, and (2) the computational cost may be high for high-dimensional models.

16.1.2   Coordinate Descent Method

Recall that G has p diagonal elements and they are updated simultaneously using the previous algorithm. With the coordinate descent algorithm, each element of G is updated conditional on all other elements of G and the other two sets of parameters β and σ^2. The updating processes for β and σ^2 remain the same as described earlier, and thus, they will not be redescribed here. We now update σ_k^2 one at a time for all $k = 1, \ldots, p$. After every element of G is updated, we have a new G denoted by $G^{(t+1)}$. Once β, σ^2, and G are updated, this only completes one cycle of the iterations. The iteration will continue until the sequence roughly converges to a constant value for each parameter. There is no inner loop for G in the iteration process because, as demonstrated later, there is an explicit solution for each element of G.

We now describe the updating procedure for σ_k^2 conditional on $\theta^{(t)}$. Since σ_k^2 contributes to the log likelihood through V, we now express V as a function of σ_k^2 and $V^{(t)}$. Define

$$V_k = \sum_{k' \neq k}^{p} Z_{k'} Z_{k'}^T \sigma_{k'}^{2(t)} + Z_k Z_k^T \sigma_k^2 + I \sigma^{2(t)} \tag{16.36}$$

as matrix V with all parameters being substituted by their values at iteration t except that σ_k^2 remains the current parameter. After some manipulation, we have an alternative expression for V_k,

$$V_k = \sum_{k'=1}^{p} Z_{k'} Z_{k'}^T \sigma_{k'}^{2(t)} + I \sigma^{2(t)} + Z_k Z_k^T (\sigma_k^2 - \sigma_k^{2(t)}) \tag{16.37}$$

Define

$$V^{(t)} = \sum_{k'=1}^{p} Z_{k'} Z_{k'}^T \sigma_{k'}^{2(t)} + I \sigma^{2(t)} \tag{16.38}$$

We now have the following expression for V_k:

$$V_k = V^{(t)} + Z_k Z_k^T (\sigma_k^2 - \sigma_k^{2(t)}) \tag{16.39}$$

Now the log likelihood function for σ_k^2 is

$$L(\sigma_k^2) = -\frac{1}{2} \ln |V_k| - \frac{1}{2} (y - X\beta^{(t)})^T V_k^{-1} (y - X\beta^{(t)}) \tag{16.40}$$

Substituting V_k^{-1} and $|V_k|$ by the Woodbury matrix identities (Golub and Van Loan 1996),

$$V_k^{-1} = (V^{(t)})^{-1} - \frac{(\sigma_k^2 - \sigma_k^{2(t)})}{Z_k^T (V^{(t)})^{-1} Z_k (\sigma_k^2 - \sigma_k^{2(t)}) + 1} (V^{(t)})^{-1} Z_k Z_k^T (V^{(t)})^{-1} \tag{16.41}$$

and

$$|V_k| = \left[Z_k^T (V^{(t)})^{-1} Z_k (\sigma_k^2 - \sigma_k^{2(t)}) + 1 \right] |V^{(t)}| \tag{16.42}$$

and ignoring all terms irrelevant to σ_k^2, we get the following log likelihood:

$$L(\sigma_k^2) = +\frac{1}{2} \frac{(\sigma_k^2 - \sigma_k^{2(t)})(y - X\beta^{(t)})^T (V^{(t)})^{-1} Z_k Z_k^T (V^{(t)})^{-1} (y - X\beta^{(t)})}{Z_k^T (V^{(t)})^{-1} Z_k (\sigma_k^2 - \sigma_k^{2(t)}) + 1}$$

$$- \frac{1}{2} \ln \left[Z_k^T (V^{(t)})^{-1} Z_k (\sigma_k^2 - \sigma_k^{2(t)}) + 1 \right] \tag{16.43}$$

Let us define

$$s_k = Z_k^T (V^{(t)})^{-1} Z_k \tag{16.44}$$

and

$$h_k = Z_k^T (V^{(t)})^{-1} (y - X\beta^{(t)}) \tag{16.45}$$

The above log likelihood function is simplified into

$$L(\sigma_k^2) = -\frac{1}{2} \ln \left[s_k (\sigma_k^2 - \sigma_k^{2(t)}) + 1 \right] + \frac{1}{2} \frac{(\sigma_k^2 - \sigma_k^{2(t)}) h_k^2}{s_k (\sigma_k^2 - \sigma_k^{2(t)}) + 1} \tag{16.46}$$

The partial derivative of the likelihood with respect to σ_k^2 is

$$\frac{\partial L(\sigma_k^2)}{\partial \sigma_k^2} = -\frac{1}{2} \frac{s_k}{s_k (\sigma_k^2 - \sigma_k^{2(t)}) + 1} + \frac{1}{2} \frac{h_k^2}{s_k (\sigma_k^2 - \sigma_k^{2(t)}) + 1}$$
$$- \frac{1}{2} \frac{s_k (\sigma_k^2 - \sigma_k^{2(t)}) h_k^2}{[s_k (\sigma_k^2 - \sigma_k^{2(t)}) + 1]^2} \tag{16.47}$$

Setting $\frac{\partial L(\sigma_k^2)}{\partial \sigma_k^2} = 0$ and solving for σ_k^2 lead to

$$\sigma_k^{2(t+1)} = \sigma_k^{2(t)} + \frac{h_k^2 - s_k}{s_k^2} \tag{16.48}$$

The solution is explicit, and no inner iterations are required for updating σ_k^2. Several characteristics of this algorithm need to be noticed: (1) The parameter σ_k^2 should be started at zero; (2) if $h_k^2 < s_k$, the iteration should stop, and the solution for σ_k^2 is $\sigma_k^{2(t)}$. In other words, the iteration process for σ_k^2 is monotonically increasing; (3) the computational cost may be high due to slow convergence for large models; and (4) it can handle a very large model, say $p > n$, because the memory storage requirement is minimum.

16.1.3 Block Coordinate Descent Method

This algorithm is a compromise between the above two methods. Here we divide matrix G into several blocks, each containing more than one variance component. We now update each block of G simultaneously. The updating process for β and σ^2 remains the same. We now introduce a special algorithm for the block updating.

Let b be the size of each block and $\frac{p}{b} = m$ be the number of blocks. If m is not an integer, it should be adjusted as $m = \text{int}(\frac{p}{b}) + 1$, and the last block should have a size less than b. Let us rewrite the variance matrix as

$$V = \sum_{k=1}^{m} Z_k G_k Z_k^T + I\sigma^2 \tag{16.49}$$

where G_k is the kth block of matrix G and Z_k is an $n \times b$ submatrix of Z corresponding to the kth block. Rewrite the variance matrix again as

$$V = \sum_{k' \neq k}^{m} Z_{k'} G_{k'} Z_{k'}^T + I\sigma^2 + Z_k G_k Z_k^T \tag{16.50}$$

Define V_{-k} as matrix V with G_k removed, i.e.,

$$V_{-k} = \sum_{k' \neq k}^{m} Z_{k'} G_{k'} Z_{k'}^T + I\sigma^2 \tag{16.51}$$

which leads to

$$V = V_{-k} + Z_k G_k Z_k^T \tag{16.52}$$

This partitioning enables us to derive the log likelihood function for G_k conditional on all other values of the parameters. Using the Woodbury matrix identities, we obtain the inverse and determinant of V as

$$V^{-1} = V_{-k}^{-1} - V_{-k}^{-1} Z_k (Z_k^T V_{-k}^{-1} Z_k + G_k^{-1})^{-1} Z_k^T V_{-k}^{-1} \tag{16.53}$$

and

$$|V| = |V_{-k} + Z_k G_k Z_k^T| = |Z_k^T V_{-k}^{-1} Z_k + G_k^{-1}||V_{-k}||G_k| \tag{16.54}$$

The log likelihood function relevant to G_k is

$$L(G_k) = +\frac{1}{2}(y - X\beta)^T V_{-k}^{-1} Z_k (Z_k^T V_{-k}^{-1} Z_k + G_k^{-1})^{-1} Z_k^T V_{-k}^{-1}(y - X\beta)$$

$$-\frac{1}{2}\ln|Z_k^T V_{-k}^{-1} Z_k + G_k^{-1}| - \frac{1}{2}\ln|G_k| \tag{16.55}$$

Define

$$h_k = Z_k^T V_{-k}^{-1}(y - X\beta) \tag{16.56}$$

and

$$s_k = Z_k^T V_{-k}^{-1} Z_k \tag{16.57}$$

Let

$$\psi_k = Z_k^T V_{-k}^{-1} Z_k + G_k^{-1} = s_k + G_k^{-1} \tag{16.58}$$

The above log likelihood function is simplified into

$$
L(G_k) = -\frac{1}{2}\ln|s_k + G_k^{-1}| - \frac{1}{2}\ln|G_k| + \frac{1}{2}\mathrm{tr}\left[h_k h_k^T (s_k + G_k^{-1})^{-1}\right]
$$

$$
= -\frac{1}{2}\ln|\psi_k| - \frac{1}{2}\ln|G_k| + \frac{1}{2}\mathrm{tr}(h_k h_k^T \psi_k^{-1}) \tag{16.59}
$$

The partial derivative of this log likelihood function with respect to G_k is

$$
\frac{\partial L(G_k)}{\partial G_k} = \frac{1}{2}G_k^{-1}\psi_k^{-1}G_k^{-1} + \frac{1}{2}G_k^{-1}\psi_k^{-1}h_k h_k^T \psi_k^{-1}G_k^{-1} - \frac{1}{2}G_k^{-1}
$$

$$
= \frac{1}{2}G_k^{-1}\left(\psi_k^{-1} + \psi_k^{-1}h_k h_k^T \psi_k^{-1}\right)G_k^{-1} - \frac{1}{2}G_k^{-1} \tag{16.60}
$$

Setting $\frac{\partial L(G_k)}{\partial G_k} = 0$, we get the following equation:

$$
G_k = \psi_k^{-1} + \psi_k^{-1}h_k h_k^T \psi_k^{-1} \tag{16.61}
$$

Again, this is an implicit equation in terms of G_k because ψ_k is a function of G_k. An explicit equation can be found by further manipulation of the above equation, as we did before for the simultaneous updating algorithm. The explicit equation is

$$
G_k = s_k^{-1}h_k h_k^T s_k^{-1} - s_k^{-1} = s_k^{-1}(h_k h_k^T - s_k)s_k^{-1} \tag{16.62}
$$

which has exactly the same form as (16.34), except that a subscript k has been added to each symbol. The matrix needs to be diagonalized because G_k is diagonal. Note that if the size of each block is one, the above equation has the following scalar form:

$$
G_k = \frac{h_k^2 - s_k}{s_k^2} \tag{16.63}
$$

which is the coordinate descent algorithm. It differs from (16.48) because s_k and h_k are defined differently. Equation (16.63) is the same as that given by Tipping (2001).

This algorithm requires repeated calculation of s_k and h_k, and it can be costly because each one needs calculation of V_{-k}^{-1}. We now introduce an efficient method to calculate s_k and h_k that does not require V_{-k}^{-1}. Let

$$
V_{-k} = V - Z_k G_k Z_k^T \tag{16.64}
$$

so that

$$
V_{-k}^{-1} = V^{-1} - V^{-1}Z_k(Z_k^T V^{-1}Z_k - G_k^{-1})^{-1}Z_k^T V^{-1} \tag{16.65}
$$

Pre- and post-multiplying the above equation by Z_k and Z_k^T lead to

$$
Z_k^T V_{-k}^{-1} Z_k = Z_k^T V^{-1}Z_k - Z_k^T V^{-1}Z_k(Z_k^T V^{-1}Z_k - G_k^{-1})^{-1}Z_k^T V^{-1}Z_k \tag{16.66}
$$

Pre- and post-multiplying (16.65) by Z_k and $\tilde{y} = y - X\beta$ yield

$$Z_k^T V_{-k}^{-1} \tilde{y} = Z_k^T V^{-1} \tilde{y} - Z_k^T V^{-1} Z_k (Z_k^T V^{-1} Z_k - G_k^{-1})^{-1} Z_k^T V^{-1} \tilde{y} \quad (16.67)$$

Let us define

$$H_k = Z_k^T V^{-1}(y - X\beta) \quad (16.68)$$

and

$$S_k = Z_k^T V^{-1} Z_k \quad (16.69)$$

which only involve inverse of V. Let us also define

$$\Psi_k = Z_k^T V^{-1} Z_k - G_k^{-1} = S_k - G_k^{-1} \quad (16.70)$$

Recall that h_k and s_k are defined in (16.56) and (16.57), respectively. After substitutions, we have the following equations:

$$s_k = S_k - S_k \Psi_k^{-1} S_k = S_k (S_k^{-1} - \Psi_k^{-1}) S_k \quad (16.71)$$

and

$$h_k = H_k - S_k \Psi_k^{-1} H_k = S_k (S_k^{-1} - \Psi_k^{-1}) H_k \quad (16.72)$$

where

$$\Psi_k^{-1} = (G_k S_k - I)^{-1} G_k = G_k (S_k G_k - I)^{-1} \quad (16.73)$$

The block coordinate descent algorithm is a general approach. When the size of each block is one, the method becomes the coordinate descent algorithm. When the block size equals p, the method becomes the simultaneous updating algorithm. In any particular situation, there is an optimal block size, which can maximize the computational speed.

16.1.4 Bayesian Estimates of QTL Effects

The three algorithms introduced above are used for estimating the prior variance components. We now treat the estimated G as the true prior matrix and provide a Bayesian estimate of γ. This Bayesian estimate is called the best linear unbiased predictor (BLUP) if G is a preselected matrix. Since G is estimated from the data, the BLUP of γ is now called the empirical Bayesian estimate. The classical Henderson's mixed model equation (Henderson 1950) can be used. An alternative form of the BLUP and the variance matrix of the BLUP are

$$\hat{\gamma} = \hat{G} Z^T \hat{V}^{-1}(y - X\hat{\beta}) \quad (16.74)$$

and

$$\text{var}(\hat{\gamma}) = \hat{G} - \hat{G} Z^T \hat{V}^{-1} Z \hat{G} \quad (16.75)$$

respectively.

For an individual QTL effect γ_k, the BLUP and the variance of the BLUP are

$$\hat{\gamma}_k = \hat{\sigma}_k^2 Z_k^T \hat{V}^{-1}(y - X\hat{\beta}) \tag{16.76}$$

and

$$\text{var}(\hat{\gamma}_k) = \hat{\sigma}_k^2 (1 - Z_k^T \hat{V}^{-1} Z_k \hat{\sigma}_k^2) \tag{16.77}$$

respectively. Let $S_{\hat{\gamma}_k} = \sqrt{\text{var}(\hat{\gamma}_k)}$ be the standard error of the BLUP. A t-test statistic is

$$t_k = \frac{\hat{\gamma}_k}{S_{\hat{\gamma}_k}} \tag{16.78}$$

which can be plotted against the genome location to produce a visual presentation of the QTL effects. One may convert the t-test statistic into the Wald-test statistic,

$$W_k = \frac{\hat{\gamma}_k^2}{S_{\hat{\gamma}_k}^2} \tag{16.79}$$

In genomic data analysis, it is common to use the LOD score test statistics,

$$\text{LOD}_k = \frac{W_k}{2\ln(10)} \tag{16.80}$$

16.2   Hierarchical Mixed Model

The classical mixed model approach may not work for extremely large models because the degree of shrinkage may not be sufficiently strong. We now incorporate a hyperprior distribution for σ_k^2. Since σ_k^2 is already a prior variance, assigning a prior to a prior involves multiple levels of prior assignments. This approach is called hierarchical prior assignment. When applied to the mixed model analysis, the model becomes a hierarchical mixed model. The scaled inverse chi-square prior (a special case of the inverse gamma prior) is most commonly used for σ_k^2. An alternative prior is the exponential prior (also called the Lasso prior). Both priors will be discussed in this section. The hierarchical prior only affects estimation of G, and therefore, updating β and σ^2 and the BLUP of γ remain the same as described in the previous section.

16.2.1   Inverse Chi-Square Prior

Assigning an independent inverse chi-square to each σ_k^2,

$$p(\sigma_k^2) = \text{Inv} - \chi^2(\sigma_k^2|\tau, \omega) \propto \left(\sigma_k^2\right)^{-(\tau+2)/2} \exp\left(-\frac{1}{2}\frac{\omega}{\sigma_k^2}\right) \tag{16.81}$$

The joint prior is

$$p(G) = \prod_{k=1}^{p} p(\sigma_k^2) \propto |G|^{-(\tau+2)/2} \exp\left[-\frac{\omega}{2} \text{tr}(G^{-1})\right] \qquad (16.82)$$

Let $\xi = \{\tau, \omega\}$ be the vector of hyperparameters of the prior distribution. The log prior is

$$L(\xi) = -\frac{1}{2}(\tau + 2) \ln |G| - \frac{\omega}{2} \text{tr}(G^{-1}) \qquad (16.83)$$

which is also called the log likelihood function for the hyperparameters. Combining the log likelihood and the log prior, we obtain the log posterior $\Phi(\theta) = L(\theta) + L(\xi)$, which expands as

$$\Phi(\theta) = -\frac{1}{2} \ln |V| - \frac{1}{2}(y - X\beta)^T V^{-1}(y - X\beta)$$

$$- \frac{1}{2}(\tau + 2) \ln |G| - \frac{\omega}{2} \text{tr}(G^{-1}) \qquad (16.84)$$

The solution of θ is obtained by maximizing the log posterior function. Therefore, the estimate of the parameter vector is called the maximum a posteriori (MAP) estimate of θ. The domain of the parameter is $\theta \in \Omega$ where Ω includes $\sigma_k^2 \geq 0$ for all $k = 1, \ldots, p$. It is important to note that $\sigma_k^2 = 0$ is allowed in order to generate a sparse model. Similar to the classical mixed model, we also adopt three different updating procedures for G.

Simultaneous Updating

The log posterior expressed as a function of G is

$$\Phi(G) = -\frac{1}{2} \ln |\psi| - \frac{1}{2} \ln |G| + \frac{1}{2} \text{tr}(qq^T \psi^{-1})$$

$$- \frac{1}{2}(\tau + 2) \ln |G| - \frac{1}{2} \omega \text{tr}(G^{-1}) \qquad (16.85)$$

The partial derivative of $\Phi(G)$ with respect to G is

$$\frac{\partial \Phi(G)}{\partial G} = \frac{1}{2} G^{-1}(\psi^{-1} + \psi^{-1} hh^T \psi^{-1} + \omega I) G^{-1} - \frac{1}{2}(\tau + 3) G^{-1} \qquad (16.86)$$

Setting $\frac{\partial \Phi(G)}{\partial G} = 0$ leads to

$$(\tau + 3) G^{-1} = G^{-1}(\psi^{-1} + \psi^{-1} hh^T \psi^{-1} + \omega I) G^{-1} \qquad (16.87)$$

Further manipulation of this equation yields

$$G = \frac{1}{\tau + 3}(\psi^{-1} + \psi^{-1}hh^T\psi^{-1} + \omega I) \tag{16.88}$$

This is an implicit equation for G because ψ is a function of G. Unfortunately, an explicit solution is hard to derive unless $(\tau, \omega) = (-2, 0)$, which is the classic mixed model. Therefore, we must use (16.88) as an iterative equation to find the solution iteratively. This iteration process is nested with an outer iteration process that involves updating β and σ^2. Therefore, the iteration using (16.88) is an inner iteration process. Let $G^{(r)}$ be the value of G at the rth inner iteration, and recall that

$$(\psi^{(r)})^{-1} = G^{(r)}(sG^{(r)} + I)^{-1} = (G^{(r)}s + I)^{-1}G^{(r)} \tag{16.89}$$

The inner iterative equation for G is

$$G^{(r+1)} = \frac{1}{\tau + 3}\text{diag}[(\psi^{(r)})^{-1} + (\psi^{(r)})^{-1}hh^T(\psi^{(r)})^{-1} + \omega I] \tag{16.90}$$

Coordinate Descent Algorithm

In a previous section, we defined

$$V^{(t)} = \sum_{k=1}^{p} Z_k Z_k^T \sigma_k^{2(t)} + I\sigma^{2(t)} \tag{16.91}$$

and

$$V = V^{(t)} + Z_k Z_k^T (\sigma_k^2 - \sigma_k^{2(t)}) \tag{16.92}$$

The log posterior expressed as a function of σ_k^2 is

$$\Phi(\sigma_k^2) = -\frac{1}{2}\ln\left[s_k(\sigma_k^2 - \sigma_k^{2(t)}) + 1\right] + \frac{1}{2}\frac{(\sigma_k^2 - \sigma_k^{2(t)})h_k^2}{s_k(\sigma_k^2 - \sigma_k^{2(t)}) + 1}$$
$$-\frac{1}{2}(\tau + 2)\ln(\sigma_k^2) - \frac{1}{2}\frac{\omega}{\sigma_k^2} \tag{16.93}$$

where $s_k = Z_k^T(V^{(t)})^{-1}Z_k$ and $h_k = Z_k^T(V^{(t)})^{-1}(y - X\beta^{(t)})$. Xu (2007) stopped at this stage and used the simplex algorithm (Nelder and Mead 1965) to update σ_k^2. Here we go beyond that by deriving an explicit updating equation. The partial derivative of the log posterior with respect to σ_k^2 is complicated (not shown here). However, when we set $\partial\Phi(\sigma_k^2)/\partial\sigma_k^2 = 0$, many complicated terms are canceled out. The equation is cubic in terms of σ_k^2, as given below:

$$c_3(\sigma_k^2)^3 + c_2(\sigma_k^2)^2 + c_1\sigma_k^2 + c_0 = 0 \tag{16.94}$$

where

$$c_3 = -(\tau + 3)s_k^2$$

$$c_2 = -(2\tau + 5)s_k(1 - s_k\sigma_k^{2(t)}) + h_k^2 + \omega s_k^2$$

$$c_1 = -(\tau + 2)(1 - s_k\sigma_k^{2(t)})^2 + 2(1 - s_k\sigma_k^{2(t)})\omega s_k$$

$$c_0 = (1 - s_k\sigma_k^{2(t)})^2\omega \tag{16.95}$$

The cubic equation has three solutions. Only one solution gives the global maximum of the log posterior. The MAP estimate of σ_k^2 takes the largest real number if that number is equal or greater than zero. Otherwise, we set $\sigma_k^2 = 0$. The PolyRoot function in SAS/IML (SAS Institute 2008a) package returns the three solutions using the four coefficients of the polynomial as the input data.

If we set $(\tau, \omega) = (\tau, 0)$, the polynomial equation becomes quadratic,

$$a(\sigma_k^2)^2 + b\sigma_k^2 + c = 0 \tag{16.96}$$

where

$$a = -(\tau + 3)s_k^2$$

$$b = -(2\tau + 5)s_k(1 - s_k\sigma_k^{2(t)}) + h_k^2$$

$$c = -(\tau + 2)(1 - s_k\sigma_k^{2(t)})^2 \tag{16.97}$$

There are two solutions for the quadratic equation, but the solution that maximizes the log posterior is

$$\sigma_k^{2(t+1)} = \frac{-b - \sqrt{b^2 - 4ac}}{2a} \tag{16.98}$$

The solution exists only if $b > 0$ and $b^2 - 4ac \geq 0$. When the condition does not hold, we set $\sigma_k^2 = 0$. Note that when $\omega = 0$, the global solution for σ_k^2 is always zero. The one given in (16.98) is actually a local solution. In this situation, we must take the local solution.

The special case, $(\tau, \omega) = (-2, 0)$, corresponds to a uniform prior for σ_k^2. The solution under this uniform prior is

$$\sigma_k^{2(t+1)} = -\frac{b}{a} = \sigma_k^{2(t)} + \frac{q_k^2 - s_k}{s_k^2} \tag{16.99}$$

This is equivalent to the result in the classical mixed model analysis, as seen in (16.48). Since we start with $\sigma_k^{2(0)} = 0$, the convergence sequence must be of monotonic increase for σ_k^2. Therefore, $q_k^2 - s_k \geq 0$; otherwise, we set $\sigma_k^{2(t+1)} = \sigma_k^{2(t)} = 0$.

Block Coordinate Descent

Definitions and notations are the same as the ones given in the corresponding section in the classical mixed model analysis. We are interested in updating the kth subset of G, denoted by G_k, where each G_k contains b elements and the total number of block is m.

The log likelihood function relevant to G_k is

$$L(G_k) = -\frac{1}{2}\ln|\psi_k| - \frac{1}{2}\ln|G_k| + \frac{1}{2}\mathrm{tr}(h_k h_k^T \psi_k^{-1}) \tag{16.100}$$

The log prior is

$$L(\xi) = -\frac{1}{2}(\tau + 2)\ln|G_k| - \frac{1}{2}\omega\mathrm{tr}(G_k^{-1}) \tag{16.101}$$

The log posterior is

$$\Phi(G_k) = -\frac{1}{2}\ln|\psi_k| - \frac{1}{2}\ln|G_k| + \frac{1}{2}\mathrm{tr}(h_k h_k^T \psi_k^{-1})$$

$$- \frac{1}{2}(\tau + 2)\ln|G_k| - \frac{1}{2}\omega\mathrm{tr}(G_k^{-1}) \tag{16.102}$$

The partial derivative of this log posterior with respect to G_k is

$$\frac{\partial\Phi(G_k)}{\partial G_k} = \frac{1}{2}G_k^{-1}\left(\psi_k^{-1} + \psi_k^{-1}h_k h_k^T \psi_k^{-1} + \omega I\right)G_k^{-1} - \frac{1}{2}(\tau + 3)G_k^{-1} \tag{16.103}$$

Setting $\frac{\partial\Phi(G_k)}{\partial G_k} = 0$, we get the following equation:

$$G_k = \frac{1}{\tau + 3}\mathrm{diag}\left(\psi_k^{-1} + \psi_k^{-1}h_k h_k^T \psi_k^{-1} + \omega I\right) \tag{16.104}$$

Because ψ_k is a function of G_k, the above equation is iterative. The inner iterations should converge before we can proceed to updating β and σ^2. Again, to avoid G_k^{-1}, we must use the following approach to calculate ψ_k^{-1}:

$$\psi_k^{-1} = G_k(s_k G_k + I)^{-1} = (G_k s_k + I)^{-1}G_k \tag{16.105}$$

16.2.2 Exponential Prior

The λ Prior

The exponential prior for σ_k^2 is

$$p(\sigma_k^2|\lambda) = \frac{\lambda^2}{2} \exp\left(-\frac{\lambda^2}{2}\sigma_k^2\right) \tag{16.106}$$

where $\lambda > 0$ is the hyperparameter of the exponential prior. The joint prior for G is

$$p(G|\lambda) = \frac{\lambda^2}{2} \exp\left[-\frac{\lambda^2}{2}\text{tr}(G)\right] \tag{16.107}$$

Although the three algorithms for updating G described earlier can all be customized to handle the exponential prior, we only demonstrate the coordinate descent approach here for the exponential prior. In a previous section, we introduced s_k and h_k in two different forms. In the coordinate descent method, we defined

$$s_k = Z_k^T (V^{(t)})^{-1} Z_k$$
$$h_k = Z_k^T (V^{(t)})^{-1} (y - X\beta) \tag{16.108}$$

In the block coordinate descent method, they were defined as

$$s_k = Z_k^T V_{-1}^{-1} Z_k$$
$$h_k = Z_k^T V_{-k}^{-1} (y - X\beta) \tag{16.109}$$

The first definition, (16.108), is introduced here mainly because it was used in the original publication of the empirical Bayesian mapping (Xu 2007). Here, we use the second definition of s_k and h_k, (16.109), for the coordinate descent algorithm. The log posterior is

$$\Phi(\sigma_k^2) = -\frac{1}{2}\ln(s_k\sigma_k^2 + 1) + \frac{1}{2}\frac{\sigma_k^2 h_k^2}{s_k\sigma_k^2 + 1} - \frac{1}{2}\lambda^2\sigma_k^2 \tag{16.110}$$

The partial derivative of this log posterior with respect to σ_k^2 is

$$\frac{\partial\Phi(\sigma_k^2)}{\partial\sigma_k^2} = -\frac{1}{2}\frac{s_k}{s_k\sigma_k^2 + 1} + \frac{1}{2}\frac{h_k^2}{(s_k\sigma_k^2 + 1)^2} - \frac{1}{2}\lambda^2 \tag{16.111}$$

Setting $\frac{\partial\Phi(\sigma_k^2)}{\partial\sigma_k^2} = 0$ leads to

$$-\lambda^2 s_k^2 (\sigma_k^2)^2 - (s_k^2 + 2\lambda^2 s_k)\sigma_k^2 + (h_k^2 - s_k - \lambda^2) = 0 \tag{16.112}$$

This is a quadratic equation with the three coefficients defined as

$$a = -\lambda^2 s_k^2$$

$$b = -(s_k^2 + 2\lambda^2 s_k)$$

$$c = h_k^2 - s_k - \lambda^2 \tag{16.113}$$

The solution is

$$\sigma_k^2 = \frac{-b - \sqrt{b^2 - 4ac}}{2a} \tag{16.114}$$

If $b^2 - 4ac < 0$, we simply set $\sigma_k^2 = 0$. The exponential prior appears to be much simpler than the scaled inverse chi-square prior.

The (a, b) Prior

This prior was introduced by Cai et al. (2011) to alleviate the problem of choosing an inappropriate value for λ. Here, we add another level to the hierarchical model at which we assign a Gamma$(\lambda | a, b)$ prior to λ with a shape parameter $a > 0$ and an inverse scale parameter $b > 0$. Cai et al. (2011) modified the λ exponential prior for σ_k^2 as

$$p(\sigma_k^2 | \lambda) = \lambda \exp(-\lambda \sigma_k^2) \tag{16.115}$$

The (a, b) prior is obtained by

$$p(\sigma_k^2 | a, b) = \int_0^\infty p(\sigma_k^2 | \lambda) p(\lambda | a, b) d\lambda = \frac{a}{b(\frac{\sigma_k^2}{b} + 1)^{(a+1)}} \tag{16.116}$$

The λ prior has been removed and replaced by the (a, b) prior. We can now preselect the values of (a, b) for this prior. The log posterior for σ_k^2 is

$$\Phi(\sigma_k^2) = -\frac{1}{2} \ln(s_k \sigma_k^2 + 1) + \frac{1}{2} \frac{\sigma_k^2 h_k^2}{s_k \sigma_k^2 + 1} - (a+1) \ln \frac{\sigma_k^2 + b}{b} \tag{16.117}$$

The partial derivative is

$$\frac{\partial \Phi(\sigma_k^2)}{\partial \sigma_k^2} = -\frac{1}{2} \frac{s_k}{s_k \sigma_k^2 + 1} + \frac{1}{2} \frac{h_k^2}{(s_k \sigma_k^2 + 1)^2} - \frac{(a+1)b}{\sigma_k^2 + b} \tag{16.118}$$

Setting $\frac{\partial \Phi(\sigma_k^2)}{\partial \sigma_k^2} = 0$ yields the following quadratic equation:

$$c_1 (\sigma_k^2)^2 + c_2 \sigma_k^2 + c_3 = 0 \tag{16.119}$$

where

$$c_1 = -[2(a + 1)b + 1]s_k^2$$

$$c_2 = -[4(a + 1) + s_k]s_k b + (h_k^2 - s_k)$$

$$c_3 = -[2(a + 1) - (h_k^2 - s_k)]b \tag{16.120}$$

The solution is

$$\sigma_k^2 = \frac{-c_2 - \sqrt{c_2^2 - 4c_1c_3}}{2c_1} \tag{16.121}$$

16.2.3 Dealing with Sparse Models

In every cycle of the iteration, we need to calculate V^{-1}. This can be costly if the sample size is large. If the shrinkage is strong, most σ_k^2 will be zero, and the model can be very sparse. Let p_r be the number of nonzero σ_k^2. Let us define G_r as the subset of matrix G corresponding to the nonzero elements. Let Z_r be the corresponding subset of Z. When $p_r < n$, we can use the following equation to calculate V^{-1}:

$$V^{-1} = \frac{1}{\sigma^2}I - \frac{1}{\sigma^2}Z_r\left(Z_r^T Z_r \frac{1}{\sigma^2} + G_r^{-1}\right)^{-1}Z_r^T \frac{1}{\sigma^2} \tag{16.122}$$

This equation only requires inverting a matrix of $p_r \times p_r$. The cost saving can be substantial if p_r is substantially less than n. Of course, if $n > p_r$, it is more efficient to invert V directly. The computer program should have a switch to choose this special algorithm if $p_r < n$ and invert V directly otherwise.

16.3 Infinitesimal Model for Whole Genome Sequence Data

Although the Bayesian method can handle high-density markers, the "high" cannot be infinite. Ultimately, there is still a limit on the number of markers because sample size is always finite. The actual limit of the marker density depends on the sample size; larger sample size is required to handle more dense markers. In the genome era, whole gene sequences will soon be available for many species, making the number of markers (SNPs) virtually infinite. How to take advantage of the whole genome sequence information to identify genes responsible for the genetic variation of complex traits is a great challenge to statisticians. Two approaches may be taken to deal with such a situation, (1) data trimming prior to the Bayesian analysis and (2) new model development. The first approach does not involve new statistical methods and is more realistic and easy to implement. The second one, however, may require a new model that is conceptually very different from all the linear models currently available in statistics. Both approaches will be discussed here in this section.

16.3.1   Data Trimming

This approach only requires data preparation and does not involve new methods. Since the marker density is too high, we can trim the markers prior to the Bayesian analysis. If two or more markers (SNPs) co-segregate in a population, i.e., they have exactly the same segregation patterns, only one of them should be used. A program is needed to select all markers with unique segregation pattern to make sure that no two markers co-segregate in the population. Depending on the sample size, this may eliminate most SNPs in the dataset because most of them may be redundant. The total number of uniquely segregated markers may still be over the limit that a model can handle. In this case, we may increase the stringency of marker trimming so that selected markers differ by at least two or more individuals, say ϱ. This number ϱ can be adjusted by trial and error until the number of selected markers is manageable. There might be some information loss if ϱ is set too high. This marker trimming approach may have already been practiced in genomic data analysis. Theoretical work is definitely lacking, and further investigation is required for the optimal ϱ for a given sample size n.

16.3.2   Concept of Continuous Genome

Let us reintroduce the linear multiple QTL model for the phenotypic value of individual j,

$$y_j = \beta + \sum_{k=1}^{p} Z_{jk} \gamma_k + \varepsilon_j \qquad (16.123)$$

where p is the number of markers included in the model. For whole genome sequence data, $p \approx \infty$, and thus, the above model becomes

$$y_j = \beta + \sum_{k=1}^{\infty} Z_{jk} \gamma_k + \varepsilon_j \qquad (16.124)$$

This is just a conceptual model because nobody can estimate parameters in the model. In mathematics, the limit of summation is integral. Therefore, we may propose the following continuous genome model by replacing the summation by integration:

$$y_j = \beta + \int_0^L Z_j(\lambda) \gamma(\lambda) \mathrm{d}\lambda + \varepsilon_j \qquad (16.125)$$

where λ is the genome location ranging from 0 to L (the size of the genome), $Z_j(\lambda)$ is the genotype indicator variable for individual j at location λ, and $\gamma(\lambda)$ is the genetic effect expressed as a function of λ. Note that $Z_j(\lambda)$ is known because, at any given location, we can observe the SNP genotypes for all individuals in the mapping population. The function of genetic effect, $\gamma(\lambda)$, however, is unknown, and estimating this function is the ultimate goal of the whole genome sequence analysis. This model is called the continuous genome model. Since the entire genome is a collection of C chromosomes, e.g., $C = 23$ for humans, this continuous genome model is more precisely expressed by

$$y_j = \beta + \sum_{t=1}^{C} \int_0^{L_t} Z_j(\lambda)\gamma(\lambda)d\lambda + \varepsilon_j \tag{16.126}$$

where L_t is the size of chromosome t for $t = 1,\ldots,C$. Conceptually, we can handle infinite number of SNPs using this continuous model. The greatest challenge is how to estimate $\gamma(\lambda)$ given the phenotypes y_j and the genotypes $Z_j(\lambda)$ of all individuals. This continuous genome model may open a new area in genome data analysis.

Prior to the development of a full theory of the infinitesimal mode, we may use numerical integration to solve this problem. Let us go back to (16.125) where only one chromosome (or a single continuous genome) is considered. Let us divide the genome into p equal distance intervals and define

$$\Delta\lambda = \frac{L}{p} \tag{16.127}$$

as the length of the interval. A numerical presentation of the integral in (16.125) is

$$y_j = \beta + \sum_{k=1}^{p} Z_j(\lambda_k)\gamma(\lambda_k)\Delta\lambda + \varepsilon_j \tag{16.128}$$

where λ_k is the middle point of the kth interval. It appears that this model still has not taken advantage of the infinite SNPs because $Z_j(\lambda_k)$ is still a value from a single point of genome. We now revise this model so that

$$y_j = \beta + \sum_{k=1}^{p} \bar{Z}_{jk}\tilde{\gamma}_k + \varepsilon_j \tag{16.129}$$

where

$$\bar{Z}_{jk} = \frac{\int_{\lambda_k - \frac{1}{2}\Delta\lambda}^{\lambda_k + \frac{1}{2}\Delta\lambda} Z_j(\lambda)d\lambda}{\int_{\lambda_k - \frac{1}{2}\Delta\lambda}^{\lambda_k + \frac{1}{2}\Delta\lambda} d\lambda} \tag{16.130}$$

is the mean Z_j for the interval covering middle point λ_k and

$$\tilde{\gamma}_k = \gamma(\lambda_k)\Delta\lambda \tag{16.131}$$

is the sum of all effects within that interval. In real data analysis, $\bar{Z}_{jk}$ simply takes the average Z_j for all markers genotyped in that interval. Comparing (16.129) with (16.123), we can see that Z_{jk} is replaced by the mean $\bar{Z}_{jk}$ and γ_k is replaced by the sum $\tilde{\gamma}_k$. We can now choose p according to the sample size n; a larger sample size allows a higher value of p, which determines the resolution. Therefore, large sample sizes lead to higher resolution of QTL mapping.

16.4   Example: Simulated Data

This example demonstrates the empirical Bayes estimates of QTL effects using a simulated dataset (Xu 2007). A BC population of 600 individuals was simulated for a single large chromosome of 1,800 cM. This giant chromosome was covered by 121 evenly spaced markers. Nine of the markers overlapped with QTL of the main effects, and 13 out of the $C_{121}^2 = 7,260$ possible marker pairs had interaction effects. In genetics terminology, interaction effects between loci are called epistatic effects (Xu 2007). Although we have not dealt with epistatic effects yet, the empirical Bayesian method developed here can be applied to epistatic effect detection without any modification at all. The only difference of the epistatic model from the pure additive model is the increase of model dimension. The true effects of the markers and marker pairs are shown in Figs. 16.1 and 16.2, respectively. The true population mean and residual variance were 5.0 and 10.0, respectively. The fixed effect vector contains only the population mean. The genetic variance contributed by all markers (including main and epistatic effects) was approximately 90.0.

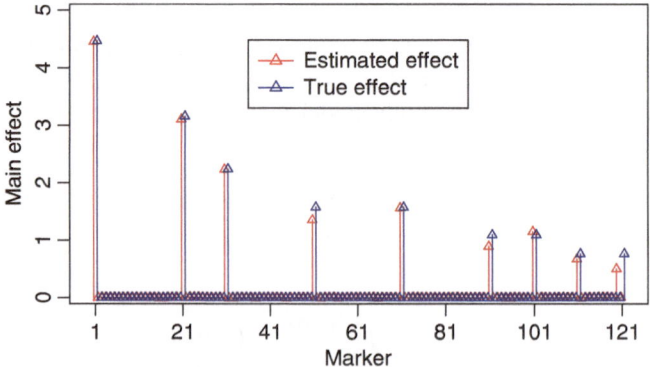

Fig. 16.1 Empirical Bayesian estimates of QTL main effects (*red*) compared with the true QTL main effects (*blue*) for the simulated data

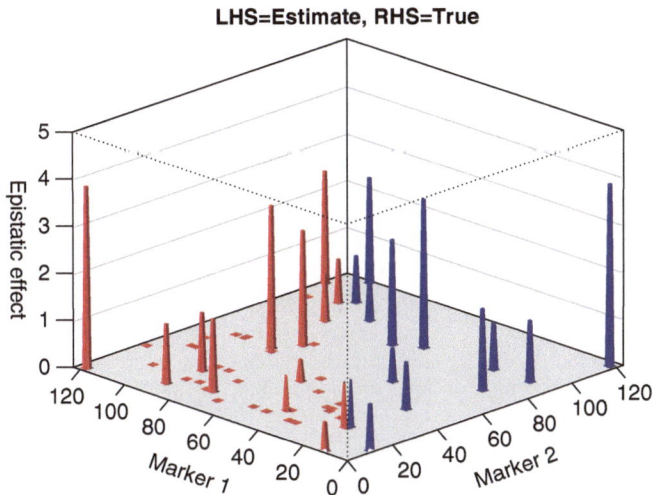

Fig. 16.2 Empirical Bayesian estimates of QTL epistatic effects (*red, left*) compared with the true QTL epistatic effects (*blue, right*) for the simulated data

The genetic contribution from the covariance terms was about 6.0, leading to a phenotypic variance of about 106.0. The theoretical proportion of the phenotypic variance contributed by an individual QTL was simply defined as the squared effect divided by the phenotypic variance. In the simulation experiment, the proportion of contribution from an individual QTL varied from 0.6 % to 20 %, whereas the proportion of contribution from a pair of QTL ranged from 0.5 % to 15 %. Some of the markers had main effects only, while others had both the main and the epistatic effects. Many of the markers with epistatic effects had no main effects. The epistatic model contained a total of $121(121 + 1)/2 = 7381$ effects, about 12 times as large as the sample size. The program and the simulated data are available on request from the author (Xu 2007). The hyperparameters were chosen as $(\tau, \omega) = (-1, 0.0005)$. Our past experience indicated that other values of the hyperparameters do not shrink the parameters properly (Xu 2003; ter Braak et al. 2005). The estimated QTL main effects along with the true values for the 121 markers are plotted in Fig. 16.1. It is clear that the estimated effects are very close to the true values. The estimated epistatic effects along with the true values are presented in Fig. 16.2 where the true values are given in the right-hand side of the 3D plot and the estimated values are given in the left-hand side of the 3D plot. The estimated and true epistatic effects are almost identical both in effect and in location. No other methods have ever been able to generate such a good result.

FIG. 15.4 ...

The ...

Part III
Microarray Data Analysis

Chapter 17
Microarray Differential Expression Analysis

Gene expression is the process by which mRNA, and eventually protein, is synthesized from the DNA template of each gene. The level of gene expression can be measured by a particular technology, called microarray technology (Schena et al. 1995), in which we can measure the expression of thousands of different RNA molecules at a given time point in the life of an organism, a tissue, or a cell. Comparisons of the levels of RNA molecules can be used to decipher the thousands of processes going on simultaneously in living organisms. Also, comparing healthy and diseased cells can yield vital information on the causes of the disease. The microarray technology has been successfully applied to several biological problems, and as arrays become more easily accessible to researchers, the popularity of these kinds of experiments will increase. The demand for good statistical analysis regimens and tools tailored for microarray data analysis will increase as the popularity of microarrays grows. The future will likely bring many new microarray applications, each with its own demands for specialized statistical analysis. Starting from this chapter, we will learn some of the basic statistical methods for microarray data analysis.

17.1 Data Preparation

First, we need to understand the format of microarray data. The data are usually arranged in a matrix form, with rows representing genes and the columns representing tissue samples. The expressing level of the ith gene in the jth sample is denoted by $y_{ij}, \forall i = 1, \ldots, N; j = 1, \ldots, M$, where N is the number of genes and M is the number of tissue samples, as shown in the sample data (Table 17.1). Each data point is assumed to have been properly transformed and normalized. Data preparation includes such data transformation and normalization.

S. Xu, *Principles of Statistical Genomics*, DOI 10.1007/978-0-387-70807-2_17,
© Springer Science+Business Media, LLC 2013

Table 17.1 A sample dataset of microarray gene expression

Gene	Sample 1	Sample 2	Sample 3	Sample 4	Sample 5	...	Sample M
1	6.73293	7.02988	6.85878	6.94274	5.00395	...	7.80914
2	4.40305	5.06069	4.95442	4.83628	6.37724	...	5.76519
3	6.58147	6.66364	6.88531	6.90465	8.32814	...	5.38998
4	4.40183	5.14982	5.21494	5.11018	6.61258	...	4.18205
5	7.03632	6.55607	7.06851	6.79872	5.2776	...	5.48147
6	5.36317	5.90971	5.46848	5.76832	5.86363	...	5.36598
7	5.3303	5.22467	4.77238	5.06765	7.97398	...	5.18403
8	6.08199	6.32077	6.12796	6.11744	6.78004	...	4.53367
9	5.83802	5.45788	5.55721	5.8354	4.69592	...	5.38174
10	7.23807	7.10562	6.92991	7.03077	5.30231	...	5.94411
11	6.11014	5.56375	5.85536	6.09334	5.92345	...	5.76738
...	...	...	...	...	...	...	...
N	8.03291	7.84023	7.65354	8.10648	5.43285	...	6.05514

17.1.1   Data Transformation

Most statistical methods require specific statistical models and distributions of the errors. Linear model is the most commonly used model for microarray data analysis. Normal distribution is most frequently assumed for the errors. However, data points in the original form usually cannot be described by a linear model, and the errors may not follow a normal distribution. Therefore, some type of data transformation is usually recommended before the data analysis. Let w_{ij} be the original expression level of the ith gene in the jth condition. The transformed value can be expressed as

$$y_{ij}^* = f(w_{ij}) \qquad (17.1)$$

where $f(.)$ represents any monotonic function. All the statistical methods to be described in the book are actually performed on the transformed data y_{ij}^*, not the raw data w_{ij}. There are many different ways to transform the data. We will list a few most commonly used ones as examples.

17.1.1.1   Logarithmic Transformation

By far the most common transformation applied to microarray readings is the logarithmic transformation

$$y_{ij}^* = \log(w_{ij}) \qquad (17.2)$$

The base of the logarithm may be 2, 10, or the natural logarithmic constant e. The choice of base is largely a matter of convenient interpretation. The log transformation tends to provide values that are approximately normally distributed and for which conventional linear regression and ANOVA models are appropriate.

Square Root Transformation

The square root transformation

$$y_{ij}^* = \sqrt{w_{ij}} \tag{17.3}$$

is a variance-stabilizing transformation. In other words, if the variance is proportional to the mean, the square root transformation will correct this to some degree so that the variance of the transformed values is independent of the mean of the transformed values.

Box–Cox Transformation Family

The two transformations described above are members of the Box–Cox family of transformations (Box and Cox 1964). This family is defined as

$$y_{ij}^* = \frac{w_{ij}^\delta - 1}{\delta} \tag{17.4}$$

where δ is the parameter of transformation chosen by the investigator. The square root transformation corresponds to $\delta = \frac{1}{2}$. The logarithmic transformation corresponds to the limit of this equation when $\delta \longrightarrow 0$. The case where $\delta = 1$ corresponds to taking no transformation at all, except for a change in the origin. The Box–Cox family provides a range of transformations that may be examined to see which value of δ yields transformed values with the desired statistical properties.

17.1.2   Data Normalization

The laboratory preparation of each biological specimen on a microarray slide introduces an arbitrary scale or dilution factor that is common to expression readings for all genes. We usually correct the readings for the scale factor and other variations using a process called normalization. The purpose of normalization is to minimize extraneous variation in the measured gene expression levels of hybridized mRNA samples so that biological differences (differential expression) can be more easily distinguished.

Practical experience has shown that, in addition to array effects, other extraneous sources of variation may be present that cloud differential gene expression if not taken into account. These sources of variations are called systematic errors, e.g., the effects of dye colors. These systematic errors will mask the true biological differences and should be removed before the data are analyzed.

Normalization Across Genes

The normalization is done by subtracting the mean expression of all genes from the individual gene expression. The normalized data point is

$$y_{ij} = y_{ij}^* - \overline{y}_{.j}^* \tag{17.5}$$

where $\overline{y}_{.j}^* = \frac{1}{N} \sum_{i=1}^{N} y_{ij}^*$ is the average expression of all genes in sample j. All subsequent statistical analysis should be performed on the normalized values y_{ij}.

Normalization Across Tissue Samples

The normalization is done by subtracting the mean expression of all tissue samples from the individual gene expression. The normalized data point is

$$y_{ij} = y_{ij}^* - \overline{y}_{i.}^* \tag{17.6}$$

where $\overline{y}_{i.}^* = \frac{1}{M} \sum_{j=1}^{M} y_{ij}^*$ is the average expression of all samples for gene i. All subsequent statistical analysis should be performed on the normalized values y_{ij}.

Normalization Across both Genes and Samples

The normalized data point is expressed as the deviation of the unnormalized value from the mean of all genes in the particular sample and the mean of all samples for the particular gene,

$$y_{ij} = y_{ij}^* - \overline{y}_{i.}^* - \overline{y}_{.j}^* + \overline{y}_{..}^* \tag{17.7}$$

where $\overline{y}_{..}^* = \frac{1}{MN} \sum_{i=1}^{N} \sum_{j=1}^{M} y_{ij}^*$ is the overall mean of gene expression across all genes and samples.

Normalization via Analysis of Covariance

Analysis of covariance is an approach to removing the influence of systematic errors on the effects of treatments in an ANOVA. If we ignore any systematic errors (effects), we may write a linear model to describe the expression of the ith gene in the jth treatment,

$$y_{ijk}^* = \mu + \alpha_i + \beta_j + \gamma_{ij} + \epsilon_{ijk} \tag{17.8}$$

where subscript k represents the kth replicate of gene i in treatment j. In this model, μ is the grand mean, α_i is the effect of the ith gene, β_j is the effect of the jth treatment, ϵ_{ijk} is the error term, and γ_{ij} is the interaction effect between the ith gene and the jth treatment. The interaction term, γ_{ij}, actually reflects the differential expression of gene i. A sufficiently large γ_{ij} indicates that the ith gene expresses differently from one treatment level to another. It is γ_{ij} that we are interested in. By formulating the above linear (ANOVA) model, we can separate γ_{ij} (the interesting part of the model) from the other effects (the parts that are not interesting to us). If we estimate the non-interesting effects and remove them from y_{ijk}^*, we have

$$y_{ijk} = y_{ijk}^* - \widehat{\mu} - \widehat{\alpha}_i - \widehat{\beta}_j = \gamma_{ij} + \epsilon_{ijk} \qquad (17.9)$$

The non-interesting effects can be estimated first and then removed from the model. This is a generalized normalization process. It has normalized the gene expression across both the genes and the treatments. The method of estimation for the non-interesting effects can be the usual least-squares method from a simple two-way ANOVA or the mixed model approach (Wolfinger et al. 2001) by treating the gene and treatment effects as fixed and the interaction effects as random. One advantage of using the mixed model approach for normalization over the previous method is that the method can handle unbalanced data.

More importantly, we can incorporate systematic effects into the ANOVA linear model as covariates and perform covariance analysis to remove the extraneous systematic errors. For example, the dye color asymmetry has led to the use of microarray study designs in which arrays are produced in pairs with the colors in one array reversed relative to the colors in the second array in order to compensate for the color differences. These are called reversed-color designs. In this kind of design, the color effects, which are not of our interest, should be included in the model. Therefore, the modified model incorporating the color effects appears

$$y_{ijlk}^* = \mu + \xi_l + \alpha_i + \beta_j + (\xi\alpha)_{li} + (\xi\beta)_{lj} + \gamma_{ij} + \epsilon_{ijlk} \qquad (17.10)$$

where $\xi_l, \forall l = 1, 2$ is the color effect, $(\xi\alpha)_{li}$ is the interaction of the lth color with the ith gene, and $(\xi\beta)_{lj}$ is the interaction of the lth color with the jth treatment. If we do not include these effects in the model, they will be absorbed by the error term. With the above model, they can be estimated and removed from the analysis, as shown below:

$$y_{ijlk} = y_{ijlk}^* - \widehat{\mu} - \widehat{\xi}_l - \widehat{\alpha}_i - \widehat{\beta}_j - \widehat{(\xi\alpha)}_{li} - \widehat{(\xi\beta)}_{lj} = \gamma_{ij} + \epsilon_{ijlk}, \qquad (17.11)$$

leaving only γ_{ij} in the model. Therefore, analysis of covariance is a generalized normalization approach to removing all non-interesting effects.

Once the data are properly transformed and normalized, they are ready to be analyzed using any of the statistical methods described in subsequent sections.

17.2 F-test and T-test

Assume that we collect tissues from M_1 mice affected by some particular disease and tissues from M_2 mice that have the same diseases but treated with a newly developed drug. The tissue of each mouse is microarrayed, and the expression of N genes are measured. The purpose of the experiment is to find which genes have different levels of expression between the untreated and treated groups of mice, i.e., to find genes responding to the drug treatment. The data matrix has N rows and $M = M_1 + M_2$ columns. However, each data point (gene expression level) is better denoted by variable y with three subscripts, y_{ijk}, where $i = 1, \ldots, N$ indexes genes, $j = 1, \ldots, P$ indexes the level of treatment ($P = 2$ in the case of two levels of treatment), and $k = 1, \ldots, M/P$ indexes the replication within each treatment group. The number of replicates within each treatment group is assumed to be $M_j = M/P$ for all $j = 1, \ldots, P$. If the data are not balanced, i.e., $M_1 \neq M_2$, the replication index k should be subscripted with j so that $k_j = 1, \ldots, M_j$. The t-test statistic of differential expression for the ith gene is

$$t_i = \frac{|\overline{y}_{i1.} - \overline{y}_{i2.}|}{s_{\overline{y}_{i1.} - \overline{y}_{i2.}}}, \tag{17.12}$$

where

$$\overline{y}_{ij.} = \frac{1}{M_j} \sum_{k=1}^{M_j} y_{ijk}, \quad \forall j = 1, 2 \tag{17.13}$$

$$s_{\overline{y}_{i1.} - \overline{y}_{i2.}} = \sqrt{s_{i1}^2/M_1 + s_{i2}^2/M_2} \tag{17.14}$$

and

$$s_{ij}^2 = \frac{1}{M_j - 1} \sum_{k=1}^{M_j} (y_{ijk} - \overline{y}_{ij.})^2, \forall j = 1, 2. \tag{17.15}$$

One can rank t_i across all $i = 1, \ldots, N$ and select all genes with $t_i > t_{\mathrm{d}f,1-\alpha}$ as differentially expressed genes, where $t_{\mathrm{d}f,1-\alpha}$ is the critical value chosen by the investigator, $\mathrm{d}f = M_1 + M_2 - 2$ is the degrees of freedom, and $0 < \alpha < 1$ is a probability that controls the type I error rate of the experiment (discussed later).

The t-test only applies to situations where there are two levels of treatment. For multiple levels of treatment, an F-test must be used. Let P be the number of levels of treatment. For example, if there are four groups of mice with the first group being the untreated group and the remaining three groups being treated with three different doses of a particular drug, then $P = 4$. The F-test statistic for the ith gene is calculated from the ANOVA table (see Table 17.2), where

$$SS_T = \sum_{j=1}^{P} M_j (\overline{y}_{ij.} - \overline{y}_{i..})^2 \tag{17.16}$$

Table 17.2 Analysis of variance (ANOVA) table for differential gene expression analysis

Variation	df	SS	MS	F-test statistic
Treatment	$df_T = P - 1$	SS_T	$MS_T = \frac{SS_T}{df_T}$	$\frac{MS_T}{MS_E}$
Error	$df_E = \sum_{j=1}^{P}(M_j - 1)$	SS_E	$MS_E = \frac{SS_E}{df_E}$	

and

$$SS_E = \sum_{j=1}^{P}\sum_{k=1}^{M_j}(y_{ijk} - \overline{y}_{ij.})^2 \tag{17.17}$$

The F-test statistic for the ith gene is defined as

$$F_i = \frac{MS_T}{MS_E} \tag{17.18}$$

The critical value used to declare significance is $F_{df_T, df_E, 1-\alpha}$, which is the $1 - \alpha$ quantile of the F-distribution with degrees of freedom df_T and df_E. All genes with $F_i > F_{df_T, df_E, 1-\alpha}$ are called significant. Again, the value of α is chosen by the investigator (discussed later).

17.3 Type I Error and False Discovery Rate

In the previous section, the critical value for a test statistic used to select the list of significant genes is denoted by $t_{df, 1-\alpha}$ for the t-test or $F_{df_T, df_E, 1-\alpha}$ for the F-test. The α value is called the type I error. A small α means a large critical value and thus generates a short list of significant genes, while a large α will produce a long list of significant genes. Therefore, type I error determines the list of significant genes. Let H_0 denote the null hypothesis that a gene is not differentially expressed and H_1 denote the alternative hypothesis that the gene is differentially expressed. When we perform a statistical test on a particular gene, we may make errors if the sample size is small. There are two types of errors we can make. If a gene is not differentially expressed but our test statistic is greater than the chosen critical value, we will make the type I error whose probability is denoted by α as mentioned before. On the other hand, if a gene is differentially expressed but our test statistic is less than the chosen critical value, we will make the type II error whose probability is denoted by β. In other words, we will make the type I error if H_0 is true but H_1 is accepted and make the type II error if H_1 is true but H_0 is accepted. The type I and type II errors are also called false-positive and false-negative errors, respectively. These two errors are negatively related, i.e., a high type I error leads to a low type II error.

The probability that H_1 is accepted while H_1 is indeed true is called the statistical power. Therefore, the statistical power is simply $\omega = 1 - \beta$. A gene can be differentially expressed or not differentially expressed. So, if we use an indicator variable h to denote the true status of the gene, we have $h = H_0$ or $h = H_1$. After the statistical test, the gene will have one of two outcomes, significant or not. Let $\hat{h}$ be the outcome of the statistical test of the gene, then $\hat{h} = H_0$ if H_0 is accepted and $\hat{h} = H_1$ if H_1 is accepted. We can now define the type I error as

$$\alpha = \Pr(\hat{h} = H_1 | h = H_0) \tag{17.19}$$

and the type II error as

$$\beta = \Pr(\hat{h} = H_0 | h = H_1). \tag{17.20}$$

The statistical power is

$$\omega = \Pr(\hat{h} = H_1 | h = H_1). \tag{17.21}$$

So far, we only discussed the type I error rate for a single test. In microarray data analysis, we have to test the differential expression for every gene. Therefore, a microarray experiment with N genes involves N hypothesis tests. The type I error for a single test needs to be adjusted to control the type I error of the entire experiment. The Bonferroni correction introduced in Sect. 8.5 of Chap. 8 applies here. If the experiment-wise type I error is γ, the nominal type I error rate after correcting for multiple tests should be

$$\alpha = \frac{\gamma}{N} \tag{17.22}$$

Assume that a microarray experiment involves $N = 1,000$ genes collected from $M_1 = 10$ cases and $M_2 = 10$ controls. A t-test statistic has been calculated for each of the N genes. We want to find the critical value for the t-test to compare so that the experiment-wise type I error is controlled at $\gamma = 0.05$. First, we need to find out the nominal type I error rate using the Bonferroni correction,

$$\alpha = \frac{\gamma}{N} = \frac{0.05}{1,000} = 0.00005$$

If the p-values are already given in the differential expression analysis, one may simply compare the gene-specific p-values to 0.00005, declaring significance if the p-value is less than 0.00005. If the p-values are not given, we need to find the critical value used to declare significance using

$$t_{\mathrm{df}, 1-\alpha/2} = t_{18, 0.999975} = 5.2879056 \tag{17.23}$$

Any genes with t-test statistics greater than 5.2879 will be declared as significant. Bonferroni correction usually provides a very conservative result of tests, i.e., reports less genes than the actual number of significant genes. In practice, Bonferroni is rarely used because of the conservativeness; instead, people often use permutation test to draw an empirical critical value. This will be discussed in the next section.

Another related performance measure in multiple testing is the so-called false discovery rate (FDR) developed by Benjamini and Hochberg (1995). The FDR measure looks at error controls from a different perspective. Instead of conditioning on the true but unknown state of whether a gene is differentially expressed or not, the FDR is defined as the probability that the test statistic indicates that the gene is differentially expressed but in fact it is not. This probability, denoted by δ, is

$$\delta = \Pr(h = H_0|\hat{h} = H_1) \tag{17.24}$$

We can control the value of FDR (δ) and use the fixed FDR value to select the list of significant genes.

17.4 Selection of Differentially Expressed Genes

Data transformation is to ensure that the residual errors of gene expressions follow a normal distribution. A normal distribution will make the t-test or F-test statistic follow the expected t- or F-distribution under the null model. For most microarray data, transformation can only improve the normality and rarely make the residual errors perfectly normal. Therefore, the critical value for a test statistic drawn from the expected t- or F-distribution is problematic. In addition, the multiple test adjustment using the Bonferroni correction is far too conservative when N is very large. Therefore, the optimal way of finding the critical value perhaps relies on some empirical methods that are data dependent. In other words, different datasets should have different critical values, reflecting the particular nature of the data.

17.4.1 Permutation Test

Permutation test is a way to generate the distribution of the test statistic under the null model. In differential expression analysis, the null model is that no genes are differentially expressed. Consider M tissue samples with M_1 samples being the control and M_2 being the treatment for $M_1 + M_2 = M$. Each sample is labeled as 0 for the control and 1 for the treatment. If the tissue samples and the labels are randomly shuffled, the association between the gene expressions and the labels will be destroyed. The distribution of the test statistic will mimic the distribution under the null model.

The total number of ways of shuffling is

$$T = \frac{M!}{M_1!M_2!}, \tag{17.25}$$

equivalent to the number of ways of randomly sampling M_1 or M_2 items from a total of M items. When M is large, T may be extremely large, making the permutation analysis very difficult. In practice, one may only use a proportion of the reshuffled samples to draw the null distribution of the test statistic.

This type of random shuffling may not generate the null distribution accurately because there is a chance that all M_1 tissue samples in a reshuffled dataset are actually from the control and M_2 samples from the treatment. If some genes are indeed differentially expressed, then the test statistics of these genes are not drawn from the null distribution. Tusher et al. (2001) proposed a balanced shuffling approach that can avoid this problem. In the balanced random shuffling, each group in the reshuffled dataset contains the samples from the original groups in proportion. Take the case-control experiment for example; in the reshuffled dataset, the M_1 controlled group should contain $\frac{M_1}{M_1+M_2} M_1$ samples from the original control group and $\frac{M_2}{M_1+M_2} M_1$ samples from the original treatment group. Similarly, the M_2 treatment group in the reshuffled dataset should contain $\frac{M_1}{M_1+M_2} M_2$ samples from the original control groups and $\frac{M_2}{M_1+M_2} M_2$ from the original treatment group. Under this restriction, we guarantee that the test statistics for all genes are sampled from the null distribution. The balanced shuffling is not easy to conduct if any one of $\frac{M_1}{M_1+M_2} M_1, \frac{M_2}{M_1+M_2} M_1, \frac{M_1}{M_1+M_2} M_2$, and $\frac{M_2}{M_1+M_2} M_2$ is not an integer. It is convenient to conduct the balanced shuffling if M_1 is an even number and $M_1 = M_2$. In this case, the total number of reshuffled datasets will be

$$T = \frac{M_1!}{(\frac{1}{2}M_1)!(\frac{1}{2}M_1)!} \frac{M_2!}{(\frac{1}{2}M_2)!(\frac{1}{2}M_2)!}. \tag{17.26}$$

For example, assume that $M = 8$ and $M_1 = M_2 = 4$; the total number of reshuffled datasets without restriction is $\frac{8!}{4!4!} = 70$, while the number of reshuffled datasets with the balance restriction is $\frac{4!}{2!2!}\frac{4!}{2!2!} = 36$.

For the kth reshuffled dataset, for $k = 1, \ldots, T$, the F-test statistics for the N genes are ranked in descending order so that

$$F_{(1)}^k > F_{(2)}^k > \cdots > F_{(N)}^k, \tag{17.27}$$

where $F_{(j)}^k$ is the jth largest F-test statistic of the kth reshuffled dataset. Let $\bar{F}_{(i)} = \frac{1}{T}\sum_{k=1}^T F_{(j)}^k$ be the average of the ith largest F-test statistic across the reshuffled datasets so that

$$\bar{F}_{(1)} > \bar{F}_{(2)} > \cdots > \bar{F}_{(N)}. \tag{17.28}$$

The empirical critical value drawn from this permutation test is $\bar{F}_{(C)}$, where C is chosen such that $\frac{C}{N} = \gamma$ and γ is a preset experiment-wise type I error rate. Let

$$F_{(1)} > F_{(2)} > \cdots > F_{(N)} \tag{17.29}$$

be the list of ranked F-test statistics calculated from the original dataset. All genes with a ranked $F_{(i)} > \bar{F}_{(C)}$ are selected as significant genes.

Table 17.3 The rank of 20
hypothetical genes

Gene	Ranking(i)	$F_{(i)}$	$\bar{F}_{(i)}$
6	1	43.6478	5.9448
4	2	17.3289	5.1476
14	3	9.8718	4.7502
9	4	6.7659	4.4627
12	5	5.9085	4.2743
1	6	5.4049	4.1099
13	7	5.1551	3.9715
15	8	4.8471	3.8872
11	9	4.4245	3.7793
7	10	4.0889	3.6800
10	11	4.0834	3.5990
3	12	4.0557	3.5371
8	13	3.9786	3.4796
17	14	3.9667	3.4209
18	15	3.9480	3.3653
5	16	3.9219	3.3272
16	17	3.9102	3.2711
19	18	3.9101	3.2242
20	19	3.8748	3.1811
2	20	3.8736	3.1357

Table 17.3 gives an example of 20 hypothetical genes and their test statistics (ranked). The table also provides the ranked average test statistics obtained from a permutation test.

Assume that we want to control the experimental type I error rate at $\gamma = 2/20 = 0.10$. Therefore, $C = 2$ and $\bar{F}_{(2)} = 5.1476$ is the critical value. Since $F_{(i)} > \bar{F}_{(2)}$ for $i = 1, \ldots, 7$, seven genes are selected as differentially expressed. The list of the significant genes is $\{6, 4, 14, 9, 12, 1, 13\}$. The permutation test also allows us to estimate the empirical FDR. Since seven genes are detected, among which two genes are expected to be false positive, the empirical FDR is $\delta = 2/7 = 0.2857$.

17.4.2 Selecting Genes by Controlling FDR

The permutation test given in the above example shows that when we set $\gamma = 0.10$, the empirical FDR is $\delta = 0.2857$. This suggests a way to select significant genes by controlling the FDR rather than the type I error rate. Let $\gamma_{(i)} = i/N$, for $i = 1, \ldots, N$, denote the type I error rate when $\bar{F}_{(i)}$ is used as the critical value. The list of significant genes includes all genes with $F_{(i')} > \bar{F}_{(i)}$, for $i' = 1, \ldots, S_{(i)}$, where $S_{(i)}$ is the largest i' such that $F_{(i')} > \bar{F}_{(i)}$. The number of significant genes under $\gamma_{(i)}$ is then $S_{(i)}$. Therefore, the empirical FDR under $\gamma_{(i)}$ is $\delta_{(i)} = i/S_{(i)}$. The $\gamma_{(i)}$ and $\delta_{(i)}$ values for the 20 hypothetical genes are listed in Table 17.4.

Table 17.4 Empirical FDR of the 20 hypothetical genes

Gene	Ranking (i)	$F_{(i)}$	$\bar{F}_{(i)}$	$\gamma_{(i)}$	$S_{(i)}$	$\delta_{(i)}$
6	1	43.6478	5.9448	0.05	4	0.2500
4	2	17.3289	5.1476	0.10	7	0.2857
14	3	9.8718	4.7502	0.15	8	0.3750
9	4	6.7659	4.4627	0.20	8	0.5000
12	5	5.9085	4.2743	0.25	9	0.5556
1	6	5.4049	4.1099	0.30	9	0.6667
13	7	5.1551	3.9715	0.35	13	0.5833
15	8	4.8471	3.8872	0.40	18	0.4444
11	9	4.4245	3.7793	0.45	20	0.4500
7	10	4.0889	3.6800	0.50	20	0.5000
10	11	4.0834	3.5990	0.55	20	0.5500
3	12	4.0557	3.5371	0.60	20	0.6000
8	13	3.9786	3.4796	0.65	20	0.6500
17	14	3.9667	3.4209	0.70	20	0.7000
18	15	3.9480	3.3653	0.75	20	0.7500
5	16	3.9219	3.3272	0.80	20	0.8000
16	17	3.9102	3.2711	0.85	20	0.8500
19	18	3.9101	3.2242	0.90	20	0.9000
20	19	3.8748	3.1811	0.95	20	0.9500
2	20	3.8736	3.1357	1.00	20	1.0000

If we want to set the FDR at $\delta = 0.375$, we will select eight significant genes. The type I error corresponding to this FDR is $\gamma = 0.15$ with a critical value of $\bar{F}_3 = 4.7502$. There are eight genes with test statistics larger than 4.7502. The list of the eight significant genes is $\{6, 4, 14, 9, 12, 1, 13, 15\}$.

The empirical method for selecting significant genes by controlling the FDR may not work for some datasets. The problem is that the relationship between $\gamma_{(i)}$ and $\delta_{(i)}$ may not be monotonic, leading to multiple values of γ corresponding to the same δ value. For example, both $\gamma_{(4)}$ and $\gamma_{(10)}$ correspond to $\delta = 0.50$. Therefore, the empirical method by controlling FDR is not recommended.

Although it is hard to select genes using the exact FDR control, Benjamini and Hochberg (1995) suggest to control FDR, not at δ but at $< \delta$. This approach does not require permutation test. It only requires calculation of the p-values corresponding to the test statistics. Genes are then ranked in ascending order based on their p-values. Significant genes under FDR $< \delta$ are selected in the following steps:

1. Rank genes based on the p-values in ascending order,

$$p_{(1)} \le p_{(2)} \le \cdots \le p_{(N)}$$

Let $g_{(i)}$ be the gene corresponding to p-value $p_{(i)}$.

2. Let $i_{\max}$ be the largest i for which

Table 17.5 False discovery rate of the 20 hypothetical genes ($\delta = 0.05$)

Gene	Ranking	$p_{(i)}$	$\frac{i}{N}$	$\frac{i}{N}\delta$
1	1	0.00036084	0.05	0.0025
7	2	0.00172817	0.10	0.0050
9	3	0.01443942	0.15	0.0075
15	4	0.02210477	0.20	0.0100
4	5	0.02354999	0.25	0.0125
3	6	0.03488643	0.30	0.0150
6	7	0.03488643	0.35	0.0175
5	8	0.03906270	0.40	0.0200
8	9	0.03927667	0.45	0.0225
10	10	0.04195647	0.50	0.0250
2	11	0.04934124	0.55	0.0275
13	12	0.10373490	0.60	0.0300
14	13	0.17910006	0.65	0.0325
20	14	0.28077512	0.70	0.0350
16	15	0.34277900	0.75	0.0375
19	16	0.35010501	0.80	0.0400
12	17	0.36536933	0.85	0.0425
11	18	0.40129684	0.90	0.0450
18	19	0.47260844	0.95	0.0475
17	20	0.48037560	1.00	0.0500

$$p_{(i)} \le \frac{i}{N}\delta$$

3. Declare significance for gene $g_{(i)}$ $\forall i = 1, \ldots, i_{\max}$. The nominal type I error rate is $\alpha = p_{(i_{\max})}$.

Table 17.5 shows an example with $N = 20$ genes under FDR $< \delta = 0.05$. It shows that $i_{\max} = 2$. Therefore, two genes ($g_{(1)} = 1$ and $g_{(2)} = 7$) are selected as significant with a nominal type I error of $\alpha = 0.00172817$.

Note that the FDR control is not a new statistical method for parameter estimation; rather, it is simply a different way provided by statisticians for biologists to decide the cutoff point for the "significant" genes.

17.4.3 Problems of the Previous Methods

The simple t-test or F-test described above is not optimal for differential expression analysis when the sample size is small. The current microarray technology is still not sufficiently effective to allow investigators to microarray a large number of tissue samples in a single microarray experiment. Therefore, microarray data are in general conducted with a very small sample size, although many genes can be measured from each tissue sample. When the sample size is small, the t-test or F-test statistics are not stable. Although both the numerator (the estimated difference between the control and the treatment) and the denominator (the estimated standard error of the difference) are subject to large estimation errors, the error in the denominator is more sensitive to the small sample size. One solution is to modify the estimation of the denominator to make it less sensitive to the small sample size. This can be done by sharing information between different genes when estimating the denominator of the t-test statistics. The genes are measured simultaneously within the same tissues. Therefore, the test statistics of the genes are correlated. This information has not been incorporated into the estimation of the standard error of the expression difference between the control and the treatment in the simple t- or F-test statistic.

17.4.4   Regularized T-test

Baldi and Long (2001) developed a Bayesian method to test differentially expressed genes. Instead of using the observed standard error of the control-treatment difference as the denominator for the t-test, they used a Bayesian method to estimate this standard error. The Bayesian estimate for the error variance is a weighted average of the observed variance and a prior variance set by the investigator. This modified t-test is called the regularized t-test. Similar idea has been proposed by Efron et al. (2001) and Tusher et al. (2001), who added a constant to the observed standard error as a new denominator to modify the calculated t-test statistic. The modified t-test in Efron et al. (2001) and Tusher et al. (2001) is

$$t_i = \frac{|\overline{y}_{i1.} - \overline{y}_{i2.}|}{s_{\overline{y}_{i1.} - \overline{y}_{i2.}} + s_0},\tag{17.30}$$

where s_0 is a constant added to the denominators of the t-tests for all genes. This constant is analogous to the prior standard deviation of Baldi and Long (2001). The constant s_0 is often chosen in such a way that it depends on the entire data of the microarray experiment. Because the value of s_0 depends on the entire data, information sharing occurs between genes. Efron et al. (2001) ranked the observed standard deviation across genes and select the 95 percentile of this empirical distribution as the value of s_0.

The regularized t-test of Baldi and Long (2001) has been implemented in a software package called Cyber-T (www.genomics.uci.edu/software.html). The significance analysis of microarrays of Tusher et al. (2001) has been implemented in a software called SAM (http://www-stat-class.standford.edu/SAM/SAMSevervlet).

17.5   General Linear Model

The t-test or the regularized t-test method only applies to two levels of a treatment, i.e., the case-control study. When the treatment has more levels, only two levels are considered at a time. In this section, we introduce a general linear model that can handle differential expression analysis with an arbitrary number of treatment levels. In addition, all genes are analyzed simultaneously under a single general linear model so that information of data is shared among genes. This will automatically generate a more accurate estimation of the error variance for each gene. The method was developed by Smyth (2004) who also provided a program named Limma (Linear Models for Microarray Data) along with the method.

Under the general linear model framework, all equations are written in matrix forms. It is more convenient to denote the microarray dataset by a matrix with rows and columns flipped (from the original dataset) so that each row represents a sample and each column represents a gene. This transposed dataset is now an $M \times N$ matrix, where M is still the number of samples and N still represents the number of genes. Note that the data matrix is assumed to have been properly transformed and normalized prior to the data analysis. Let $y_j = \{y_{1j}, \ldots, y_{Mj}\}^T$ denote the jth column of the data matrix, i.e., it stores the expressions of the jth gene for all the M samples. We now use the following linear model to describe y_j:

$$y_j = \beta + Z\gamma_j + \epsilon_j, \forall j = 1, \ldots, N \tag{17.31}$$

where β is an $M \times 1$ vector for the mean expressions across all genes, γ_j is a $P \times 1$ vector ($P < M$) of latent variables for P different groups of tissue samples, i.e., the effects for P groups, and ϵ_j is an $M \times 1$ vector for the residual errors. Finally, Z is an $M \times P$ design matrix. For example, if there are $M = 6$ tissue samples and $P = 3$ types of tissues and assume that the first two samples are from the first type of tissue, the second two samples from the second type of tissue, and the last two samples from the third type of tissue, then a class variable is denoted by a vector $G = \{1, 1, 2, 2, 3, 3\}$ and the design matrix $Z = \text{design}(G)$ has the form of

$$Z = \begin{bmatrix} 1 & 0 & 0 \\ 1 & 0 & 0 \\ 0 & 1 & 0 \\ 0 & 1 & 0 \\ 0 & 0 & 1 \\ 0 & 0 & 1 \end{bmatrix} \tag{17.32}$$

Detailed view of the linear model is

$$\begin{bmatrix} y_{1j} \\ y_{2j} \\ y_{3j} \\ y_{4j} \\ y_{5j} \\ y_{6j} \end{bmatrix} = \begin{bmatrix} \beta_1 \\ \beta_2 \\ \beta_3 \\ \beta_4 \\ \beta_5 \\ \beta_6 \end{bmatrix} + \begin{bmatrix} 1 & 0 & 0 \\ 1 & 0 & 0 \\ 0 & 1 & 0 \\ 0 & 1 & 0 \\ 0 & 0 & 1 \\ 0 & 0 & 1 \end{bmatrix} \begin{bmatrix} \gamma_{1j} \\ \gamma_{2j} \\ \gamma_{3j} \end{bmatrix} + \begin{bmatrix} \epsilon_{1j} \\ \epsilon_{2j} \\ \epsilon_{3j} \\ \epsilon_{4j} \\ \epsilon_{5j} \\ \epsilon_{6j} \end{bmatrix} \tag{17.33}$$

Note that β_i for $i = 1,\ldots, M$ is the mean of expression levels for all genes measured in the ith tissue sample. Assume that ϵ_j has a multivariate $N(0, I\sigma_j^2)$ distribution. We now present two methods for the general linear model (GLM) analysis.

17.5.1   Fixed Model Approach

The fixed model approach of the GLM is simply an alternative way to perform the F-test (described early), except that the GLM provides more flexible way for the F-test. Under the fixed model framework, information sharing only occurs with the mean expressions (vector β). In fact, vector β may be removed before the analysis in the normalization step. In that case, the model is simply represented by $y_j = Z\gamma_j + \epsilon_j$, where vector y_j has been normalized. The GLM analysis is so general that we can estimate β simultaneously along with γ_j and other parameters. Here we emphasize simultaneous analysis, and thus, all genes are included in the same model. The log likelihood function for gene j is

$$L_j(\beta, \gamma_j, \sigma_j^2) = -\frac{1}{2}\ln(\sigma_j^2) - \frac{1}{2\sigma_j^2}(y_j - \beta - Z\gamma_j)^T(y_j - \beta - Z\gamma_j) \quad (17.34)$$

Assume that all genes are independent (this assumption may often be violated); the overall log likelihood function is

$$L(\beta, \gamma, \psi) = -\frac{1}{2}\sum_{j=1}^{N}\ln(\sigma_j^2) - \sum_{j=1}^{N}\frac{1}{2\sigma_j^2}(y_j - \beta - Z\gamma_j)^T(y_j - \beta - Z\gamma_j)$$

$$(17.35)$$

where $\gamma = \{\gamma_j\}_{j=1}^{N}$ and $\psi = \left\{\sigma_j^2\right\}_{j=1}^{N}$. The maximum likelihood estimates of β and γ can be obtained explicitly. However, using the following iterative approach simplifies the estimation:

$$\beta = \frac{1}{N}\sum_{j=1}^{N}(y_j - Z\gamma_j)$$

$$\gamma_j = (Z^T Z)^{-1}Z^T(y_j - \beta), \forall j = 1,\ldots, N \quad (17.36)$$

The iteration process converges quickly to produce the MLE of β and γ, denoted by $\hat{\beta}$ and $\hat{\gamma}$. The MLE of σ_j^2 is

$$\hat{\sigma}_j^2 = \frac{1}{M}(y_j - \hat{\beta} - Z\hat{\gamma}_j)^T(y_j - \hat{\beta} - Z\hat{\gamma}_j), \forall j = 1,\ldots, N \quad (17.37)$$

The variance of the MLE of γ_j can be approximated using

$$\text{var}(\hat{\gamma}_j) = (Z^T Z)^{-1} \hat{\sigma}_j^2, \forall j = 1, \ldots, N \tag{17.38}$$

The next step is to perform statistical tests for differentially expressed genes. For three treatment groups, there are two orthogonal linear contrasts, which can be expressed as

$$\alpha_j = L^T \gamma_j = \begin{bmatrix} 1 & -\frac{1}{2} & -\frac{1}{2} \\ 0 & 1 & -1 \end{bmatrix} \begin{bmatrix} \gamma_{1j} \\ \gamma_{2j} \\ \gamma_{3j} \end{bmatrix} = \begin{bmatrix} \gamma_{1j} - \frac{1}{2}\gamma_{2j} - \frac{1}{2}\gamma_{3j} \\ \gamma_{2j} - \gamma_{3j} \end{bmatrix} \tag{17.39}$$

where

$$L = \begin{bmatrix} 1 & 0 \\ -\frac{1}{2} & 1 \\ -\frac{1}{2} & -1 \end{bmatrix} \tag{17.40}$$

The null hypothesis for $H_0 : \alpha_j = 0$ can be tested using

$$F_j = \frac{1}{r} \hat{\alpha}_j^T \text{var}^{-1}(\hat{\alpha}_j) \hat{\alpha}_j \tag{17.41}$$

where $r = r(L) = 2$ is the rank of matrix L and

$$\text{var}(\hat{\alpha}_j) = \text{var}(L^T \hat{\gamma}_j) = L^T \text{var}(\hat{\gamma}_j) L = L^T (Z^T Z)^{-1} L \hat{\sigma}_j^2 \tag{17.42}$$

Therefore,

$$F_j = \frac{1}{r} \hat{\alpha}_j^T \text{var}^{-1}(\hat{\alpha}_j) \hat{\alpha}_j = \frac{1}{r\hat{\sigma}_j^2} \hat{\gamma}_j^T L (L^T (Z^T Z)^{-1} L)^{-1} L^T \hat{\gamma}_j \tag{17.43}$$

There is not much advantage of this GLM-generated F-test over the F-test described earlier, except that one can choose different linear contrast matrix L to perform different biologically meaningful tests.

As mentioned earlier, information sharing is not obvious for the fixed model approach of F-test. The problem for the sensitivity of $\hat{\sigma}_j^2$ to small sample size remains unsolved. A slight modification can resolve this problem. Let us now make the assumption of

$$\epsilon_j \sim N(0, I\sigma^2), \forall j = 1, \ldots, N \tag{17.44}$$

which states that all genes share a common residual error variance. The MLE estimate of σ^2 becomes

$$\hat{\sigma}^2 = \frac{1}{MN} \sum_{j=1}^{N} (y_j - \hat{\beta} - Z\hat{\gamma}_j)^T (y_j - \hat{\beta} - Z\hat{\gamma}_j) \tag{17.45}$$

After this slight modification, the F-test statistic is

$$F_j = \frac{1}{r\hat{\sigma}^2} \hat{\gamma}_j^T L (L^T (Z^T Z)^{-1} L)^{-1} L^T \hat{\gamma}_j \tag{17.46}$$

This is a much more robust test statistic. However, it is questionable for the validity of the assumption of common residual error variance across all genes.

17.5.2   Random Model Approach

We now modify the model to allow all other non-interesting effects to be included in the model, which is

$$y_j = X\beta + Z\gamma_j + \epsilon_j \tag{17.47}$$

where X is an $M \times q$ design matrix and β is a $q \times 1$ vector. Under the fixed model approach, we simply assume that $X = I$ (an identity matrix). Now this assumption has been relaxed, although the fixed model approach can handle $X \neq I$ equally well. We now still assume $\epsilon_j \sim N(0, I\sigma_j^2), \forall j = 1, \dots, N$. In addition, we make another assumption, $\gamma_j \sim N(0, \Pi), \forall j = 1, \dots, N$, where Π is a $p \times p$ positive definite matrix, an unknown matrix subject to estimation. With this assumption, the model becomes a random model or mixed model considering β being fixed effects. The variance–covariance matrix Π is shared by all genes, which is what we call the information sharing among genes. Under the mixed model framework, we estimate parameters $\theta = \{\beta, \Pi, \psi\}$ using the ML method, while $\gamma = \{\gamma_j\}$ are predicted rather than estimated because they are no longer called parameters. We may use a two-step approach to predict γ. The first step is to estimate the parameters, and the second step is to predict γ given the estimated parameters and the data. Vector y_j now follows a multivariate normal distribution

$$y_j \sim N\left(X\beta, Z\Pi Z^T + I\sigma_j^2\right) \tag{17.48}$$

The log likelihood function for gene j is

$$L_j(\theta) = -\frac{1}{2} \ln |Z\Pi Z^T + I\sigma_j^2|$$
$$- \frac{1}{2}(y_j - X\beta)^T \left(Z\Pi Z^T + I\sigma_j^2\right)^{-1} (y_j - X\beta) \tag{17.49}$$

which leads to an overall log likelihood function of

$$L(\theta) = \sum_{j=1}^{N} L_j(\theta) \tag{17.50}$$

The two-step approach can be combined into a single step but with an iterative mechanism for the solution. The EM algorithm is such an algorithm with explicit expression of the solution in each iterative cycle. In the E-step, we predict γ_j conditional on the parameters and the data,

$$\hat{\gamma}_j = E(\gamma_j) = \left(\Pi^{-1}\sigma_j^2 + Z^T Z\right)^{-1} Z^T (y_j - X\beta) \tag{17.51}$$

and

$$\Sigma_j = \text{var}(\hat{\gamma}_j) = \left(\Pi^{-1}\sigma_j^2 + Z^T Z\right)^{-1} \sigma_j^2 \tag{17.52}$$

These allow us to calculate

$$E(\gamma_j \gamma_j^T) = \text{var}(\gamma_j) + E(\gamma_j)E(\gamma_j^T) = \Sigma_j + \hat{\gamma}_j \hat{\gamma}_j^T \tag{17.53}$$

which is required in the M-step. In the M-step, we calculate the parameter values using the quantities obtained in the E-step,

$$\beta = \frac{1}{N}(X^T X)^{-1} X^T \sum_{j=1}^{N} (y_j - Z\hat{\gamma}_j)$$

$$\Pi = \frac{1}{N} \sum_{j=1}^{N} E(\gamma_j \gamma_j^T) = \frac{1}{N} \sum_{j=1}^{N} (\Sigma_j + \hat{\gamma}_j \hat{\gamma}_j^T)$$

$$\sigma_j^2 = \frac{1}{M} E[(y_j - X\beta - Z\gamma_j)^T (y_j - X\beta - Z\gamma_j)] \tag{17.54}$$

where the residual variance can be further expressed as

$$\sigma_j^2 = \frac{1}{M} y_j^T (y_j - X\beta - Z\hat{\gamma}_j) \tag{17.55}$$

The E-step and M-step are alternated until a certain criterion of convergence is reached. Once the EM iteration converges, the predicted γ_j and the variance of the prediction are obtained as shown below:

$$\hat{\gamma}_j = \left(\hat{\Pi}^{-1}\hat{\sigma}_j^2 + Z^T Z\right)^{-1} Z^T (y_j - X\hat{\beta}) \tag{17.56}$$

and

$$\Sigma_j = \left(\hat{\Pi}^{-1}\hat{\sigma}_j^2 + Z^T Z\right)^{-1} \hat{\sigma}_j^2 \tag{17.57}$$

The F-test for $H_0 : L^T \gamma = 0$ is given by

$$
\begin{aligned}
F_j &= \frac{1}{r} \hat{\gamma}_j^T L (L^T \Sigma_j L)^{-1} L^T \hat{\gamma}_j \\
&= \frac{1}{r \hat{\sigma}_j^2} \hat{\gamma}_j^T L \left[L^T \left(\hat{\Pi}^{-1} \hat{\sigma}_j^2 + Z^T Z \right)^{-1} L \right]^{-1} L^T \hat{\gamma}_j
\end{aligned}
\qquad (17.58)
$$

We now compare the F-tests under the random model and that under the fixed model,

$$
F_j(\text{random model}) = \frac{1}{r \hat{\sigma}_j^2} \hat{\gamma}_j^T L \left[L^T \left(\hat{\Pi}^{-1} \hat{\sigma}_j^2 + Z^T Z \right)^{-1} L \right]^{-1} L^T \hat{\gamma}_j
$$

$$
F_j(\text{fixed model}) = \frac{1}{r \hat{\sigma}_j^2} \hat{\gamma}_j^T L (L^T (Z^T Z)^{-1} L)^{-1} L^T \hat{\gamma}_j
\qquad (17.59)
$$

The difference is obvious; an extra term $\hat{\Pi}^{-1} \hat{\sigma}_j^2$ occurs in the random model F-test. This illustrates the information sharing across genes. Further manipulation on the variance–covariance matrix of the predicted γ_j, we get

$$
\Sigma_j = \left(\hat{\Pi}^{-1} \hat{\sigma}_j^2 + Z^T Z \right)^{-1} \hat{\sigma}_j^2 = \left(\hat{\Pi}^{-1} + \frac{1}{\hat{\sigma}_j^2} Z^T Z \right)^{-1}
\qquad (17.60)
$$

The first term $\hat{\Pi}^{-1}$ is shared for all genes, and the second term $Z^T Z / \hat{\sigma}_j^2$ is gene specific. This idea is the same as the regularized t-test except that the shrinkage factor $\hat{\Pi}^{-1}$ is a matrix and it is estimated from the data rather than chosen by the investigator a priori.

Chapter 18
Hierarchical Clustering of Microarray Data

Sometimes a microarray experiment is aimed at identifying different groups of genes. Genes within a group show similar expression patterns across samples, while different groups of genes show different expression patterns. There are sufficient evidences showing that genes with similar expression patterns across samples tend to share similar functions (Eisen et al. 1998). Therefore, clustering genes into different groups may provide a clue to uncover gene functions. Biologists are more interested in identifying gene functions by clustering genes into different groups based on differential expression patterns across samples. Medical professionals, on the other hand, may be more interested in clustering specimens (tissue samples) into different clusters based on differential expression patterns across genes for the purpose of disease diagnosis. The assumption is that different tissues (e.g., cancer vs normal) tend to have different expression patterns across genes. In this chapter, we will learn two important methods of cluster analysis, UPGMA and NJ. Both methods belong to a general hierarchical clustering system because the final result of each method is represented by a treelike structure or phylogeny in terms of evolutionary studies.

18.1 Distance Matrix

A number of distance measures are available for comparing proximities of expression vectors in the gene space. The most popular measurement of distance is the Euclidean distance. The Euclidean distance between gene i and j is defined as

$$d_{ij} = \left[\frac{1}{M} \sum_{k=1}^{M} (y_{ki} - y_{kj})^2 \right]^{\frac{1}{2}} \tag{18.1}$$

S. Xu, *Principles of Statistical Genomics*, DOI 10.1007/978-0-387-70807-2_18,
© Springer Science+Business Media, LLC 2013

These distances are stored in a symmetric matrix with the diagonals equal to zero

$$D = \begin{bmatrix} 0 & d_{12} & \cdots & d_{1N} \\ d_{21} & 0 & \cdots & d_{2N} \\ \vdots & \vdots & \ddots & \vdots \\ d_{N1} & d_{N2} & \cdots & 0 \end{bmatrix} \tag{18.2}$$

This property of a distance matrix is represented by $d_{ij} = 0$ for $i = j$ and $d_{ij} = d_{ji}$ for $i, j = 1, \ldots, N$. This distance matrix is the input data for the cluster analysis methods described in this chapter.

18.2   UPGMA

UPGMA is an abbreviation of a method called unweighted pair group method using arithmetic mean (a very ugly name). It is also called the average distance method. UPGMA is the simplest method for cluster analysis. The method was originally developed by Sokal and Michener (1958) for constructing a dendrogram (or phenogram) as shown in Fig. 18.1. However, it is used most often by evolutionary biologists to construct phylogenetic trees. Eisen et al. (1998) first applied this method to cluster genes. We now use a simple example of five genes to demonstrate the UPGMA algorithm. The example was obtained from Swofford et al. (1996) for phylogenetic reconstruction. In the terminology of phylogeny, each gene is called an OTU (operational taxonomic unit). In this example, the five "genes" are actually five bacterium species or five OTUs. They are *Bacillus subtilis* (Bsu), *Bacillus stearothermophilus* (Bst), *Lactobacillus viridescens* (Lvi), *Acholeplasma modicum* (Amo), and *Micrococcus luteus* (Mlu). These five species are represented by OTUs 1, 2, 3, 4, and 5, respectively. Since we are dealing with genes rather than OTUs, we will no longer use the term OTU hereafter. The original distance matrix of the five genes and distance matrices of reduced dimension are given in Table 18.1.

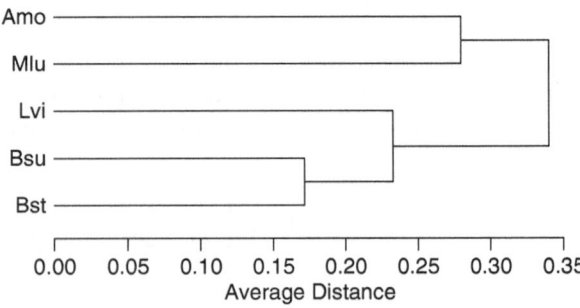

Fig. 18.1 UPGMA tree for five bacterium strains (genes). The data were obtained from Swofford et al. (1996)

Table 18.1 The original distance matrix of five genes and subsequent distance matrices of reduced dimension in the UPGMA clustering process

Gene	1	2	3	4	5
1	–	0.1715	0.2147	0.3091	0.2326
2	d_{12}	–	0.2991	0.3399	0.2058
3	d_{13}	d_{23}	–	0.2795	0.3943
4	d_{14}	d_{24}	d_{34}	–	0.4289
5	d_{15}	d_{25}	d_{35}	d_{45}	–

Gene	(1,2)	3	4	5
(1,2)	–	0.2569	0.3245	0.2192
3	$d_{(1,2)3}$	–	0.2795	0.3943
4	$d_{(1,2)4}$	d_{34}	–	0.4289
5	$d_{(1,2)5}$	d_{35}	d_{45}	–

Gene	((1,2),5)	3	4
((1,2),5)	–	0.3027	0.3593
3	$d_{((1,2),5)3}$	–	0.2795
4	$d_{((1,2),5)4}$	d_{34}	–

Gene	((1,2),5)	(3,4)
((1,2),5)	–	0.3310
(3,4)	$d_{((1,2),5)(3,4)}$	–

The UPGMA algorithm starts with the full matrix for the five genes (5×5 distance matrix). The actual values of the distance matrix are listed in the upper-right triangle and the symbols of the distances are given in the lower-left triangle (see the first distance matrix in Table 18.1). First, we find two genes with the smallest distance, which are genes 1 and 2. We then merge the two genes as a single "gene," denoted by gene (1,2). We now calculate the average distance of "gene" (1,2) with the remaining genes using the following equations:

$$d_{(1,2)3} = \frac{1}{2}(d_{13} + d_{23}) = \frac{1}{2}(0.2147 + 0.2991) = 0.2569$$

$$d_{(1,2)4} = \frac{1}{2}(d_{14} + d_{24}) = \frac{1}{2}(0.3091 + 0.3399) = 0.3245$$

$$d_{(1,2)5} = \frac{1}{2}(d_{15} + d_{25}) = \frac{1}{2}(0.2326 + 0.2058) = 0.2192 \qquad (18.3)$$

We now have reduced the number of genes by one. The new distance matrix with four genes is given in the second block in Table 18.1. We treat "gene" (1,2) as a single gene and start merging the next two closest genes. The smallest distance occurs between "gene" (1,2) and gene 5. Therefore, we combine (1,2) and 5 as a single "gene" denoted by gene $((1,2),5)$. The new distances of this cluster with the remaining genes are

$$d_{((1,2),5)3} = \frac{1}{3}(d_{13} + d_{23} + d_{35}) = \frac{1}{3}(0.2147 + 0.2991 + 0.3943) = 0.3027$$

$$d_{((1,2),5)4} = \frac{1}{3}(d_{14} + d_{24} + d_{45}) = \frac{1}{3}(0.3091 + 0.3399 + 0.4289) = 0.3593$$

$$(18.4)$$

We now construct a new reduced distance matrix with three genes, which is given in the third block of Table 18.1. The smallest distance occurs between genes 3 and 4, which are merged into a new gene denoted by "gene" $(3,4)$. The average distance between clusters $((1,2),5)$ and $(3,4)$ is

$$d_{((1,2),5)(3,4)} = \frac{1}{3 \times 2}(d_{13} + d_{23} + d_{35} + d_{14} + d_{24} + d_{45})$$

$$= \frac{1}{6}(0.2147 + 0.2991 + 0.3943 + 0.3091 + 0.3399 + 0.4289)$$

$$= 0.3310 \tag{18.5}$$

At this moment, we have two clusters, $((1,2),5)$ and $(3,4)$. The last step is to merge the two clusters together to complete the clustering analysis. The final cluster contains all the five genes represented by $(((1,2),5),(3,4))$. The dendrogram for this clustering result is given in Fig. 18.1.

The general formula for calculating the distance between cluster X and cluster Y is

$$d_{XY} = \frac{1}{n_X n_Y} \sum_{i=1}^{n_X} \sum_{j=1}^{n_Y} d_{ij} \tag{18.6}$$

where n_X and n_Y are the numbers of genes in clusters X and Y, respectively, and i and j index the genes contained in clusters X and Y. A cluster can be a single gene or multiple genes that are joined in a previous step of the UPGMA analysis. UPGMA is only one of many different methods available for constructing a phylogenetic tree. Many software packages are available to perform cluster analysis and draw phylogenetic trees, e.g., the cluster and tree procedures in SAS (SAS Institute 2008b).

18.3   Neighbor Joining

The neighbor-joining (NJ) method was developed by Saitou and Nei (1987) for reconstructing phylogenies. It is so far the most popular method for constructing evolutionary trees. It can be used to cluster genes, although we have not seen a single

Fig. 18.2 Neighbor-joining tree for five bacterium strains (genes). Data were obtained from Swofford et al. (1996)

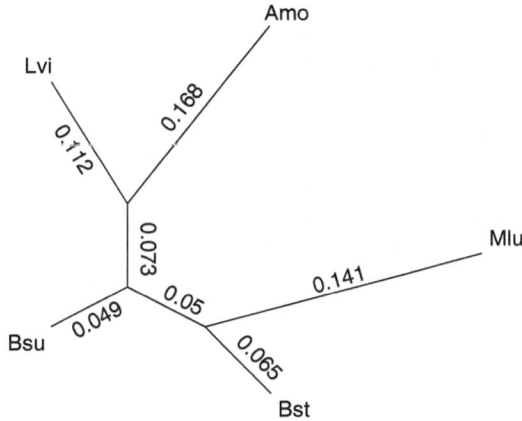

publication that uses NJ to cluster genes based on expression levels. The method is more comprehensive than the UPGMA method. The advantage of NJ over UPGMA is that it does not assume constant evolutionary rate across lineages (the molecular clock assumption). The final result of the clustering remains a bifurcating tree (unrooted tree), but different branches are allowed to have different lengths. The method takes the same input distance matrix as that used in the UPGMA method, but it uses a different criterion to combine genes. We will first introduce the idea of neighbor joining and then describe the NJ algorithm in detail. For the same distance matrix of the five genes used in the section of UPGMA, the neighbor-joining tree is illustrated in Fig. 18.2. For a set of genes, neighbors are defined as a pair of genes that join together before they connect with other genes. For example, in the NJ tree (Fig. 18.2), genes Bst and Mlu are neighbors because they join each other first before joining the remaining three genes. The NJ algorithm will find all neighbors in sequence to form a final bifurcating tree.

18.3.1 Principle of Neighbor Joining

The algorithm starts with a star phylogeny defined in Fig. 18.3a, where the internal node of the star phylogeny is denoted by Y. There is only one internal node for a star phylogeny. The length of this star phylogeny is defined as the sum of the lengths of all branches. For N genes, the length of the star phylogeny is

$$S_N = \sum_{k=1}^{N} L_{kY} = \frac{1}{N-1} \sum_{i<j}^{N} d_{ij} \qquad (18.7)$$

Fig. 18.3 Star phylogeny of five genes and decomposition of the star phylogeny. (**a**) The star phylogeny of five genes; (**b**) a decomposed phylogeny with 3 and 4 being chosen as a pair of neighbors and forming a node X. The remaining three genes, 1, 2, and 5, and node X form a new star phylogeny with four genes. Node X is a synthetic "new" gene consisting of 3 and 4

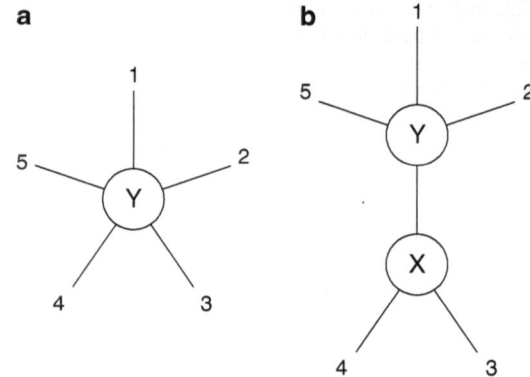

where L_{kY} is the length of the branch connecting gene k to node Y. The above relationship is due to the fact that the distance between gene i and gene j is

$$d_{ij} = L_{iY} + L_{jY} \tag{18.8}$$

which leads to

$$\sum_{i<j}^{N} d_{ij} = \sum_{i<j}^{N} (L_{iY} + L_{jY}) = (N-1) \sum_{k=1}^{N} L_{kY} = (N-1)S_N \tag{18.9}$$

When adding up all the distances, each branch is counted $N-1$ times, which explains why the total length of the star phylogeny is the sum of all the distances divided by $N-1$. From the star phylogeny, the NJ algorithm tries to find a pair of neighbors to decompose the star phylogeny. For a star phylogeny with N genes, there are $N(N-1)/2$ possible pairs of neighbors for consideration. Let us assume that genes 3 and 4 are candidate pair of neighbors. The decomposed phylogeny is demonstrated in Fig. 18.3b, where there is one more internal node, denoted by X. Node X is a new "gene," which replaces genes 3 and 4. The new gene (node X) and the remaining genes form a new star phylogeny with the number of genes reduced by one. Let the length of the new star phylogeny (X and the remaining genes) be denoted by S_{N-1}. The length of the decomposed phylogeny (Fig. 18.3b) is

$$S_{3,4} = S_{N-1} + d_{34} \tag{18.10}$$

The notation $S_{i,j}$ represents the length of the phylogeny that is decomposed from the star phylogeny of N genes with i and j chosen as the pair of neighbors. The question is now turned into how to calculate S_{N-1}. Let us further partition Fig. 18.3b into two starlike phylogenies as shown in Fig. 18.4.

Fig. 18.4 Two starlike phylogenies obtained from the decomposed phylogeny given in Fig. 18.3b. (**a**) Star phylogeny with four genes {1,2,3,5}, (**b**) star phylogeny with four genes {1,2,4,5}

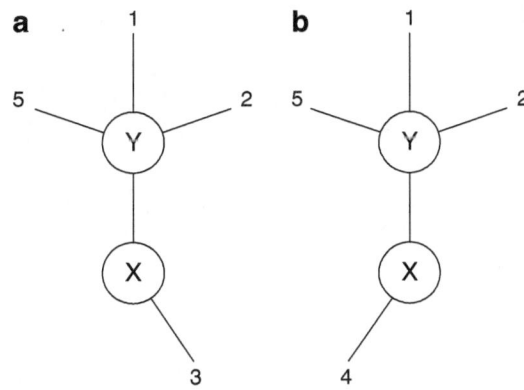

From Fig. 18.4a, we obtain the length of this star phylogeny by

$$S_{N-1} + L_{3X} = \frac{1}{N-2} \sum_{i<j, i\neq 4, j\neq 4}^{N} d_{ij} = \frac{1}{N-2} \left(\sum_{i<j}^{N} d_{ij} - \sum_{k=1}^{N} d_{4k} \right) \qquad (18.11)$$

Similarly, we obtain the length of the star phylogeny shown in Fig. 18.4b by

$$S_{N-1} + L_{4X} = \frac{1}{N-2} \sum_{i<j, i\neq 3, j\neq 3}^{N} d_{ij} = \frac{1}{N-2} \left(\sum_{i<j}^{N} d_{ij} - \sum_{k=1}^{N} d_{3k} \right) \qquad (18.12)$$

Adding the two equations together, we obtain

$$2S_{N-1} + d_{34} = \frac{1}{N-2} \left(2\sum_{i<j}^{N} d_{ij} - \sum_{k=1}^{N} d_{4k} - \sum_{k=1}^{N} d_{3k} \right) \qquad (18.13)$$

Note that $L_{3X} + L_{4X} = d_{34}$. The right-hand side of the above equation involves distances only and thus is computable from the data. Solving for S_{N-1} yields

$$S_{N-1} = \frac{1}{2(N-2)} \left(2\sum_{i<j}^{N} d_{ij} - \sum_{k=1}^{N} d_{4k} - \sum_{k=1}^{N} d_{3k} \right) - \frac{1}{2}d_{34} \qquad (18.14)$$

Substituting (18.14) into (18.10), we get

$$S_{3,4} = \frac{1}{2(N-2)} \left(2\sum_{i<j}^{N} d_{ij} - \sum_{k=1}^{N} d_{3k} - \sum_{k=1}^{N} d_{4k} \right) + \frac{1}{2}d_{34} \qquad (18.15)$$

Rearrangement of the above equation yields

$$S_{3,4} = \frac{1}{2}d_{34} - \frac{1}{2(N-2)}\left(\sum_{k=1}^{N} d_{3k} + \sum_{k=1}^{N} d_{4k}\right) + \frac{1}{N-2}\sum_{i<j}^{N} d_{i,j} \qquad (18.16)$$

In general, the total length of a phylogeny with N genes and one internal branch that separates genes p and q from the remaining genes (see Fig. 18.3b) can be expressed as

$$S_{p,q} = \frac{1}{2}d_{pq} - \frac{1}{2(N-2)}\left(\sum_{k=1}^{N} d_{pk} + \sum_{k=1}^{N} d_{qk}\right) + \frac{1}{N-2}\sum_{i<j}^{N} d_{ij} \qquad (18.17)$$

Understanding how to calculate the length of a phylogeny decomposed from a starlike phylogeny (one internal branch and two internal nodes) is the key to perform the NJ method for tree reconstruction. For N genes, there are a total of $N(N-1)/2$ possible ways to choose a potential pair of neighbors, and all of them have to be evaluated. The pair of neighbors that have the minimum length of the tree topology, i.e., the minimum $S_{p,q}$ value, is considered as neighbors and joined together at this stage. Once genes p and q are joined, they are represented by an internal node X. This node is considered as a new "gene," which, together with the remaining genes, forms a new star phylogeny with a total number of $N-1$ genes ($N-2$ individual genes plus a combined "gene" X). We then calculate the pairwise distances between all the $N-1$ genes to form a new distance matrix. The distances between node X (formed from joining p and q) and a remaining gene, say gene k, for $k \neq p, q$, are

$$d_{kX} = \frac{1}{2}(d_{pk} + d_{qk} - d_{pq}) \qquad (18.18)$$

The derivation of this equation is given below, assuming that $p = 3$ and $q = 4$. From Fig. 18.3b, we can see that

$$d_{3k} = L_{3X} + d_{kX}$$
$$d_{4k} = L_{4X} + d_{kX} \qquad (18.19)$$

Adding these two equations, we get

$$d_{3k} + d_{4k} = d_{34} + 2d_{kX} \qquad (18.20)$$

Rearrangement of this equation leads to

$$d_{kX} = \frac{1}{2}(d_{3k} + d_{4k} - d_{34}) \qquad (18.21)$$

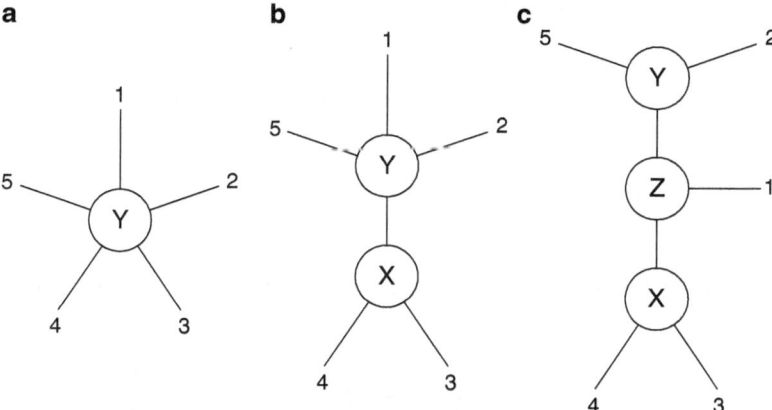

Fig. 18.5 The recurrent process of neighbor joining. (**a**) The star phylogeny, (**b**) the decomposed phylogeny with 3 and 4 being chosen as a pair of neighbors and forming a node X, (**c**) the final resolved phylogeny with 1 and node X being combined into node Z, which is connected with 2 and 5 by node Y. Note that Y always is treated as the center of a star phylogeny. In this case, (**a**) is a star phylogeny with five genes (1, 2, 3, 4, 5), (**b**) is a start phylogeny with four genes (1, 2, 5 , X), (**c**) is a star phylogeny with three genes (2, 5, Z)

The distances between pairs of genes that exclude p and q remain the same as the distances in the original distance matrix. This new distance matrix (with dimension $N - 1$) is used as the input data for further decomposition of the star phylogeny. The process continues until $N - 1 = 3$ because such a star phylogeny cannot be further decomposed. The entire process of the NJ algorithm for the sample data of five bacterium strains (see the section of UPGMA) is illustrated in Fig. 18.5, where Fig. 18.5a shows the original star phylogeny with $N = 5$ genes, Fig. 18.5b shows a decomposed phylogeny with one internal branch and two internal nodes, and Fig. 18.5c shows the final resolved phylogeny with two internal branches and three internal nodes.

The original NJ algorithm developed by Saitou and Nei (1987) evaluates the branch lengths sequentially during the tree building process. For example, when genes 3 and 4 are joined, the branch lengths L_{3X} and L_{4X} are estimated from the phylogeny given in Fig. 18.3b by the following process. Adding (18.11) to (18.12) and subtracting (18.12) from (18.11), we generate the following two simultaneous equations:

$$L_{3X} + L_{4X} = d_{34}$$

$$L_{3X} - L_{4X} = \frac{1}{N-2} \left(\sum_{k=1}^{N} d_{3k} - \sum_{k=1}^{N} d_{4k} \right) \qquad (18.22)$$

Solving the two equations for L_{3X} and L_{4X}, we obtain

$$L_{3X} = \frac{1}{2}d_{34} + \frac{1}{2(N-2)}\left(\sum_{k=1}^{N}d_{3k} - \sum_{k=1}^{N}d_{4k}\right)$$

$$L_{4X} = \frac{1}{2}d_{34} + \frac{1}{2(N-2)}\left(\sum_{k=1}^{N}d_{4k} - \sum_{k=1}^{N}d_{3k}\right) \qquad (18.23)$$

The general formula for estimating the lengths of the two branches coming from node X is

$$L_{pX} = \frac{1}{2}d_{pq} + \frac{1}{2(N-2)}\left(\sum_{k=1}^{N}d_{pk} - \sum_{k=1}^{N}d_{qk}\right)$$

$$L_{qX} = \frac{1}{2}d_{pq} + \frac{1}{2(N-2)}\left(\sum_{k=1}^{N}d_{qk} - \sum_{k=1}^{N}d_{pk}\right) \qquad (18.24)$$

18.3.2 Computational Algorithm

The original NJ algorithm of Saitou and Nei (1987) calculates $S_{p,q}, \forall p < q$ for all possible pairs of N genes. The pair of genes that have the minimum $S_{p,q}$ (see (18.17)) is chosen as the neighbors and is thus subject to merging. Note that the last term of (18.17) is invariant with respect to p and q. It is the sum of all the distances (upper triangle of the distance matrix with N genes) divided by $N - 2$. Computationally, calculating this term for all possible pairs of genes is simply a waste of time because it is a constant regardless of which pair of genes is evaluated. Therefore, Studier and Keppler (1988) modified the clustering criterion by deleting this common term. Swofford et al. (1996) further revised Studier and Keppler's (1988) modification to define the clustering criterion by

$$d_{pq}^{*} = d_{pq} - \frac{1}{N-2}\left(\sum_{k=1}^{N}d_{pk} + \sum_{k=1}^{N}d_{qk}\right) \qquad (18.25)$$

The relationship between d_{pq}^{*} and $S_{p,q}$ is

$$d_{pq}^{*} = 2S_{p,q} - \frac{1}{N-2}\sum_{i<j}^{N}d_{ij} \qquad (18.26)$$

The reverse relationship is

$$S_{p,q} = \frac{1}{2}d_{pq}^* + \frac{1}{N-2}\sum_{i<j}^{N} d_{ij} \tag{18.27}$$

This simple linear relationship means that if d_{pq}^* is minimum across all possible pairs of genes, then $S_{p,q}$ should also be minimum. Therefore, the clustering result will be identical regardless whether d_{pq}^* or $S_{p,q}$ is used. Let $r_p = \sum_{k=1}^{N} d_{pk}$ and $r_q = \sum_{k=1}^{N} d_{qk}$, which are called the net divergences of genes p and q from other genes, respectively (Swofford et al. 1996). Equation (18.25) can be expressed by

$$d_{pq}^* = d_{pq} - \frac{r_p + r_q}{N-2} \tag{18.28}$$

Swofford et al. (1996) called d_{ij}^* the rate-corrected distance. Upon replacement of the original distance matrix by the rate-corrected distance matrix, the modified NJ algorithm is identical to the UPGMA algorithm. The modified NJ algorithm is now summarized as follows:

1. For N genes, calculate the net divergence of gene j using

$$r_j = \sum_{k=1}^{N} d_{jk}, \forall j = 1, \ldots, N \tag{18.29}$$

2. Calculate the rate-corrected distance between p and q using

$$d_{pq}^* = d_{pq} - \frac{r_p + r_q}{N-2}, \forall p \neq q \tag{18.30}$$

3. Find the minimum d_{pq}^* for all $p < q$, and join p and q to form an internal node X, which replaces p and q.
4. Calculate the distances between X and each remaining gene using

$$d_{kX} = \frac{1}{2}(d_{pk} + d_{qk} - d_{pq}), \forall k \neq p, q \tag{18.31}$$

5. Remove the distance from p to node X and the distance from q to node X, and construct a distance matrix with $N - 1$ genes (genes p and q are joined as a combined "gene").
6. Go back to Step 1 recurrently unless the star phylogeny is completely resolved.

The recurrent process continues until a fully resolved bifurcating tree is generated. Table 18.2 shows the detailed process of the NJ algorithm for the five strain bacterium example. When $N = 5$, the smallest rate-corrected distance occurs between genes 3 and 4. Therefore, genes 3 and 4 form a new "gene" denoted by node X. In the next step with 4 genes, the smallest rate-corrected distance occurs between gene 1 and node X. However, it also occurs between genes 2 and 5. In such

Table 18.2 The distance and rate-corrected distance matrices of five genes and subsequent distance matrices of reduced dimension in the NJ clustering process

Gene	1	2	3	4	5	R
1	–	0.1715	0.2147	0.3091	0.2326	0.9279
2	−0.4766	–	0.2991	0.3399	0.2058	1.0163
3	−0.4905	−0.4356	–	0.2795	0.3943	1.1876
4	−0.4527	−0.4514	−0.5689	–	0.4289	1.3574
5	−0.4972	−0.5535	−0.4221	−0.4441	–	1.2616

Gene	1	2	5	X(3,4)	R
1	–	0.1751	0.2326	0.1222	0.5263
2	−0.3701	–	0.2058	0.1797	0.5571
5	−0.3856	−0.4278	–	0.2719	0.7103
X(3,4)	−0.4278	−0.3856	−0.3701	–	0.5739

Gene	2	5	Z(1,(3,4))	R
2	–	0.2.58	0.1146	0.3204
5	−0.5116	–	0.1912	0.3970
Z(1,(3,4))	−0.5116	−0.5116	–	0.3058

Gene	5	Y(2,(1,(3,4)))
5	–	0.1412
Y(2,(1,(3,4)))	–	–

The upper-right triangle gives the distance matrix, and the lower-left triangle gives the rate-corrected distance matrix

a tie situation, one of them is arbitrarily chosen. We decided to join gene 1 and node X to form a new node Z. The next step involves 3 genes, and thus, we can stop here. However, continuing to reduce the dimension allows us to calculate the lengths of the three branches emerging from node Y.

At each step, two branches from a new node are generated. The lengths of the two branches are estimated immediately after they emerge. In the first step, branches L_{3X} and L_{4X} emerged from node X, and the estimated lengths of the two branches are

$$L_{3X} = \frac{1}{2}d_{34} + \frac{r_3 - r_4}{2(N-2)} = \frac{1}{2} \times 0.2795 + \frac{1.1876 - 1.3574}{2 \times (5-2)} = 0.1114$$

$$L_{4X} = \frac{1}{2}d_{34} + \frac{r_4 - r_3}{2(N-2)} = \frac{1}{2} \times 0.2795 + \frac{1.3574 - 1.1876}{2 \times (5-2)} = 0.1681$$

In the second step, branches L_{1Z} and L_{XZ} emerged, whose lengths are estimated from

$$L_{1Z} = \frac{1}{2}d_{1X} + \frac{r_1 - r_X}{2(4-2)} = \frac{1}{2} \times 0.1222 + \frac{0.5263 - 0.5739}{2 \times (4-2)} = 0.0492$$

$$L_{XZ} = \frac{1}{2}d_{1X} + \frac{r_X - r_1}{2(4-2)} = \frac{1}{2} \times 0.1222 + \frac{0.5739 - 0.5263}{2 \times (4-2)} = 0.0730$$

The next recurrent step provides estimated lengths of branches L_{2Y} and L_{ZY}, which are

$$L_{2Y} = \frac{1}{2}d_{2Z} + \frac{r_2 - r_Z}{2(3-2)} = \frac{1}{2} \times 0.1146 + \frac{0.3204 - 0.3058}{2 \times (3-2)} = 0.0646$$

$$L_{ZY} = \frac{1}{2}d_{2Z} + \frac{r_Z - r_2}{2(3-2)} = \frac{1}{2} \times 0.1146 + \frac{0.3058 - 0.3204}{2 \times (3-2)} = 0.0500$$

The last step gives the estimated length of L_{5Y}, which is

$$L_{5Y} = d_{5Y} = 0.1412$$

The lengths of all branches obtained above are illustrated in Fig. 18.2.

Saitou and Nei (1987) proved that the estimated branch lengths given in each step are the least-squares estimates of the decomposed star phylogeny at that step. For example, L_{3X} and L_{4X} are estimated from the first step. They are the least-squares estimates of the decomposed star phylogeny given in Fig. 18.3b. We can see that these estimated branches (obtained from each step) are not the least-squares estimates of the branches from the final resolved phylogeny (Fig. 18.2). The differences, however, are very small. Now let us reestimate all the branches from the final phylogeny using the least-squares method. We can define each distance as the sum of several branches plus a random error, as shown below.

$$
\begin{bmatrix} d_{12} \\ d_{13} \\ d_{14} \\ d_{15} \\ d_{23} \\ d_{24} \\ d_{25} \\ d_{34} \\ d_{35} \\ d_{45} \end{bmatrix}
=
\begin{bmatrix}
1\ 0\ 1\ 1\ 0\ 0\ 0 \\
0\ 0\ 0\ 1\ 1\ 1\ 0 \\
0\ 0\ 0\ 1\ 1\ 0\ 1 \\
0\ 1\ 1\ 1\ 0\ 0\ 0 \\
1\ 0\ 1\ 0\ 1\ 1\ 0 \\
1\ 0\ 1\ 0\ 1\ 0\ 1 \\
1\ 1\ 0\ 0\ 0\ 0\ 0 \\
0\ 0\ 0\ 0\ 0\ 1\ 1 \\
0\ 1\ 1\ 0\ 1\ 1\ 0 \\
0\ 1\ 1\ 0\ 1\ 0\ 1
\end{bmatrix}
\begin{bmatrix} L_{2Y} \\ L_{5Y} \\ L_{YZ} \\ L_{1Z} \\ L_{XZ} \\ L_{3X} \\ L_{4X} \end{bmatrix}
+
\begin{bmatrix} e_{12} \\ e_{13} \\ e_{14} \\ e_{15} \\ e_{23} \\ e_{24} \\ e_{25} \\ e_{34} \\ e_{35} \\ e_{45} \end{bmatrix}
\qquad (18.32)
$$

In a compact matrix notation for the above equations, we have

$$d = AL + e \tag{18.33}$$

where matrix A contains values of zeros and ones. The least-squares estimate of vector L is

$$L = (A^T A)^{-1} A^T d \tag{18.34}$$

The estimated branch lengths for the bacterium example are

$$\hat{L} = \begin{bmatrix} L_{2Y} \\ L_{5Y} \\ L_{YZ} \\ L_{1Z} \\ L_{XZ} \\ L_{3X} \\ L_{4X} \end{bmatrix} = \begin{bmatrix} 0.0620 \\ 0.1438 \\ 0.0500 \\ 0.0492 \\ 0.0730 \\ 0.1115 \\ 0.1681 \end{bmatrix} \tag{18.35}$$

The total length of the final phylogeny is

$$\hat{L}_{(5,(2,(1,(3,4))))} = L_{2Y} + L_{5Y} + L_{YZ} + L_{1Z} + L_{XZ} + L_{3X} + L_{4X} = 0.6574 \tag{18.36}$$

The lengths of branches obtained from the original NJ algorithm are

$$\tilde{L} = \begin{bmatrix} L_{2Y} \\ L_{5Y} \\ L_{YZ} \\ L_{1Z} \\ L_{XZ} \\ L_{3X} \\ L_{4X} \end{bmatrix} = \begin{bmatrix} 0.0646 \\ 0.1412 \\ 0.0500 \\ 0.0492 \\ 0.0730 \\ 0.1114 \\ 0.1681 \end{bmatrix} \tag{18.37}$$

The total length of the final phylogeny calculated from the above estimated branch lengths is

$$\tilde{L}_{(5,(2,(1,(3,4))))} = L_{2Y} + L_{5Y} + L_{YZ} + L_{1Z} + L_{XZ} + L_{3X} + L_{4X} = 0.6575 \tag{18.38}$$

We can see that $\hat{L}_{(5,(2,(1,(3,4))))}$ is remarkably close to $\tilde{L}_{(5,(2,(1,(3,4))))}$. Therefore, we can use either way to calculate the lengths of the branches in the final resolved phylogeny. If the number of genes is large, calculating the branch lengths with the recurrent process may be time consuming. Therefore, it is recommended to calculate the branch lengths after the tree is fully resolved using the least-squares method given in (18.34).

18.4 Other Methods

There are many other methods available for phylogeny reconstruction. The UPGMA and NJ methods described previously belong to a class of methods called distance matrix-based method because the input data is a distance matrix. These methods are computationally efficient and thus can handle a relatively large number of genes. Another class of methods are called sequence data (or raw data)-based methods. These methods include the parsimony method (Sober 1983), the maximum likelihood method (Felsenstein 1981a,b), and the Bayesian method (Huelsenbeck et al. 2001). Because the sequence data-based methods use the raw data without conversion to distance data, these methods may capture more information from the data. Therefore, the sequence data-based methods have a better chance to recover the true phylogeny. The tradeoff is that these methods are computationally more demanding than the distance matrix-based methods, especially the maximum likelihood and Bayesian methods, which are very time consuming. Therefore, sequence data-based methods can only handle a small number of genes. In microarray data analysis, we usually have to deal with 10,000 or 20,000 genes simultaneously. Therefore, the sequence data-based methods are not the first choice for microarray data analysis.

The parsimony method (Sober 1983) uses particular criteria to define the optimality of a phylogeny. For example, the length of a tree can be used as a criterion of optimality. The length of a tree equals the sum of the lengths of all branches. The length of a branch, however, is defined as the number of "mutations" (changes in nucleotides of a DNA sequence or changes in amino acids of a protein sequence) occurred in that lineage during evolution. A parsimony tree is the one that has the minimum length. The maximum likelihood method adopts an evolutionary model for mutation and calculates the likelihood function for each possible tree. The one with the maximum likelihood value is the maximum likelihood tree. So, the criterion in the ML method is the likelihood function, rather than the length of a tree. Investigators can choose different evolutionary models for different data to construct the likelihood function for a tree. While evaluating the likelihood function of a tree, we also estimate parameters involved in the model. Therefore, the ML method is a kind of model-based method. The Bayesian method is also a model-based method and thus requires a model and parameters to define the model. In addition, the parameters are assigned a prior distribution (another model to describe the distribution of the parameters). The tree topology is considered as one of the parameters, and thus, a prior distribution of a topology can be assigned. Combining the data (sequence or other raw data) and the evolutionary model, we can build the likelihood function. Combining the likelihood function with the prior distribution of the parameters, we obtain the joint distribution of the data and the parameters.

From the joint distribution, the posterior distribution of the parameters given the data can be derived, from which realized parameter values are sampled, including the tree topology. The topology that has the maximum posterior probability is the Bayesian tree.

The parsimony and ML methods require an algorithm to find the optimal tree. The safest algorithm is the exhaustive search where all possible tree topologies have to be evaluated to find the optimal topology. The branch and bound algorithm used in map construction described in Chap. 2 is better than the exhaustive search method because most often a small percentage of the trees are evaluated and the algorithm guarantees to find the "best" tree. The Bayesian method does not require evaluation of all possible trees; instead, it samples a topology at each iteration of the Bayesian sampling process. If the model is correct, the true topology will be sampled more often than other tree topologies. The topologies that are close in appearance to the true topology will also have high chances to be sampled. Because both the ML and the Bayesian methods require estimating (or sampling) parameters, they are computationally more demanding than the parsimony method. The Bayesian method is the most time-consuming method because it requires repeated sampling of parameters (including the tree topology). To make sure that the posterior sample represents the true posterior distribution, the posterior sample must be sufficiently large.

For the microarray gene clustering analysis, ML and Bayesian methods may not be the choice at all. The parsimony method may be possibly considered if the number of genes does not exceed 100. Therefore, UPGMA and NJ may be the only methods for consideration in the hierarchical cluster analysis of expressed genes. Unfortunately, even though UPGMA and NJ are computationally efficient, they still cannot handle more than a thousand genes simultaneously. The large number of genes may be prescreened before conducting the cluster analysis. For example, majority of the genes may not be differentially expressed at all, i.e., their expression levels are constant across the samples. These genes should be eliminated. Alternatively, investigators may have some prior knowledge of the functions of some genes, and clustering those genes only may allow the investigators to answer some specific biological questions.

18.5 Bootstrap Confidence

The final clustering result (the tree) does not have a confidence support. A common property of the hierarchical clustering methods is that they all generate a bifurcating tree. In a bifurcating tree, each internal branch separates the genes into two clusters. A confidence support can be put on each internal branch to indicate the confidence of the separation of the two clusters. A statistical resampling method developed by Efron (1979), called the bootstrap method, can be applied here to construct the confidence support for each internal branch (Felsenstein 1985).

First, we need to generate bootstrap samples using a random number generator. Recall that the data matrix is an $M \times N$ matrix, where M is the number of tissue samples and N is the number of genes. If we denote the gene expression data for the ith tissue by a row vector

$$y_i = [y_{i1}, y_{i2}, \ldots, y_{iN}] \tag{18.39}$$

The entire dataset can be expressed by an $M \times N$ matrix

$$y = \begin{bmatrix} y_1 \\ y_2 \\ \vdots \\ y_M \end{bmatrix} \tag{18.40}$$

A bootstrap sample contains a randomly selected M rows of matrix y with replacement to form a new $M \times N$ matrix, denoted by

$$y^{(k)} = \begin{bmatrix} y_{(1)} \\ y_{(2)} \\ \vdots \\ y_{(M)} \end{bmatrix} \tag{18.41}$$

where $y_{(i)}$ is the ith row of the bootstrap data matrix and it is selected randomly from the original data matrix with replacement. Because of the replacement sampling, some rows of the original data matrix may appear in the bootstrap sample many times, and some may not be present at all. We use a superscript $^{(k)}$ to indicate that $y^{(k)}$ is the kth bootstrap sample because we need to repeat the sampling process many times (say $S = 100$ times). For each bootstrap sample (dataset), we do the clustering analysis and draw a tree. After all the S bootstrap samples have been analyzed, we count the number of times that a particular internal branch has been preserved in the bootstrap samples. For example, assume that the internal branch separating genes $\{3, 4\}$ from genes $\{1, 2, 5\}$ shown in the final NJ tree (Fig. 18.5c) is the current internal branch for evaluation. From the $S = 100$ bootstrap trees, we count how many trees that separate genes $\{3,4\}$ from genes $\{1,2,5\}$ (branch connecting Y and Z is reserved). This number divided by $S = 100$ is the confidence of this internal branch, regardless how the genes within each of the bifurcating groups are organized. Each internal branch should be evaluated for the confidence. The confidence measurement is a percentage and thus has a value between 0 and 1. A tree with high confidences for most internal branches is more reliable than a tree with low confidences for most of the internal branches.

Chapter 19
Model-Based Clustering of Microarray Data

19.1 Cluster Analysis with the K-means Method

The K-means method (MacQueen 1967; Hartigan 1975; Hartigan and Wong 1979) is not a model-based clustering method. It is a data-partitioning method that divides the entire data into K disjoint groups. The idea is similar to the multivariate Gaussian mixture model analysis (a model-based method). Both deal with the problem of data partitioning and thus are in clear contrast to the hierarchical clustering methods described in Chap. 18. The K-means method is simple but very useful in microarray data analysis. As a result, it earns a section here in this chapter, even though it is not a model-based clustering method.

In the contest of microarray data analysis, the aim of the K-means algorithm is to divide all N genes in M dimensions into K clusters so that the within-cluster sum of squares is minimum. It is not practical to require that the solution has minimal within-cluster sum of squares against all partitions, except when N and M are small and $K = 2$. We seek, instead, a "local" optimum, a solution such that no movement of a gene from one cluster to another will reduce the within-cluster sum of squares.

The algorithm requires a data matrix y with N genes in M dimensions (i.e., y is an $M \times N$ matrix) and a matrix of K initial cluster centers (center is also called centroid) denoted by an $M \times K$ matrix μ. The number of genes in cluster k is denoted by N_k for $k = 1, \ldots, K$ and $\sum_{k=1}^{K} N_k = N$. Let $d(j, k)$ be the Euclidean distance between gene j and the centroid of cluster k. The general procedure is to search for a K-partitioning of local minimum within-cluster sum of squares by moving genes from one cluster to another. The within-cluster sum of squares is defined as

$$Q = \sum_{k=1}^{K} \sum_{j=1}^{N_k} d^2(j, k) \tag{19.1}$$

S. Xu, *Principles of Statistical Genomics*, DOI 10.1007/978-0-387-70807-2_19,
© Springer Science+Business Media, LLC 2013

where the Euclidean distance $d(j, k)$ between gene j and the centroid of cluster k is defined as

$$d(j, k) = \|y_j - \mu_k\| = \sqrt{(y_j - \mu_k)^T (y_j - \mu_k)} \qquad (19.2)$$

Note that y_j is an $M \times 1$ vector of the expression levels for gene j and μ_k is the centroid (an $M \times 1$ vector) of cluster k.

The K-means algorithm can be summarized as:

1. Choose an initial centroid matrix $\mu = \{\mu_1^{(0)}, \ldots, \mu_K^{(0)}\}$.
2. Calculate $d(j, k)$ for all $j = 1, \ldots, N$ and $k = 1, \ldots, K$.
3. Assign gene j into cluster k if $d(j, k) = \min\{d(j, 1), \ldots, d(j, K)\}$.
4. When all genes have been assigned, recalculate $\mu = \{\mu_1, \ldots, \mu_K\}$, where $\mu_k = \frac{1}{N_k} \sum_{j=1}^{N_k} y_j$ is the mean of all y_j that belong to cluster k.
5. Repeat Step 2 to Step 4 until the centroid matrix μ no longer changes. This produces a separation of the genes into K distinct clusters.

The remaining question in the K-means method is how to determine K and the initial centroid matrix μ. The number of clusters is set by the investigator a priori. Since the K-means algorithm is computationally fast, one can perform the method using different K values and select the one that produces the most "meaningful" result. Alternatively, the investigator may already have some idea about how many clusters the data should fall into, say K_0. Only a few different K values around K_0 may be evaluated. The initial value of the centroid matrix can be chosen arbitrarily. Since the method only finds a local optimum, different initial values of μ may generate different results. Therefore, multiple initial values of μ should be tried to make sure that the optimum obtained is close to the global optimum. One approach to choosing the initial value of μ is to randomly partition the data into K clusters and use the mean value of each cluster to form the centroid matrix μ. Another approach is to handpick K genes which appear to be very "different" and use the expression levels of these genes as the centroids. The K-means method that updates the centroids constantly based on the expression levels of the genes included in each cluster is called unsupervised K-means. The final centroid matrix μ is entirely determined by the data.

In some applications of the K-means method, the entire set of genes are divided into two categories. In one category, the genes are already assigned a priori into K clusters based on prior knowledge of the investigators. The other category contains genes whose cluster identities are not known. The purpose of the K-means analysis is to assign each of the genes in category two to one of the K clusters defined by genes in category one. This type of analysis is called supervised cluster analysis. Genes in category one are called the training sample, while genes in category two are called the testing sample. The training sample is only used to provide the initial centroid matrix $\mu^{(0)}$. Let S_k be the number of genes in cluster k of the training sample for $k = 1, \ldots, K$ and $\sum_{k=1}^{K} S_k = S$, where S is the total number of genes

Fig. 19.1 K-means clustering analysis for $N = 2,000$ genes with $K = 5$ clusters each having $N_k = 400$ genes

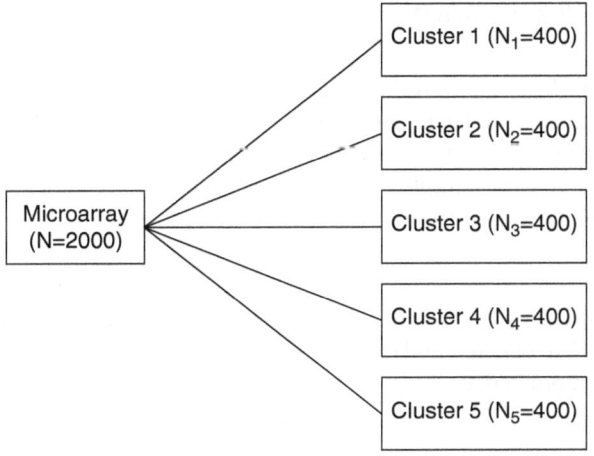

in the training sample. In the supervised K-means analysis, the updated centroid matrix is calculated using

$$\mu_k = \frac{1}{N_k + S_k}\left(S_k\mu^{(0)} + \sum_{j=1}^{N_k} y_j\right) \tag{19.3}$$

where j now indexes all genes in cluster k of the testing sample. The new μ_k will never replace $\mu^{(0)}$, but it is updated constantly using the above equation. The iteration process stops when further moving genes from one cluster to another does not change the centroid. Figure 19.1 illustrates schematically the result of a K-means clustering analysis for $N = 2,000$ genes with $K = 5$ clusters each having $N_k = 400$ genes.

In contrast to the hierarchical clustering analysis, which can only handle a few hundred genes at a time, the K-means method, along with the model-based methods to be introduced later, can handle almost unlimited number of genes.

19.2 Cluster Analysis Under Gaussian Mixture

In contrast to the hierarchical clustering methods, the model-based clustering analysis classifies genes into distinct clusters. Among clusters, genes show different patterns of expression, but within clusters, genes share the same patterns of expression. The method requires specific statistical models to fit the data, and the errors of fitness are supposed to follow some specific distributions. Theory and method of Gaussian mixture can be found in McLachlan and Peel (2000) and Fraley and Raftery (2002). The Gaussian mixture models have been widely applied to microarray data analysis (Ghosh and Chinnaiyan 2002; McLachlan et al. 2002; Pan et al. 2002; Ouyang et al. 2004; Qu and Xu 2004, 2006; McNicholas and Murphy 2010).

19.2.1 *Multivariate Gaussian Distribution*

Before we introduce the mixture model of microarray data analysis, we will review the basic definition of multivariate normal distribution, which is also called multivariate Gaussian distribution. Let us denote the expression of the jth gene across M samples by a column vector y_j. Assume that y_j follows a multivariate Gaussian distribution denoted by

$$y_j \sim N(\mu, \Sigma) \tag{19.4}$$

where μ is an $M \times 1$ vector of means (not the centroid matrix appearing in the K-means method) and Σ is an $M \times M$ symmetrical and positive definite matrix expressed as

$$\Sigma = \begin{bmatrix} \sigma_{11} & \sigma_{12} & \cdots & \sigma_{1M} \\ \sigma_{12} & \sigma_{22} & \cdots & \sigma_{2M} \\ \vdots & \vdots & \ddots & \vdots \\ \sigma_{1M} & \sigma_{2M} & \cdots & \sigma_{MM} \end{bmatrix} \tag{19.5}$$

The multivariate normal density is

$$\phi(y_j | \mu, \Sigma) = \frac{1}{(2\pi)^{M/2} |\Sigma|^{1/2}} \exp\left[-\frac{1}{2} (y_j - \mu)^T \Sigma^{-1} (y_j - \mu) \right] \tag{19.6}$$

To estimate the parameters, $\theta = \{\mu, \Sigma\}$, we construct the following log likelihood function:

$$L(\theta) = -\frac{N}{2} \log |\Sigma| - \frac{1}{2} \sum_{j=1}^{N} (y_j - \mu)^T \Sigma^{-1} (y_j - \mu) \tag{19.7}$$

The parameters can be estimated using the maximum likelihood method by setting

$$\frac{\partial}{\partial \mu} L(\theta) = \frac{\partial}{\partial \Sigma} L(\theta) = 0 \tag{19.8}$$

and solving for θ. This requires rules for matrix derivatives (beyond the scope of this class), but the final result for μ is

$$\frac{\partial}{\partial \mu} L(\theta) = -\frac{1}{2} \frac{\partial}{\partial \mu} \sum_{j=1}^{N} (y_j - \mu)^T \Sigma^{-1} (y_j - \mu)$$

$$= \Sigma^{-1} \sum_{j=1}^{N} (y_j - \mu)$$

$$= \Sigma^{-1} \sum_{j=1}^{N} y_j - N \Sigma^{-1} \mu \tag{19.9}$$

Setting $\frac{\partial}{\partial \mu} L(\theta) = 0$ and solving for μ lead to

$$\hat{\mu} = \frac{1}{N} \sum_{j=1}^{N} y_j \tag{19.10}$$

The partial derivative of the log likelihood function with respect to Σ is

$$\frac{\partial}{\partial \Sigma} L(\theta) = -\frac{N}{2} \frac{\partial}{\partial \Sigma} \log |\Sigma| - \frac{1}{2} \frac{\partial}{\partial \Sigma} \sum_{j=1}^{N} (y_j - \mu)^T \Sigma^{-1} (y_j - \mu)$$

$$= -\frac{N}{2} \frac{1}{|\Sigma|} \frac{\partial}{\partial \Sigma} |\Sigma| - \frac{1}{2} \frac{\partial}{\partial \Sigma} \mathrm{tr} \left[\Sigma^{-1} \sum_{j=1}^{N} (y_j - \mu)(y_j - \mu)^T \right]$$

$$= -\frac{N}{2} \Sigma^{-1} + \frac{1}{2} \Sigma^{-1} \sum_{j=1}^{N} (y_j - \mu)(y_i - \mu)^T \Sigma^{-1} \tag{19.11}$$

Setting $\frac{\partial}{\partial \Sigma} L(\theta) = 0$ and solving for Σ yield

$$\hat{\Sigma} = \frac{1}{N} \sum_{j=1}^{N} (y_j - \hat{\mu})(y_j - \hat{\mu})^T \tag{19.12}$$

The MLE of the parameters can be solved explicitly without resorting to any iterative algorithm.

19.2.2   Mixture Distribution

Assume that these genes are sampled from K multivariate normal distributions (clusters), but we do not know which gene comes from which cluster. The problem becomes a mixture model problem. Let $G_j = k$ be the cluster index for the jth gene, i.e., if the jth gene is from the kth cluster, then $G_j = k, \forall k = 1, \ldots, K$. Let us redefine $\mu = \{\mu_1, \ldots, \mu_K\}$ as an $M \times K$ matrix (similar to the centroid in the K-means analysis), where μ_k is an $M \times 1$ vector for the mean of cluster k. Let us define Σ_k as the covariance matrix of cluster k. Given that the jth gene is from the kth cluster, the density of y_j is multivariate normal with mean μ_k and variance matrix Σ_k, i.e., $y_j | G_j = k \sim N(\mu_k, \Sigma_k), \forall k = 1, \ldots, K$. The conditional density of y_j given $G_j = k$ is

$$\phi_k (y_j | \mu_k, \Sigma_k) = \frac{1}{(2\pi)^{M/2} |\Sigma_k|^{1/2}} \exp \left[-\frac{1}{2} (y_j - \mu_k)^T \Sigma_k^{-1} (y_j - \mu_k) \right] \tag{19.13}$$

Let $\pi_k > 0$ be the proportion of genes belonging to the kth cluster, where $\sum_{k=1}^{K} \pi_k = 1$. These proportions are also called the mixing proportions. They are treated as the prior probability that a randomly sampled gene is from the kth cluster. The density of the mixture distribution is

$$f(y_j|\theta) = \sum_{k=1}^{K} \pi_k \phi_k(y_j|\mu_k, \Sigma_k) \tag{19.14}$$

where

$$\theta = \{\pi_1, \ldots, \pi_K, \mu_1, \ldots, \mu_K, \Sigma_1, \ldots, \Sigma_K\} \tag{19.15}$$

is the vector of parameters. The overall log likelihood function is

$$L(\theta) = \sum_{j=1}^{N} \ln f(y_j|\theta) \tag{19.16}$$

The likelihood function for the mixture model is messy, and thus, no explicit solution for θ is available. Therefore, we resort to a numerical algorithm for estimating the MLE of θ.

19.2.3   The EM Algorithm

The EM algorithm for parameter estimation largely remains the same as that in the segregation analysis except that the dimension of parameters becomes multivariate. Instead of re-deriving the EM algorithm, we simply provide detailed steps of the EM algorithm. With the EM algorithm, we assign each gene into one of the K clusters with a certain probability denoted by

$$\pi_{jk} = \Pr(G_j = k|y_j, \theta) \tag{19.17}$$

This probability is called the posterior probability of $G_j = k$ given the data and the parameter values. The EM algorithm starts with some initial values of all unknown parameters and iteratively updates each parameter conditional on the parameter values in the previous round of iteration. The iteration process is summarized as follows:

1. Set $t = 0$ and initialize all parameters $\theta^{(t)}$, including

$$\pi_k^{(t)} = 1/K, \forall k = 1, \ldots, K \tag{19.18}$$

2. Update the posterior probabilities of cluster assignments,

$$\pi_{jk}^{(t)} = \frac{\pi_k^{(t)} \phi_k(y_j | \mu_k^{(t)}, \Sigma_k^{(t)})}{\sum_{k'=1}^{K} \pi_{k'}^{(t)} \phi_{k'}(y_j | \mu_{k'}^{(t)}, \Sigma_{k'}^{(t)})} \tag{19.19}$$

for all $j = 1, \ldots, N$ and $k = 1, \ldots, K$.

3. Update the mean vectors by

$$\mu_k^{(t+1)} = \left(N \pi_k^{(t)}\right)^{-1} \sum_{j=1}^{N} \pi_{jk}^{(t)} y_j, \forall k = 1, \ldots, K \tag{19.20}$$

4. Update the covariance matrices by

$$\Sigma_k^{(t+1)} = \left(N \pi_k^{(t)}\right)^{-1} \sum_{j=1}^{N} \pi_{jk}^{(t)} \left(y_j - \mu_k^{(t+1)}\right) \left(y_j - \mu_k^{(t+1)}\right)^{T} \tag{19.21}$$

for all $k = 1, \ldots, K$.

5. Update the mixing proportions,

$$\pi_k^{(t+1)} = \frac{1}{N} \sum_{j=1}^{N} \pi_{jk}^{(t)}, \forall k = 1, \ldots, K \tag{19.22}$$

6. Increment t by $t + 1$, and repeat Step 2 to Step 5 until a certain criterion of convergence is reached.

Once the EM iteration converges, the parameter values are the maximum likelihood estimates, denoted by $\hat{\theta}$. The conditional posterior probability of gene j from cluster k is denoted by $\hat{\pi}_{jk}$. Gene j will be assigned to the kth cluster if

$$\hat{\pi}_{jk} = \max\{\hat{\pi}_{j1}, \ldots, \hat{\pi}_{jK}\} \tag{19.23}$$

One can also set a cutoff point for cluster assignment. For example, assign gene j into cluster k if $\hat{\pi}_{jk} = \max\{\hat{\pi}_{j1}, \ldots, \hat{\pi}_{jK}\} > \alpha$, where $0 < \alpha < 1$ is the cutoff point. Those genes whose maximum posterior probability is less than α are claimed to be unassigned. This flexibility of the model-based method is an advantage over the K-means method.

19.2.4 Supervised Cluster Analysis

In the supervised cluster analysis, we know the functions of genes in the training sample and thus know which gene belongs to which cluster in the training sample. The purpose of the supervised cluster analysis is to assign genes in the testing sample (contains genes with unknown functions) into one of the clusters defined in the training sample. The method was developed by Qu and Xu (2004). Let S be the training sample size, which is partitioned into K clusters with the sample size for the kth cluster being S_k, for $\sum_{k=1}^{K} S_k = S$. From the training sample, we estimate $\tilde{\mu} = \{\tilde{\mu}_1, \ldots, \tilde{\mu}_K\}$ and $\widetilde{\Sigma} = \{\widetilde{\Sigma}_1, \ldots, \widetilde{\Sigma}_K\}$, where $\tilde{\mu}_k$ is an $M \times 1$ vector for the mean of cluster k and $\widetilde{\Sigma}_k$ is an $M \times M$ variance–covariance matrix for cluster k. The estimated $\tilde{\mu}$ and $\widetilde{\Sigma}$ obtained from the training sample are used to guide the cluster analysis for the testing sample. Equations (19.10) and (19.12) introduced in a previous section are used to estimate $\tilde{\mu}_k$ and $\widetilde{\Sigma}_k$, respectively, except that the sample size N in those equations is now replaced by S_k because the sample size for cluster k in the training sample is S_k. This concludes the first step of the supervised cluster analysis.

The second step of the supervised cluster analysis is to update the estimated parameters from genes in the testing sample and assign each of the N genes in the testing sample to one of the K clusters. We will again use the EM algorithm to update the parameters, which is described as follows:

1. Set $t = 0$ and initialize all parameters, including

$$\mu_k^{(t)} = \tilde{\mu}_k, \ \Sigma_k^{(t)} = \widetilde{\Sigma}_k \text{ and } \pi_k^{(t)} = \frac{S_k + N/K}{S + N}, \forall k = 1, \ldots, K$$

2. Update the posterior probabilities of cluster assignments,

$$\pi_{jk}^{(t)} = \frac{\pi_k^{(t)} \phi_k(y_j | \mu_k^{(t)}, \Sigma_k^{(t)})}{\sum_{k'=1}^{K} \pi_{k'}^{(t)} \phi_{k'}(y_j | \mu_{k'}^{(t)}, \Sigma_{k'}^{(t)})} \tag{19.24}$$

   for all $\forall j = 1, \ldots, N$ and $k = 1, \ldots, K$.
3. Update the mean vectors by

$$\mu_k^{(t+1)} = \left[S_k + \sum_{j=1}^{N} \pi_{jk}^{(t)} \right]^{-1} \left[S_k \tilde{\mu}_k + \sum_{j=1}^{N} \pi_{jk}^{(t)} y_j \right] \tag{19.25}$$

   for all $k = 1, \ldots, K$.
4. Update the variance–covariance matrices by

$$\Sigma_k^{(t+1)} = \left[S_k + \sum_{j=1}^{N} \pi_{jk}^{(t)} \right]^{-1}$$
$$\times \left[S_k \widetilde{\Sigma}_k + \sum_{j=1}^{N} \pi_{jk}^{(t)} (y_j - \mu_k^{(t+1)})(y_j - \mu_k^{(t+1)})^T \right] \tag{19.26}$$

   for all $k = 1, \ldots, K$.

5. Update the mixing proportions,

$$\pi_k^{(t+1)} = \frac{1}{S+N} \left[S_k + \sum_{j-1}^{N} \pi_{jk}^{(t)} \right], \forall k = 1, \ldots, K \qquad (19.27)$$

6. Increment t by $t + 1$ and repeat Steps 2–5 until a certain criterion of convergence is reached.

19.2.5   Semisupervised Cluster Analysis

It is possible that genes in the training sample may not cover all possible clusters in the testing sample. In other words, some genes in the testing sample may not belong to any of the clusters in the training sample. In this case, the supervised cluster analysis may be combined with the unsupervised cluster analysis, which is called the semisupervised cluster analysis. Assume that we want to classify all N genes into K clusters. Among the K clusters, $K' < K$ of them are defined in the training sample where $K - K'$ clusters occur only in the testing sample. The algorithm is identical to that of the supervised algorithm except that $S_k = 0$ for $k = K' + 1, \ldots, K$. In other words, we set each of the additional clusters that are not included in the training sample empty. No further modification is required. The values of $\tilde{\mu}_k$ and $\widetilde{\Sigma}_k$ for $k = K' + 1, \ldots, K$ are arbitrary, e.g., $\tilde{\mu}_k = 0$ and $\widetilde{\Sigma}_k = 0$, because they do not affect the EM estimates of the parameters.

19.3   Inferring the Number of Clusters

The number of clusters K is usually unknown. The EM algorithm described above is based on a fixed value of K. In the unsupervised and semisupervised methods, the K value needs to be estimated from the data. Note that in the supervised cluster analysis, K is determined exclusively by the training sample, and thus it is given. In this section, we will introduce a special method to infer K, called the Bayesian information criterion (BIC), which was proposed by Schwarz (1978), although there are many other methods available in the literature. For the unsupervised cluster analysis, the BIC for K clusters is given by

$$\Lambda(K) = -2L(\theta|K) + \dim(\theta|K)\ln(N) \qquad (19.28)$$

where

$$L(\theta|K) = \sum_{j=1}^{N} \ln \sum_{k=1}^{K} \pi_k \phi_k(y_j|\mu_k, \Sigma_k) \qquad (19.29)$$

is the log likelihood function evaluated at $\theta = \hat{\theta}$ and

$$\dim(\theta|K) = MK + \frac{1}{2}M(M+1)K + K - 1 \qquad (19.30)$$

is the dimension of parameter vector θ. Recall that

$$\theta = \{\mu_1, \ldots, \mu_K, \Sigma_1, \ldots, \Sigma_k, \pi_1 \ldots, \pi_K\}$$

where the dimension for the μ's is MK, the dimension for the Σ's is $\frac{1}{2}M(M+1)K$, and the dimension for the π's is $K-1$. The $K-1$ comes from the fact that only $K-1$ of the π's are independent parameters because $\pi_K = 1 - \sum_{k=1}^{K-1} \pi_k$. Using the BIC as the criterion of optimality, the K value that minimizes $\Lambda(K)$ is the optimal number of clusters.

The BIC for the semisupervised clustering for K clusters is also given by (19.28). The dimension of the parameters remains the same as that given in (19.30). However, the log likelihood function is different from (19.29). The correct log likelihood function for the semisupervised clustering is

$$L(\theta|K) = \sum_{j=1}^{N} \ln \sum_{k=1}^{K} \pi_k \phi_k(y_j|\mu_k, \Sigma_k)$$

$$+ \sum_{k=1}^{K} \sum_{j=1}^{S_k} \ln \phi_k(y_j^*|\mu_k, \Sigma_k) \qquad (19.31)$$

where y_j^* indicates data from the training sample. The additional information gained from the training sample appears in the second term of the right-hand side of (19.31).

19.4   Microarray Experiments with Replications

Recall that in the microarray experiment without replication,

$$y_j = \{y_{1j}, \ldots, y_{Mj}\}$$

is an $M \times 1$ vector for the expression level of gene j in all M conditions. Let us assume that each condition represents a subject (e.g., a human, an animal, or a plant). Assume that the ith subject of the experiment is replicated r_i times and the mean value of the r_i replications is entered into the dataset. In this case, the ith row and the jth column of matrix y are actually $\bar{y}_{ij}$, a mean of r_i replications. We now use a linear model to describe a single measurement y_{ij} by

$$y_{ij} = \gamma_{ij} + \epsilon_{ij} \qquad (19.32)$$

where γ_{ij} is the true expression of subject i for gene j and ϵ_{ij} is a measurement error with an assumed $N(0, \sigma^2)$ distribution. If $\bar{y}_{ij}$ is the mean of r_i replications, the model becomes

$$\bar{y}_{ij} = \gamma_{ij} + \bar{\epsilon}_{ij} \tag{19.33}$$

where $\bar{\epsilon}_{ij}$ is the mean error with an $N(0, \sigma^2/r_i)$ distribution. We now assume that vector $\gamma_j = \{\gamma_{1j}, \ldots, \gamma_{Mj}\}$ has a mixture of K multivariate Gaussian distributions with the kth component (cluster) being $\gamma_j \sim N(\mu_k, \Sigma_k)$. Now let us define $\bar{\epsilon}_j = \{\bar{\epsilon}_{1j}, \ldots, \bar{\epsilon}_{Mj}\}$ as a vector of the mean errors and $\bar{y}_j = \{\bar{y}_{1j}, \ldots, \bar{y}_{Mj}\}$ as a vector for the mean expression of gene j. We can write the following model to describe

$$\bar{y}_j = \gamma_j + \bar{\epsilon}_j \tag{19.34}$$

where $\bar{\epsilon}_j \sim N(0, D\sigma^2)$ and D is an $M \times M$ diagonal matrix,

$$D = \begin{bmatrix} \frac{1}{r_1} & 0 & \cdots & 0 \\ 0 & \frac{1}{r_2} & \cdots & 0 \\ \vdots & \vdots & \ddots & \vdots \\ 0 & 0 & \cdots & \frac{1}{r_M} \end{bmatrix} \tag{19.35}$$

This leads to a multivariate Gaussian mixture distribution for $\bar{y}_j$,

$$\bar{y}_j \sim \sum_{k=1}^{K} \pi_k N(\mu_k, \Sigma_k + D\sigma^2) \tag{19.36}$$

Comparing this distribution to the multivariate Gaussian distribution in the unreplicated experiment introduced before, we can see that there is an extra parameter σ^2 involved, which represents the variance of repeated measurement errors. The parameter vector now is defined as

$$\theta = \{\pi_1, \ldots, \pi_K, \mu_1, \ldots, \mu_K, \Sigma_1, \ldots, \Sigma_K, \sigma^2\}$$

The log likelihood function is

$$L(\theta) = \sum_{j=1}^{N} \ln \sum_{k=1}^{K} \pi_k \phi_k(\bar{y}_j | \mu_k, \Sigma_k + D\sigma^2) \tag{19.37}$$

The EM algorithm for the MLE of parameters is summarized below:

1. Set $t = 0$ and initialize all parameters $\theta^{(t)}$, including $\sigma^{2(t)}$ and

$$\pi_k^{(t)} = 1/K, \forall k = 1, \ldots, K$$

2. Update the posterior probabilities of cluster assignments,

$$\pi_{jk}^{(t)} = \frac{\pi_k^{(t)} \phi_k(\bar{y}_j | \mu_k^{(t)}, \Sigma_k^{(t)} + D\sigma^{2(t)})}{\sum_{k'=1}^{K} \pi_{k'}^{(t)} \phi_{k'}(\bar{y}_j | \mu_{k'}^{(t)}, \Sigma_{k'}^{(t)} + D\sigma^{2(t)})} \tag{19.38}$$

for all $j = 1, \ldots, N$ and $k = 1, \ldots, K$.

3. Update the mean vectors using

$$\mu_k^{(t+1)} = \left[\sum_{j=1}^{N} \pi_{jk}^{(t)} V_k^{-1} \right]^{-1} \left[\sum_{j=1}^{N} \pi_{jk}^{(t)} V_k^{-1} \bar{y}_j \right] \tag{19.39}$$

where $V_k = \Sigma_k + D\sigma^{2(t)}, \forall k = 1, \ldots, K$.

4. Update the covariance matrices by

$$\Sigma_k^{(t+1)} = \left(N\pi_k^{(t)} \right)^{-1} \sum_{j=1}^{N} \pi_{jk}^{(t)} \mathrm{E} \left[(\gamma_j - \mu_k^{(t+1)})(\gamma_j - \mu_k^{(t+1)})^T \right] \tag{19.40}$$

for all $k = 1, \ldots, K$.

5. Update σ^2 by

$$\sigma^{2(t+1)} = \frac{1}{MN} \sum_{j=1}^{N} \sum_{k=1}^{K} \pi_{jk}^{(t)} \mathrm{E} \left[(\bar{y}_j - \gamma_j)^T D^{-1} (\bar{y}_j - \gamma_j) \right] \tag{19.41}$$

6. Update the mixing proportions by

$$\pi_k^{(t+1)} = \frac{1}{N} \sum_{j=1}^{N} \pi_{jk}^{(t)}, \forall k = 1, \ldots, K \tag{19.42}$$

7. Increment t by $t + 1$ and repeat Steps 2–6 until a certain criterion of convergence is reached.

In the above EM iteration process, γ_j appears twice, all in the expected quadratic forms. The first appearance is

$$\mathrm{E}[(\gamma_j - \mu_k)(\gamma_j - \mu_k)^T] = \mathrm{E}(\gamma_j - \mu_k)\mathrm{E}(\gamma_j - \mu_k)^T + \mathrm{var}(\gamma_j - \mu_k) \tag{19.43}$$

and the second appearance is

$$\mathrm{E}\left[(\gamma_j - \bar{y}_j)^T D^{-1}(\gamma_j - \bar{y}_j)\right] = \mathrm{E}(\gamma_j - \bar{y}_j)^T D^{-1}\mathrm{E}(\gamma_j - \bar{y}_j) + \mathrm{tr}\left[D^{-1}\mathrm{var}(\gamma_j - \bar{y}_j)\right] \tag{19.44}$$

Although the cluster label k does not occur explicitly in (19.44), this equation is cluster specific. For the kth cluster, we have

$$E(\gamma_j - \bar{y}_j) = E(\gamma_j - \mu_k) + E(\mu_k - \bar{y}_j) = E(\gamma_j - \mu_k) + (\mu_k - \bar{y}_j) \quad (19.45)$$

and

$$\text{var}(\gamma_j - \bar{y}_j) = \text{var}(\gamma_j - \mu_k) + \text{var}(\mu_k - \bar{y}_j) = \text{var}(\gamma_j - \mu_k) \quad (19.46)$$

This result is due to $E(\mu_k - \bar{y}_j) = (\mu_k - \bar{y}_j)$, and $\text{var}(\mu_k - \bar{y}_j) = 0$ as $\mu_k - \bar{y}_j$ is not a function of variable γ_j. Now there are only two terms that need our attention, which are $E(\gamma_j - \mu_k)$ and $\text{var}(\gamma_j - \mu_k)$. They are the posterior mean and posterior variance of γ_j for cluster k given below:

$$E(\gamma_j - \mu_k) = \Sigma_k (\Sigma_k + D\sigma^2)^{-1}(\bar{y}_j - \mu_k) \quad (19.47)$$

and

$$\text{var}(\gamma_j - \mu_k) = \Sigma_k - \Sigma_k(\Sigma_k + D\sigma^2)^{-1}\Sigma_k \quad (19.48)$$

The optimal number of clusters can be found using the BIC criterion (see 19.29). The dimension of the parameters, however, is

$$\dim(\theta|K) = MK + \frac{1}{2}M(M+1)K + K \quad (19.49)$$

which is one shy that of the experiment without replication.

One caveat of the model-based cluster analysis is the "identifiability" problem, which occurs as two or more clusters having identical distributions (i.e., same cluster mean and the same cluster variance matrix). Several approaches can be used to solve this problem. One ad hoc approach (adopted by Qu and Xu (2006)) is to introduce a small noise vector to each μ_k at each iteration. This small perturbation will eventually separate all μ_k from each other. Another approach is to revise the model so that all clusters share the same variance matrix, i.e., $\Sigma_k = \Sigma, \forall k = 1, \ldots, K$. The most effective approach is the stochastic EM algorithm (SEM), in which the cluster label for each gene is randomly sampled from the posterior distribution rather than taking the posterior mean (Zhan et al. 2011). The SEM algorithm will be described in the next chapter (Chap. 20). Conceptually, this approach is the same as the ad hoc method, but statistically it is more rigorous and should generate the best result of clustering.

Chapter 20
Gene-Specific Analysis of Variances

In the differential expression analysis, the subjects are divided into two groups, case and control. A generalization of the differential expression analysis is the situation where the subjects are divided into multiple treatment groups. In addition, there may be multiple factors with possible interactions among different factors, e.g., the factorial design. This chapter deals with microarray data analysis under the factorial design.

20.1 General Linear Model

We now use an example to demonstrate the linear model for a factorial design with two factors and their interaction effects. Suppose that $M = 12$ subjects are microarrayed for N genes. The 12 subjects come from three age groups (A) and two genders (B). The original design is demonstrated in Table 20.1 (see next page).

S. Xu, *Principles of Statistical Genomics*, DOI 10.1007/978-0-387-70807-2_20,
© Springer Science+Business Media, LLC 2013

Table 20.1 Factorial design
with two factors and six
factor interactions

Subject	Age (A)	Gender (B)
1	Young (A1)	Male (B1)
2	Young (A1)	Male (B1)
3	Young (A1)	Female (B2)
4	Young (A1)	Female (B2)
5	Middle age (A2)	Male (B1)
6	Middle age (A2)	Male (B1)
7	Middle age (A2)	Female (B2)
8	Middle age (A2)	Female (B2)
9	Old age (A3)	Male (B1)
10	Old age (A3)	Male (B1)
11	Old age (A3)	Female (B2)
12	Old age (A3)	Female (B2)

The design matrices and the linear model are shown below for each of the N genes:

$$
\begin{bmatrix} y_1 \\ y_2 \\ y_3 \\ y_4 \\ y_5 \\ y_6 \\ y_7 \\ y_8 \\ y_9 \\ y_{10} \\ y_{11} \\ y_{12} \end{bmatrix}
=
\begin{bmatrix} 1 \\ 1 \\ 1 \\ 1 \\ 1 \\ 1 \\ 1 \\ 1 \\ 1 \\ 1 \\ 1 \\ 1 \end{bmatrix} \mu
+
\begin{bmatrix} 1 & 0 & 0 \\ 1 & 0 & 0 \\ 1 & 0 & 0 \\ 1 & 0 & 0 \\ 0 & 1 & 0 \\ 0 & 1 & 0 \\ 0 & 1 & 0 \\ 0 & 1 & 0 \\ 0 & 0 & 1 \\ 0 & 0 & 1 \\ 0 & 0 & 1 \\ 0 & 0 & 1 \end{bmatrix}
\begin{bmatrix} \alpha_1 \\ \alpha_2 \\ \alpha_3 \end{bmatrix}
+
\begin{bmatrix} 1 & 0 \\ 1 & 0 \\ 0 & 1 \\ 0 & 1 \\ 1 & 0 \\ 1 & 0 \\ 0 & 1 \\ 0 & 1 \\ 1 & 0 \\ 1 & 0 \\ 0 & 1 \\ 0 & 1 \end{bmatrix}
\begin{bmatrix} \beta_1 \\ \beta_2 \end{bmatrix}
$$

$$
+ \begin{bmatrix}
1 & 0 & 0 & 0 & 0 & 0 \\
1 & 0 & 0 & 0 & 0 & 0 \\
0 & 1 & 0 & 0 & 0 & 0 \\
0 & 1 & 0 & 0 & 0 & 0 \\
0 & 0 & 1 & 0 & 0 & 0 \\
0 & 0 & 1 & 0 & 0 & 0 \\
0 & 0 & 0 & 1 & 0 & 0 \\
0 & 0 & 0 & 1 & 0 & 0 \\
0 & 0 & 0 & 0 & 1 & 0 \\
0 & 0 & 0 & 0 & 1 & 0 \\
0 & 0 & 0 & 0 & 0 & 1 \\
0 & 0 & 0 & 0 & 0 & 1
\end{bmatrix}
\begin{bmatrix}
\gamma_{11} \\
\gamma_{12} \\
\gamma_{21} \\
\gamma_{22} \\
\gamma_{31} \\
\gamma_{32}
\end{bmatrix}
+ \begin{bmatrix}
\varepsilon_1 \\
\varepsilon_2 \\
\varepsilon_3 \\
\varepsilon_4 \\
\varepsilon_5 \\
\varepsilon_6 \\
\varepsilon_7 \\
\varepsilon_8 \\
\varepsilon_9 \\
\varepsilon_{10} \\
\varepsilon_{11} \\
\varepsilon_{12}
\end{bmatrix}
\tag{20.1}
$$

This linear model can be rewritten in a compact matrix notation as

$$
y_j = 1\mu_j + Z_\alpha \alpha_j + Z_\beta \beta_j + Z_\gamma \gamma_j + \varepsilon_j \tag{20.2}
$$

where the design matrices are not of full rank because the sum of columns of each design matrix equals 1, which is the design matrix of the population mean. This presents no problem here because the model effects, α_j, β_j, and γ_j, will be treated as random effects. The residual errors are assumed to be $N(0, I\sigma_j^2)$ with an unknown σ_j^2 for each gene. The population mean μ_j is not something of interest and is assigned a single normal prior distribution

$$
p(\mu_j) = N(\mu_j | 0, \Sigma_\mu) \tag{20.3}
$$

where Σ_μ is an unknown variance. The age and gender effects and their interactions are assigned a Gaussian mixture prior apiece, i.e,

$$
p(\alpha_j) = \pi_\alpha N(\alpha_j | 0, \Sigma_\alpha) + (1 - \pi_\alpha) N(\alpha_j | 0, \Sigma_0) \tag{20.4}
$$

where π_α is the mixing proportion, Σ_α is an unknown 3×3 variance–covariance matrix, and $\Sigma_0 = 10^{-8} \times I$ is a constant matrix with a small positive value in the diagonal. Genes classified into the α cluster are associated with the age. Similarly, we assign a Gaussian mixture to the gender effects

$$
p(\beta_j) = \pi_\beta N(\beta_j | 0, \Sigma_\beta) + (1 - \pi_\beta) N(\beta_j | 0, \Sigma_0) \tag{20.5}
$$

where $\Sigma_0 = 10^{-8} \times I$ is again a constant matrix with a dimensionality of 2×2. The interaction effects are also assigned a Gaussian mixture prior

$$
p(\gamma_j) = \pi_\gamma N(\gamma_j | 0, \Sigma_\gamma) + (1 - \pi_\gamma) N(\gamma_j | 0, \Sigma_0) \tag{20.6}
$$

The parameter vector in the above model is

$$\theta = \left\{ \pi_\alpha, \pi_\beta, \pi_\gamma, \Sigma_\mu, \Sigma_\alpha, \Sigma_\beta, \Sigma_\gamma, \sigma_1^2, \ldots, \sigma_N^2 \right\} \tag{20.7}$$

Let us define $\eta_j^\alpha = \{0, 1\}$, $\eta_j^\beta = \{0, 1\}$, and $\eta_j^\gamma = \{0, 1\}$ as the class labels (binary indicator variables) for the three effects, and they are missing values. Denote $\eta_j = [\eta_j^\alpha \ \eta_j^\beta \ \eta_j^\gamma]$ as a vector of the class labels. Given η_j, the expectation and variance matrix of y_j are $E(y_j) = 0$ and

$$\text{var}(y_j) = V_j = 1\Sigma_\mu 1^T + Z_\alpha \Theta_j^\alpha Z_\alpha^T + Z_\beta \Theta_j^\beta Z_\beta^T + Z_\gamma \Theta_j^\gamma Z_\gamma^T + I\sigma_j^2 \tag{20.8}$$

where

$$\Theta_j^\alpha = \eta_j^\alpha \Sigma_\alpha + \left(1 - \eta_j^\alpha\right) \Sigma_0$$

$$\Theta_j^\beta = \eta_j^\beta \Sigma_\beta + \left(1 - \eta_j^\beta\right) \Sigma_0$$

$$\Theta_j^\gamma = \eta_j^\gamma \Sigma_\gamma + \left(1 - \eta_j^\gamma\right) \Sigma_0 \tag{20.9}$$

are the prior variance matrices for the two factors and the factor interactions. The problem is a little more complicated than differential expression analysis. Therefore, we need a special algorithm called stochastic EM algorithm (McLachlan et al. 2002).

20.2   The SEM Algorithm

The SEM algorithm is a hybrid technology combining the EM algorithm and Monte Carlo simulation (McLachlan and Peel 2000). If η_j is known, all parameters can be estimated using the EM algorithm. The stochastic EM algorithm takes advantage of the EM property by introducing a stochastic step to sample η_j from its posterior distribution. The posterior probability of $\eta_j^\alpha = 1$ is calculated as follows:

$$\rho_j^\alpha = E(\eta_j^\alpha | \cdots) = \frac{\pi_\alpha N \left(y_j | 0, \Gamma_j^\alpha\right)}{\pi_\alpha N \left(y_j | 0, \Gamma_j^\alpha\right) + (1 - \pi_\alpha) N \left(y_j | 0, \Gamma_j^0\right)} \tag{20.10}$$

where

$$\Gamma_j^\alpha = 1\Sigma_\mu 1^T + Z_\alpha \Sigma_\alpha Z_\alpha^T + Z_\beta \Theta_j^\beta Z_\beta^T + Z_\gamma \Theta_j^\gamma Z_\gamma^T + I\sigma_j^2$$

$$\Gamma_j^0 = 1\Sigma_\mu 1^T + Z_\alpha \Sigma_0 Z_\alpha^T + Z_\beta \Theta_j^\beta Z_\beta^T + Z_\gamma \Theta_j^\gamma Z_\gamma^T + I\sigma_j^2 \tag{20.11}$$

The two variance matrices differ by one term, that is, Σ_α and Σ_0. Similarly, the posterior probability of $\eta_j^\beta = 1$ is calculated as follows:

$$\rho_j^\beta = E\left(\eta_j^\beta | \cdots\right) = \frac{\pi_\beta N\left(y_j | 0, \Gamma_j^\beta\right)}{\pi_\beta N\left(y_j | 0, \Gamma_j^\beta\right) + (1 - \pi_\beta) N\left(y_j | 0, \Gamma_j^0\right)} \tag{20.12}$$

where

$$\Gamma_j^\beta = 1 \Sigma_\mu 1^T + Z_\alpha \Theta_j^\alpha Z_\alpha^T + Z_\beta \Sigma_\beta Z_\beta^T + Z_\gamma \Theta_j^\gamma Z_\gamma^T + I \sigma_j^2$$
$$\Gamma_j^0 = 1 \Sigma_\mu 1^T + Z_\alpha \Theta_j^\alpha Z_\alpha^T + Z_\beta \Sigma_0 Z_\beta^T + Z_\gamma \Theta_j^\gamma Z_\gamma^T + I \sigma_j^2 \tag{20.13}$$

Again, the two variance matrices differ by Σ_β and Σ_0. Finally, the posterior probability of $\eta_j^\gamma = 1$ is calculated as follows:

$$\rho_j^\gamma = E\left(\eta_j^\gamma | \cdots\right) = \frac{\pi_\gamma N\left(y_j | 0, \Gamma_j^\gamma\right)}{\pi_\gamma N\left(y_j | 0, \Gamma_j^\gamma\right) + (1 - \pi_\gamma) N\left(y_j | 0, \Gamma_j^0\right)} \tag{20.14}$$

where

$$\Gamma_j^\gamma = 1 \Sigma_\mu 1^T + Z_\alpha \Theta_j^\alpha Z_\alpha^T + Z_\beta \Theta_j^\beta Z_\beta^T + Z_\gamma \Sigma_\gamma Z_\gamma^T + I \sigma_j^2$$
$$\Gamma_j^0 = 1 \Sigma_\mu 1^T + Z_\alpha \Theta_j^\alpha Z_\alpha^T + Z_\beta \Theta_j^\beta Z_\beta^T + Z_\gamma \Sigma_0 Z_\gamma^T + I \sigma_j^2 \tag{20.15}$$

Note that the two variance matrices differ by Σ_γ and Σ_0. In the stochastic step, $\eta_j = [\eta_j^\alpha \ \eta_j^\beta \ \eta_j^\gamma]$ are sampled from their Bernoulli posterior distributions. The EM steps for updating other parameters are all derived based on the sampled η_j. The SEM steps are now summarized below:

1. Initialize all parameters.
2. Sample $\eta_j = [\eta_j^\alpha \ \eta_j^\beta \ \eta_j^\gamma]$ from

$$p(\eta_j^\alpha | \cdots) = \text{Bernoulli}(\eta_j^\alpha | \rho_j^\alpha), \forall j = 1, \ldots, N$$
$$p(\eta_j^\beta | \cdots) = \text{Bernoulli}(\eta_j^\beta | \rho_j^\beta), \forall j = 1, \ldots, N$$
$$p(\eta_j^\gamma | \cdots) = \text{Bernoulli}(\eta_j^\gamma | \rho_j^\gamma), \forall j = 1, \ldots, N \tag{20.16}$$

3. Update $\pi = \begin{bmatrix} \pi_\alpha \ \pi_\beta \ \pi_\gamma \end{bmatrix}$ from

$$\pi_\alpha = \frac{1}{N} \sum_{j=1}^{N} \eta_j^\alpha$$

$$\pi_\beta = \frac{1}{N} \sum_{j=1}^{N} \eta_j^\beta$$

$$\pi_\gamma = \frac{1}{N} \sum_{j=1}^{N} \eta_j^\gamma \tag{20.17}$$

4. Calculate the posterior expectation and variance for each of $\{\mu_j \ \alpha_j \ \beta_j \ \gamma_j\}$ using

$$\hat{\mu}_j = E(\mu_j|\cdots) = \Sigma_\mu 1^T V_j^{-1} y_j, \forall j = 1,\ldots,N$$

$$\hat{\alpha}_j = E(\alpha_j|\cdots) = \Theta_j^\alpha Z_\alpha^T V_j^{-1} y_j, \forall j = 1,\ldots,N$$

$$\hat{\beta}_j = E(\beta_j|\cdots) = \Theta_j^\beta Z_\beta^T V_j^{-1} y_j, \forall j = 1,\ldots,N$$

$$\hat{\gamma}_j = E(\gamma_j|\cdots) = \Theta_j^\gamma Z_\gamma^T V_j^{-1} y_j, \forall j = 1,\ldots,N \qquad (20.18)$$

and

$$\hat{M}_j = \text{var}(\mu_j|\cdots) = \Sigma_\mu - \Sigma_\mu 1^T V_j^{-1} 1 \Sigma_\mu$$

$$\hat{A}_j = \text{var}(\alpha_j|\cdots) = \Theta_j^\alpha - \Theta_j^\alpha Z_\alpha^T V_j^{-1} Z_\alpha \Theta_j^\alpha$$

$$\hat{B}_j = \text{var}(\beta_j|\cdots) = \Theta_j^\beta - \Theta_j^\beta Z_\beta^T V_j^{-1} Z_\beta \Theta_j^\beta$$

$$\hat{G}_j = \text{var}(\gamma_j|\cdots) = \Theta_j^\gamma - \Theta_j^\gamma Z_\gamma^T V_j^{-1} Z_\gamma \Theta_j^\gamma \qquad (20.19)$$

5. Update each of the $\{\Sigma_\mu \ \Sigma_\alpha \ \Sigma_\beta \ \Sigma_\gamma\}$ using

$$\Sigma_\mu = \frac{1}{N} \sum_{j=1}^{N} E\left(\mu_j^2|\cdots\right) = \frac{1}{N} \sum_{j=1}^{N} \left(\hat{\mu}_j^2 + \hat{M}_j\right)$$

$$\Sigma_\alpha = \frac{1}{\pi_\alpha N} \sum_{j=1}^{N} \eta_j^\alpha E\left(\alpha_j \alpha_j^T|\cdots\right) = \frac{1}{\pi_\alpha N} \sum_{j=1}^{N} \eta_j^\alpha \left(\hat{\alpha}_j \hat{\alpha}_j^T + \hat{A}_j\right)$$

$$\Sigma_\beta = \frac{1}{\pi_\beta N} \sum_{j=1}^{N} \eta_j^\beta E\left(\beta_j \beta_j^T|\cdots\right) = \frac{1}{\pi_\beta N} \sum_{j=1}^{N} \eta_j^\beta \left(\hat{\beta}_j \hat{\beta}_j^T + \hat{B}_j\right)$$

$$\Sigma_\gamma = \frac{1}{\pi_\gamma N} \sum_{j=1}^{N} \eta_j^\gamma E\left(\gamma_j \gamma_j^T|\cdots\right) = \frac{1}{\pi_\gamma N} \sum_{j=1}^{N} \eta_j^\gamma \left(\hat{\gamma}_j \hat{\gamma}_j^T + \hat{G}_j\right) \quad (20.20)$$

6. The residual error variance is updated using

$$\sigma_j^2 = \frac{1}{M} y_j^T \left(y - 1\hat{\mu}_j - \eta_j^\alpha Z_\alpha \hat{\alpha}_j - \eta_j^\beta Z_\beta \hat{\beta}_j - \eta_j^\gamma Z_\gamma \hat{\gamma}_j\right), \forall j = 1,\ldots,N$$
$$(20.21)$$

7. Return to Step 2 until all parameters have converged to their stationary distributions.

   The convergence criterion for the SEM is different from that of the EM algorithm. Because of the Monte Carlo sampling, the parameters do not converge to a constant vector; rather, they converge to a stationary distribution. We need to monitor the iteration process of each parameter by looking at the trace plot (parameter values against the iteration). Once all parameters are stabilized, we can take the values from the last T iterations and calculate the average values. These average parameter values across iterations are the SEM estimates of the parameters.

20.3 Hypothesis Testing

Genes are selected based on their posterior probabilities of the cluster labels. For example, all genes with $\rho_j^\alpha \geq 0.9$ may be declared as significantly associated with age. Similar criterion may be used for declaration of association with the gender and age by gender interaction. In addition, we can also present the F-test statistics for each gene. The F-test statistics for associations with the age, gender, and age by gender interaction for gene j are

$$F_j^\alpha = \frac{1}{3}\hat{\alpha}_j^T \left(\Theta_j^\alpha - \Theta_j^\alpha Z_\alpha^T V_j^{-1} Z_\alpha \Theta_j^\alpha\right)^{-1} \hat{\alpha}_j$$

$$F_j^\beta = \frac{1}{2}\hat{\beta}_j^T \left(\Theta_j^\beta - \Theta_j^\beta Z_\beta^T V_j^{-1} Z_\beta \Theta_j^\beta\right)^{-1} \hat{\beta}_j$$

$$F_j^\gamma = \frac{1}{6}\hat{\gamma}_j^T \left(\Theta_j^\gamma - \Theta_j^\gamma Z_\gamma^T V_j^{-1} Z_\gamma \Theta_j^\gamma\right)^{-1} \hat{\gamma}_j \qquad (20.22)$$

where

$$\begin{aligned}
\Theta_j^\alpha &= \rho_j^\alpha \Sigma_\alpha + \left(1 - \rho_j^\alpha\right) \Sigma_0 \\
\Theta_j^\beta &= \rho_j^\beta \Sigma_\beta + \left(1 - \rho_j^\beta\right) \Sigma_0 \\
\Theta_j^\gamma &= \rho_j^\gamma \Sigma_\gamma + \left(1 - \rho_j^\gamma\right) \Sigma_0
\end{aligned} \qquad (20.23)$$

Note that (20.23) differ from (20.9) by replacing the missing η by the expectation ρ. An overall F-test for association of gene j with all effects can also be presented. Let us define

$$\xi_j^T = \left[\alpha_j^T \; \beta_j^T \; \hat{\gamma}_j^T\right] \qquad (20.24)$$

as the vector of all effects and

$$Z = Z_\alpha||Z_\beta||Z_\gamma \qquad (20.25)$$

as the horizontal concatenation of the three matrices that represent the design matrix for all the effects. Let

$$\Theta_j = \text{BlockDiag}\left\{\Theta_j^\alpha, \Theta_j^\beta \Theta_j^\gamma\right\} \qquad (20.26)$$

Define

$$V_j = 1\Sigma_\mu 1^T + Z\Theta_j Z^T + I\sigma_j^2$$

(20.27)

The overall F-test statistic is

$$F_j = \frac{1}{11}\hat{\xi}_j^T \left(\Theta_j - \Theta_j Z^T V_j^{-1} Z\Theta_j\right)^{-1}\hat{\xi}_j$$

(20.28)

where the denominator, 11, is the total number of effects (3 for age, 2 for gender, and 6 for age by gender interaction).

Chapter 21
Factor Analysis of Microarray Data

In differential expression analysis, the subjects are divided into two groups, the treatment group and the control group. In the gene-specific analysis of variance (ANOVA), the subjects are divided into multiple groups, each corresponding to a particular treatment. Sometimes, the subjects may not be grouped based on any criteria. They are simply selected randomly from a population for microarray analysis. The purpose of the microarray analysis is simply to find gene networks that coexpress in a system. Although the model-based cluster analysis can be applied directly to the subjects, genes are classified into several clusters based on their different expression profiles across the subjects. When the number of subjects is large, the analysis is not efficient because there are too many parameters to estimate. Factor analysis is a more efficient way to perform the model-based cluster analysis by clustering genes based on the factor loadings of some hidden factors. This chapter is focused on factor analysis of microarray expression data. It is equivalent to differential expression analysis or ANOVA without knowing the identities of case and control or the group labels of the treatments.

21.1 Background of Factor Analysis

21.1.1 Linear Model of Latent Factors

We use the model and the method given by McLachlan and Peel (2000). Let y_j be an $M \times 1$ vector for expressions of gene j measured from M subjects. The linear model involved in the factor analysis is

$$y_j = \mu + B u_j + \varepsilon_j = \mu + \sum_{l=1}^{r} B_l u_{jl} + \varepsilon_j \tag{21.1}$$

S. Xu, *Principles of Statistical Genomics*, DOI 10.1007/978-0-387-70807-2_21,
© Springer Science+Business Media, LLC 2013

where μ is an $M \times 1$ vector for the mean, B is an $M \times r$ unknown matrix called the factor loading, u_j is an $r \times 1$ vector of factors, and ε_j is an $M \times 1$ vector of the residual errors. The $r \times 1$ vector of factors u_j is unknown variables with an assumed multivariate normal distribution $N(0, I)$, where I is an $r \times r$ identity matrix. This assumption means that the r factors are independent and each one has a standardized normal distribution. The residual error vector is assumed to be distributed as $N(0, D)$ where $D = \text{diag}[d_1, d_2, \ldots, d_r]$ is a diagonal matrix. The number of factors r is arbitrary but should be less than M. This linear model of factors is the standard one commonly seen in the literature (McLachlan and Peel 2000). We will revise this model later when dealing with various computational algorithms of the factor analysis. The unknown parameters in the standard factor analysis are $\theta = \{\mu, B, D\}$. The strength of the relationship between the factors and the gene expressions is represented by the factor loading matrix B. If $B=0$ or close to 0, the factors are irrelevant to gene expressions. In factor analysis, the relationship between the factors and the gene expressions is described by the covariance structure,

$$\Gamma = B \text{var}(u_j) B^T + D = B B^T + D \tag{21.2}$$

21.1.2 EM Algorithm

The observed log likelihood function is

$$L(\theta) = -\frac{N}{2} \log |B B^T + D| - \frac{1}{2} \sum_{j=1}^{N} (y_j - \mu)^T (B B^T + D)^{-1} (y_j - \mu)$$

$$\tag{21.3}$$

However, the log likelihood function to be maximized in the EM algorithm is the expectation of the complete-data log likelihood function,

$$E[L(\theta, u)] = -\frac{1}{2} \sum_{j=1}^{N} E[(y_j - \mu - B u_j)^T D^{-1} (y_j - \mu - B u_j)]$$

$$- \frac{N}{2} \log |D| \tag{21.4}$$

where the expectation is taken with respect to factors u_j for $j = 1, \ldots, N$. The E-steps involve calculating various terms of expectations related to u_j. These include

$$\hat{u}_j = E(u_j | \cdots) = [(B^T D^{-1} B)^{-1} + I]^{-1} B^T D^{-1} (y_j - \mu) \tag{21.5}$$

and

$$\hat{S}_j = \text{var}(u_j \mid \cdots) = [(B^T D^{-1} B)^{-1} + I]^{-1} \tag{21.6}$$

An alternative expressions of these are

$$\hat{u}_j = \text{E}(u_j \mid \cdots) = B^T \Gamma^{-1}(y_j - \mu) \tag{21.7}$$

and

$$\hat{S}_j = \text{var}(u_j \mid \cdots) = I - B^T \Gamma^{-1} B \tag{21.8}$$

The M-steps involve maximization of the expected complete-data log likelihood function. The partial derivative of $\text{E}[L(\theta, u)]$ with respect to μ is

$$\frac{\partial \text{E}(L(\theta, u)]}{\partial \mu} = \sum_{j=1}^{N} D^{-1}(y_j - \mu - B\hat{u}_j) \tag{21.9}$$

Setting this partial derivative to zero and solving the equation lead to

$$\mu = \frac{1}{N} \sum_{j=1}^{N} (y_j - B\hat{u}_j) \tag{21.10}$$

where $\hat{u}_j = \text{E}(u_j \mid \cdots)$ is the conditional expectation of u_j given in (21.7). The partial derivative of $\text{E}[L(\theta, u)]$ with respect to B is

$$\frac{\partial \text{E}[L(\theta, u)]}{\partial B} = \sum_{j=1}^{N} E\left[D^{-1}(y_j - \mu - Bu_j)u_j^T \right]$$

$$= \sum_{j=1}^{N} E\left[D^{-1}(y_j - \mu)u_j^T \right] - \sum_{j=1}^{N} D^{-1} BE\left(u_j u_j^T \right) \tag{21.11}$$

Setting $\partial \text{E}[L(\theta, u)]/\partial B = 0$, we get

$$\sum_{j=1}^{N} (y_j - \mu)\text{E}(u_j^T \mid \cdots) - B \sum_{j=1}^{N} \text{E}(u_j u_j^T \mid \cdots) = 0 \tag{21.12}$$

Therefore,

$$B = \left[\sum_{j=1}^{N} (y_j - \mu)\hat{u}_j^T \right] \left[\sum_{j=1}^{N} (\hat{u}_j \hat{u}_j^T + \hat{S}_j) \right]^{-1} \tag{21.13}$$

Defining

$$\Psi = \sum_{j=1}^{N} (y_j - \mu)(y_j - \mu)^T \tag{21.14}$$

and substituting (21.7) and (21.8) into (21.13), we have

$$B = [\Psi \Gamma^{-1} B][B^T \Gamma^{-1} \Psi \Gamma^{-1} B + N(I - B^T \Gamma^{-1} B)]^{-1} \tag{21.15}$$

The last M-step is to solve matrix D by setting $\partial \mathrm{E}[L(\theta, u)]/\partial D = 0$. Let us rewrite $\mathrm{E}[L(\theta, u)]$ as

$$\mathrm{E}[L(\theta, u)] = -\frac{1}{2} \sum_{j=1}^{N} \mathrm{tr}\{D^{-1} E[(y_j - \mu - Bu_j)(y_j - \mu - Bu_j)^T]\}$$

$$-\frac{N}{2} \log |D| \tag{21.16}$$

The partial derivative of the above likelihood with respect to D is

$$\frac{\partial \mathrm{E}[L(\theta, u)]}{\partial D} = -\frac{1}{2} \sum_{j=1}^{N} \frac{\partial}{\partial D} \mathrm{tr}\{D^{-1} \mathrm{E}[(y_j - \mu - Bu_j)(y_j - \mu - Bu_j)^T]\}$$

$$-\frac{N}{2} \frac{\partial \log |D|}{\partial D} \tag{21.17}$$

where

$$\frac{\partial \log |D|}{\partial D} = D^{-1} \tag{21.18}$$

and

$$\frac{\partial}{\partial D} \mathrm{tr}\{D^{-1} \mathrm{E}[(y_j - \mu - Bu_j)(y_j - \mu - Bu_j)^T]\}$$

$$= -D^{-1} E[(y_j - \mu - Bu_j)(y_j - \mu - Bu_j)^T]D^{-1} \tag{21.19}$$

Substituting (21.18) and (21.19) into (21.17), we get

$$\frac{\partial E[L(\theta, u)]}{\partial D} = +\frac{1}{2} D^{-1} \sum_{j=1}^{N} \mathrm{E}[(y_j - \mu - Bu_j)(y_j - \mu - Bu_j)^T]D^{-1}$$

$$-\frac{N}{2} D^{-1} \tag{21.20}$$

Setting $\partial \mathrm{E}[L(\theta, u)]/\partial D = 0$ and solving for D lead to

$$D = \frac{1}{N}\text{diag}\left\{\sum_{j=1}^{N} \text{E}[(y_j - \mu - Bu_j)(y_j - \mu - Bu_j)^T]\right\} \qquad (21.21)$$

where

$$\text{E}[(y_j - \mu - Bu_j)(y_j - \mu - Bu_j)^T]$$
$$= (y_j - \mu - B\hat{u}_j)(y_j - \mu - B\hat{u}_j)^T + B\hat{S}_j B^T \qquad (21.22)$$

Therefore,

$$D = \frac{1}{N}\text{diag}\left\{\sum_{j=1}^{N} [(y_j - \mu - B\hat{u}_j)(y_j - \mu - B\hat{u}_j)^T + B\hat{S}_j B^T]\right\} \qquad (21.23)$$

The diagonal operator appears in the equation because D is a diagonal matrix. Note that the EM algorithm requires repeatedly inversion of matrix $\Gamma = BB^T + D$ and it can be time consuming if M is large. Because Γ is a highly structured matrix, we can use the following Woodbury matrix identity to invert this matrix:

$$(BB^T + D)^{-1} = D^{-1} - D^{-1}B(I + B^T D^{-1}B)^{-1}B^T D^{-1} \qquad (21.24)$$

The computing time saved can be substantial if r is much smaller than M because D is a diagonal matrix whose inverse is simply

$$D^{-1} = \text{diag}[d_1^{-1}, \ldots, d_r^{-1}] \qquad (21.25)$$

and $I + B^T D^{-1}B$ is an $r \times r$ matrix and r is small.

The EM algorithm is now summarized in the following steps:

1. Initialize all parameters.
2. Calculate the expectation and variance of u_j using (21.7) and (21.8).
3. Update μ using (21.10).
4. Update B using (21.15).
5. Update matrix D using (21.23).
6. Repeat Steps (2)–(5) until the iteration process converges to a satisfactory level.

The E-step is represented by Step (2), whereas the M-step consists of Steps (3)–(5) of the EM algorithm. The convergence criterion is determined by

$$||\theta^{(t+1)} - \theta^{(t)}|| \leq 10^{-8} \qquad (21.26)$$

where $\theta^{(t+1)}$ and $\theta^{(t)}$ are the parameter values in two consecutive iterations of the EM algorithm and the small number 10^{-8} is an arbitrary positive number set by the investigator.

21.1.3   Number of Factors

The number of factors can be determined using the Bayesian information criterion (BIC),

$$\text{BIC} = -2L(\theta) + (2M + rM)\ln(N) \tag{21.27}$$

where $L(\theta)$ is the observed log likelihood function, $\theta = \{\mu, B, D\}$ is the parameter vector, N is the number of genes, and $p = 2M + rM$ is the total number of parameters involved in the model. Note that μ and D each has a dimension of M and the dimension of B is $M \times r$, leading to a total number of free parameters $2M + M \times r$. The optimal number of factors is the one that minimizes the BIC value.

21.2   Cluster Analysis

We now perform cluster analysis on the genes based on the covariance structures to find a few coregulated networks. Let C be the total number of clusters and $\eta_j = 1, \ldots, C$ be the cluster label for gene j. A multivariate representation of η_j is

$$\delta_j = [\delta_{j1} \cdots \delta_{jC}] \tag{21.28}$$

where $\delta_{j\kappa}$ is a Bernoulli variable defined as

$$\delta_{j\kappa} = \begin{cases} 1 & \text{for } \eta_j = \kappa \\ 0 & \text{for } \eta_j \neq \kappa \end{cases} \tag{21.29}$$

We now use the stochastic expectation and maximization (SEM) algorithm to perform the cluster analysis. The SEM algorithm is largely the same as the EM algorithm except that the unknown class label for each gene is sampled from the posterior probability before each of the EM cycles starts. Let us define

$$\pi = [\pi_1 \cdots \pi_C] \tag{21.30}$$

as the mixing proportions and

$$\rho_j = [\rho_{j1} \cdots \rho_{jC}], \forall j = 1, \ldots, N \tag{21.31}$$

as the posterior probability of δ_j . Let μ_κ be the mean of cluster κ and

$$\Gamma_\kappa = B_\kappa B_\kappa^T + D_\kappa, \forall \kappa = 1, \ldots, C \tag{21.32}$$

is the cluster-specific covariance matrix. The posterior probability of $\delta_{j\kappa}$ is calculated using

$$\rho_{j\kappa} = \frac{\pi_{\kappa} N(y_j | \mu_{\kappa}, \Gamma_{\kappa})}{\sum_{\kappa'=1}^{C} \pi_{\kappa'} N(y_j | \mu_{\kappa'}, \Gamma_{\kappa'})}, \forall \kappa = 1, \ldots, C \qquad (21.33)$$

The SEM algorithm is summarized as follows:

1. Initialize all parameters.
2. Sample δ_j from

$$p(\delta_j | \cdots) = \text{Multinomial}(\delta_j | 1, \rho_j), \forall j = 1, \ldots, N \qquad (21.34)$$

3. Update π using

$$\pi_{\kappa} = \frac{1}{N} \sum_{j=1}^{N} \delta_{j\kappa}, \forall \kappa = 1, \ldots, C \qquad (21.35)$$

4. Calculate $\text{E}(u_j | \cdots)$ and $\text{var}(u_j | \cdots)$ using

$$\hat{u}_j = \text{E}(u_j | \cdots) = \sum_{\kappa=1}^{C} \delta_{j\kappa} B_{\kappa}^T \Gamma^{-1} (y_j - \mu_{\kappa}) \qquad (21.36)$$

and

$$\hat{S}_j = \text{var}(u_j | \cdots) = \sum_{\kappa=1}^{C} \delta_{j\kappa} (I - B_{\kappa}^T \Gamma_{\kappa}^{-1} B_{\kappa}) \qquad (21.37)$$

5. Update μ_{κ} using

$$\mu_{\kappa} = \frac{1}{\pi_{\kappa} N} \sum_{j=1}^{N} \delta_{j\kappa} (y_j - B_{\kappa} \hat{u}_j), \forall \kappa = 1, \ldots, C \qquad (21.38)$$

6. Update B_{κ} using

$$B_{\kappa} = \left[\sum_{j=1}^{N} \delta_{j\kappa} (y_j - \mu_{\kappa}) \hat{u}_j^T \right] \left[\sum_{j=1}^{N} \delta_{j\kappa} (\hat{u}_j \hat{u}_j^T + \hat{S}_j) \right]^{-1} \qquad (21.39)$$

for all $\kappa = 1, \ldots, C$.
7. Update D_{κ} using

$$D_{\kappa} = \frac{1}{\pi_{\kappa} N} \text{diag} \left[\sum_{j=1}^{N} \delta_{j\kappa} (SS_{j\kappa} + B_{\kappa} \hat{S}_j B_{\kappa}^T) \right] \qquad (21.40)$$

where

$$SS_{j\kappa} = (y_j - \mu_{\kappa} - B_{\kappa} \hat{u}_j)(y_j - \mu_{\kappa} - B_{\kappa} \hat{u}_j)^T \qquad (21.41)$$

8. Repeat Steps (2)–(7) until all parameters have converged to their perspective stationary distributions.

21.3　Differential Expression Analysis

A differential expression analysis can be performed even though there are no "case" and "control" labels for the subjects. We can classify each gene into one of two clusters; one is the neutral cluster in which the mean and covariance structure are μ_0 and $\Gamma_0 = D_0$, respectively. The other cluster is defined by μ_1 and $\Gamma_1 = B_1 B_1^T + D_1$. The mixture distribution for gene j is

$$p(y_j) = \pi N(y_j|\mu_1, \Gamma_1) + (1 - \pi)N(y_j|\mu_0, \Gamma_0) \qquad (21.42)$$

where π is the proportion of genes belonging to cluster 1. Let $\eta_j = \{0, 1\}$ be the cluster label for gene j. We now use the stochastic expectation and maximization (SEM) algorithm to perform the cluster analysis. Let $\rho_j = \mathrm{E}(\eta_j|\cdots)$ be the conditional posterior probability of $\eta_j = 1$, which is calculated using

$$\rho_j = \frac{\pi N(y_j|\mu_1, \Gamma_1)}{\pi N(y_j|\mu_1, \Gamma_1) + (1 - \pi)N(y_j|\mu_0, \Gamma_0)} \qquad (21.43)$$

The SEM algorithm is summarized as follows:

1. Initialize all parameters.
2. Sample η_j from

$$p(\eta_j|\cdots) = \mathrm{Bernoulli}(\eta_j|\rho_j), \forall j = 1, \ldots, N \qquad (21.44)$$

3. Update π using

$$\pi = \frac{1}{N} \sum_{j=1}^{N} \eta_j \qquad (21.45)$$

4. Calculate $\mathrm{E}(u_j|\ldots)$ and $\mathrm{var}(u_j|\ldots)$ for those genes that come from cluster 1 using

$$\hat{u}_j = \mathrm{E}(u_j|\cdots) = B_1^T \Gamma_1^{-1}(y_j - \mu_1) \qquad (21.46)$$

and

$$\hat{S}_j = \mathrm{var}(u_j|\cdots) = I - B_1^T \Gamma_1^{-1} B_1 \qquad (21.47)$$

5. Update μ_κ using

$$\mu_1 = \frac{1}{\pi N} \sum_{j=1}^{N} \eta_j (y_j - B_1 \hat{u}_j) \qquad (21.48)$$

and

$$\mu_0 = \frac{1}{(1 - \pi)N} \sum_{j=1}^{N} (1 - \eta_j)y_j \qquad (21.49)$$

6. Update B_1 using

$$B_1 = \left[\sum_{j=1}^{N} \eta_j (y_j - \mu_1) \hat{u}_j^T \right] \left[\sum_{j=1}^{N} \eta_j (\hat{u}_j \hat{u}_j^T + \hat{S}_j) \right]^{-1} \qquad (21.50)$$

7. Update D_1 and D_0 using

$$D_0 = \frac{1}{(1 - \pi)N} \text{diag} \left[\sum_{j=1}^{N} (1 - \eta_j)(y_j - \mu_0)(y_j - \mu_0)^T \right] \qquad (21.51)$$

and

$$D_1 = \frac{1}{\pi N} \text{diag} \left[\sum_{j=1}^{N} \eta_j \left(SS_j + B_1 \hat{S}_j B_1^T \right) \right] \qquad (21.52)$$

where

$$SS_j = (y_j - \mu_1 - B_1 \hat{u}_j)(y_j - \mu_1 - B_1 \hat{u}_j)^T \qquad (21.53)$$

8. Repeat Steps (2)–(7) until all parameters have converged to their perspective stationary distributions.

21.4 MCMC Algorithm

We only examine the MCMC algorithm for the differential expression analysis (the two-cluster analysis). In Bayesian analysis, we need to assign a prior distribution for each parameter. For the mixing proportion, we assign $p(\pi) = \text{Beta}(\pi | 1, 1)$. Each of the cluster means is assigned an uninformative prior, $p(\mu_0) = p(\mu_1) = 1$. The prior for B_1 is also uninformative, $p(B_1) = 1$. The only parameter left is D. We now assign each element of D a scaled inverse chi-square distribution

$$p(d_i) = \text{Inv} - \chi^2(d_i | \tau, \omega), \forall i = 1, \dots, M \qquad (21.54)$$

The posterior distribution of each variable involved in the analysis has an explicit form, and thus, the Gibbs sampler approach can be used for the MCMC analysis. The MCMC algorithm is summarized below:

1. Sample all variables from their prior distributions.
2. Sample η_j from the following Bernoulli distribution:

$$p(\eta_j | \cdots) = \text{Bernoulli}(\eta_j | \rho_j), \forall j = 1, \dots, N \qquad (21.55)$$

3. Simulate π from the following beta distribution:

$$p(\pi|\cdots) = \text{Beta}\left[\pi \,\bigg|\, 1 + \sum_{j=1}^{N} \eta_j, 1 + N - \sum_{j=1}^{N} \eta_j\right] \qquad (21.56)$$

4. Sample factors u_j in cluster 1 from the following normal distribution:

$$p(u_j|\cdots) = N(u_j|\hat{u}_j, \hat{S}_j) \qquad (21.57)$$

where

$$\hat{u}_j = \text{E}(u_j|\cdots) = B_1^T \Gamma_1^{-1}(y_j - \mu_1) \qquad (21.58)$$

and

$$\hat{S}_j = \text{var}(u_j|\cdots) = I - B_1^T \Gamma_1^{-1} B_1 \qquad (21.59)$$

5. Simulate μ_0 and μ_1 from their perspective normal distributions,

$$p(\mu_1|\cdots) = N\left[\mu_1 \,\bigg|\, \frac{1}{\pi N}\sum_{j=1}^{N}\eta_j(y_j - B_1\hat{u}_j), \frac{1}{\pi N}D_1\right] \qquad (21.60)$$

and

$$p(\mu_0|\cdots) = N\left[\mu_0 \,\bigg|\, \frac{1}{(1-\pi)N}\sum_{j=1}^{N}(1-\eta_j)y_j, \frac{1}{(1-\pi)N}D_0\right] \qquad (21.61)$$

6. Let B_{1l} be the lth column of matrix $B_1 = [B_{11}\ B_{12} \cdots B_{1r}]$, which is sampled from the following normal distribution:

$$p(B_{1l}|\cdots) = N\left[B_{1l} \,\bigg|\, \left(\sum_{j=1}^{N}\eta_j u_{jl}^2\right)^{-1}\sum_{j=1}^{N}\eta_j u_{jl}y_j^*, \left(\sum_{j=1}^{N}\eta_j u_{jl}^2\right)^{-1}D_1\right]$$

$$(21.62)$$

where u_{jl} is the lth element of vector u_j and

$$y_j^* = y_j - \mu_1 - \sum_{l'\neq l}^{r} B_{1l'}u_{jl'} \qquad (21.63)$$

is y_j adjusted by the mean and values of other factors (also called the offset of y_j).

7. Simulate matrix D_1 one element at a time using the following distribution. Let v_i be the ith diagonal element of the following $M \times M$ matrix:

$$V = \sum_{j=1}^{N}\eta_j[(y_j - \mu_1 - B_1u_j)(y_j - \mu_1 - B_1u_j)^T] \qquad (21.64)$$

Let d_i be the ith diagonal element of matrix D_1. We now simulate d_i from the following scaled inverse chi-square distribution:

$$p(d_i|\cdots) = \text{Inv} - \chi^2(d_i|\tau + \pi N, \omega + v_i), \forall i = 1, \ldots, M \qquad (21.65)$$

8. Simulate matrix D_0 one element at a time using the following distribution. Redefine v_i as the ith diagonal element of matrix

$$V = \sum_{j=1}^{N} (1 - \eta_j)(y_j - \mu_0)(y_j - \mu_0)^T \qquad (21.66)$$

and define d_i as the ith diagonal element of matrix D_0. We now simulate d_i from the following scaled inverse chi-square distribution:

$$p(d_i|\cdots) = \text{Inv} - \chi^2(d_i|\tau + (1 - \pi)N, \omega + v_i), \forall i = 1, \ldots, M \qquad (21.67)$$

9. Go back to Step (2) and repeat the sampling process until a desired length of the Markov chain is reached.

Let T be the posterior sample size and $\eta_j^{(t)}$ be the tth observation of the sampled cluster label for gene j for $t = 1, \ldots, T$. Define

$$\bar{\rho}_j = \frac{1}{T} \sum_{t=1}^{T} \eta_j^{(t)} \qquad (21.68)$$

as the posterior mean of η_j across all the posterior sample. Gene j will be classified into the differentially expressed cluster if $\bar{\rho}_j \geq 0.9$, where 0.9 is an arbitrarily set cutoff point.

Chapter 22
Classification of Tissue Samples Using Microarrays

Classification of tissue samples is an important aspect of disease diagnosis and treatment (Golub et al. 1999; Zhu and Hastie 2004; Liao and Chin 2007). Conventional diagnosis of disease has been based on examination of the morphological appearance of stained tissue specimens under light microscopy. However, this method is subjective and highly depends on trained pathologists. Microarray technology offers an alternative approach of disease diagnosis, and it can be more objective and accurate than the conventional diagnosis. This chapter introduces tissue sample classification using microarrays. Numerous studies have been reported using microarrays to classify disease states and cancer types (Lee and Lee 2003; Yeung and Bumgarner 2003; Statnikov et al. 2005; Jirapech-Umpai and Aitken 2005; Dagliyan et al. 2011). The support vector machine (SVM) is one of the methods that have been successfully applied to cancer diagnosis problem (Lee and Lee 2003). However, we will use the model-based logistic regression method developed by Zhu and Hastie (2004) as the theoretical basis to develop our own penalized logistic regression classifier. The penalized logistic regression method has an advantage of providing an estimate of the underlying probability of disease occurrence.

22.1 Logistic Regression

We will introduce the logistic regression for classification of tissue samples under two states of the disease phenotype, affected (case) and normal (control). In contrast to the microarray data analysis described in previous chapters, we now use an $M \times N$ matrix Z to denote the microarray data where M is the number of subjects and N is the number of transcripts (genes). The disease "phenotype" is denoted by an $M \times 1$ vector y with $y_j = 0$ for control and $y_j = 1$ for case where $j = 1, \ldots, M$ indexes the subjects. We first review the basic principles of logistic regression analysis and then develop a penalized regression method to select genes. Finally, we will use the

S. Xu, *Principles of Statistical Genomics*, DOI 10.1007/978-0-387-70807-2_22,
© Springer Science+Business Media, LLC 2013

selected genes to predict disease phenotypes for future samples. Before collecting more samples, we can evaluate the method using cross validation analysis to obtain some confidence about the predictability of the method.

The probability distribution for y_j is

$$p(y_j) = \mu_j^{y_j}(1 - \mu_j)^{1-y_j} \tag{22.1}$$

where

$$\mu_j = \mathrm{E}(y_j) = \frac{e^{\xi_j}}{1 + e^{\xi_j}} \tag{22.2}$$

is the expectation of y_j and ξ_j is a linear predictor described as

$$\xi_j = X_j\beta + \sum_{k=1}^{N} Z_{jk}\gamma_k = X_j\beta + Z_j\gamma \tag{22.3}$$

Note that the genes are now indexed by k rather than j. The relationship between the expectation and the linear predictor is the logit link function

$$\xi_j = \mathrm{logit}(\mu_j) = \ln\frac{\mu_j}{1 - \mu_j} = X_j\beta + \sum_{k=1}^{N} Z_{jk}\gamma_k \tag{22.4}$$

where X_j is a design matrix for some effects not related to the gene expression and γ_k is the effect of gene k on the binary disease status. The gene expression level Z_{jk} is standardized prior to the analysis, i.e., it is subtracted by the mean expression of gene k and divided by the standard deviation. Theoretically, the standardization is not required, but in practice, it is useful because the expression levels may not be in the same scale. The variation in the scale will cause computational problems for the logistic model. When there is no systematic effects other than the gene expressions, the design matrix X_j is simply equal to one, and β simply represents the intercept of the logistic regression. The logit link function is one of a few link functions we can choose. An alternative link function is the probit link function commonly used in the threshold model of quantitative genetics (Falconer and Mackay 1996; Lynch and Walsh 1998).

The traditional logistic regression is performed by maximizing the following log likelihood function:

$$L(\theta) = \sum_{j=1}^{M} \left[y_j \ln\mu_j + (1 - y_j)\ln(1 - \mu_j) \right] \tag{22.5}$$

where $\theta = \{\beta, \gamma\}$ is the parameter vector and $\gamma = \begin{bmatrix} \gamma_1 \cdots \gamma_N \end{bmatrix}^T$ is the vector of gene effects. The iteratively reweighted least-squares method for generalized linear models (Wedderburn 1974) can be used to estimated the parameters. The method requires the first- and second-order partial derivatives of the likelihood function with respect to the linear predictor. The first partial derivative of $L(\theta)$ with respect to the parameters is

$$\frac{\partial L(\theta)}{\partial \theta} = \begin{bmatrix} \frac{\partial L(\theta)}{\partial \beta} \\ \frac{\partial L(\theta)}{\partial \gamma} \end{bmatrix} = \begin{bmatrix} \sum\limits_{j=1}^{M} X_j^T (y_j - \mu_j) \\ \sum\limits_{j=1}^{M} Z_j^T (y_j - \mu_j) \end{bmatrix} \tag{22.6}$$

and the second-order partial derivative is

$$\frac{\partial^2 L(\theta)}{\partial \theta \, \partial \theta^T} = \begin{bmatrix} \frac{\partial^2 L(\theta)}{\partial \beta \partial \beta^T} & \frac{\partial^2 L(\theta)}{\partial \beta \partial \gamma^T} \\ \frac{\partial^2 L(\theta)}{\partial \gamma \partial \beta^T} & \frac{\partial^2 L(\theta)}{\partial \gamma \partial \gamma^T} \end{bmatrix} = \begin{bmatrix} -\sum\limits_{j=1}^{M} X_j^T W_j(\theta) X_j & -\sum\limits_{j=1}^{M} X_j^T W_j(\theta) Z_j \\ -\sum\limits_{j=1}^{M} Z_j^T W_j(\theta) X_j & -\sum\limits_{j=1}^{M} Z_j^T W_j(\theta) Z_j \end{bmatrix} \tag{22.7}$$

where

$$W_j(\theta) = \mu_j(1 - \mu_j), \forall j = 1, \ldots, M \tag{22.8}$$

is the variance of the data point. The intercept β and gene effect γ can be estimated sequentially or jointly using the Newton–Raphson iteration. For the sequential approach, β is estimated conditional on γ, and then γ is estimated conditional on β. The Newton–Raphson iteration for β is

$$\beta^{(t+1)} = \beta^{(t)} + \left[\sum_{j=1}^{M} X_j^T W_j(\theta^{(t)}) X_j \right]^{-1} \left[\sum_{j=1}^{M} X_j^T (y_j - \mu_j) \right]$$

$$= \left[\sum_{j=1}^{M} X_j^T W_j(\theta^{(t)}) X_j^T \right]^{-1} \left[\sum_{j=1}^{M} X_j^T W_j(\theta^{(t)}) y_j^* \right] \tag{22.9}$$

where

$$y_j^* = X_j \beta^{(t)} + W_j^{-1}(\theta^{(t)})(y_j - \mu_j) \tag{22.10}$$

is called the adjusted response or pseudodata. It is written in the form of weighted regression analysis on the pseudodata. The corresponding Newton–Raphson iteration for γ conditional on β is

$$\gamma^{(t+1)} = \gamma^{(t)} + \left[\sum_{j=1}^{M} Z_j^T W_j(\theta^{(t)}) Z_j \right]^{-1} \left[\sum_{j=1}^{M} Z_j^T (y_j - \mu_j) \right]$$

$$= \left[\sum_{j=1}^{M} Z_j^T W_j(\theta^{(t)}) Z_j^T \right]^{-1} \left[\sum_{j=1}^{M} Z_j^T W_j(\theta^{(t)}) y_j^* \right] \tag{22.11}$$

where

$$y_j^* = Z_j \gamma^{(t)} + W_j^{-1}(\theta^{(t)})(y_j - \mu_j) \tag{22.12}$$

is the pseudodata corresponding to γ.

Once the parameters are estimated, denoted by $\hat{\theta} = \{\hat{\beta}, \hat{\gamma}\}$, the disease outcome can be predicted using

$$\hat{\mu}_j = \frac{e^{X_j \hat{\beta} + Z_j \hat{\gamma}}}{1 + e^{X_j \hat{\beta} + Z_j \hat{\gamma}}} \tag{22.13}$$

which is the probability of $y_j = 1$ given the expression levels for N genes. Of course, the probability of $y_j = 0$ is

$$1 - \hat{\mu}_j = \frac{1}{1 + e^{X_j \hat{\beta} + Z_j \hat{\gamma}}} \tag{22.14}$$

The traditional logistic regression analysis applied to microarray data has two problems: (1) The sample size M is usually small, e.g., $M < 200$ is typical in the current microarray experiments. (2) The number of genes is usually large, e.g., $N > 20,000$ is common. With the traditional regression analysis, we cannot handle all the N genes simultaneously because the model will be overfit and thus will loose the desired predictability. Therefore, a preliminary treatment of the data is required prior to the analysis. Among the N genes, majority of the genes will not be differentially expressed. These genes can be detected by visual inspection of the expression profiles (the plot of the gene expression against the subjects). These genes can be immediately eliminated in the preprocessing stage. The genes that cannot be eliminated in this stage will be subject to further analysis. The following discussion assumes that all the N genes have survived the preprocessing and thus N may be in the order of a few hundred to one or two thousand. Even with a few hundred genes, we still have to face the overfitting problem because the sample size M is not sufficiently large. Therefore, we perform a penalized logistic regression analysis as described in the following section.

22.2   Penalized Logistic Regression

The penalized logistic regression is a generalization of the Bayesian shrinkage posterior mode estimation for quantitative traits to the binary traits (disease outcome). The penalized log likelihood function is defined as

$$\psi(\theta) = \sum_{j=1}^{M} \left[y_j \ln \mu_j + (1 - y_j) \ln(1 - \mu_j) \right] + \sum_{k=1}^{N} \ln N(\gamma_k | 0, \tau_k) \tag{22.15}$$

where the quadratic penalty, also called the L_2 penalty, is used and it is equivalent to assigning each regression coefficient to a normal prior. Other penalties can also be used, e.g., the L_1 penalty used in the Lasso regression analysis (Tibshirani 1996; Friedman et al. 2010). The penalty involves τ_k whose values need to be determined prior to the data analysis. We now introduce a higher level prior to each τ_k and include the log density of the hyperprior in the penalty also. The scaled inverse chi-square distribution is chosen as the hyperprior. We now revise the parameter vector to include the τ_k, i.e., $\theta = \{\beta, \gamma, \tau\}$ where $\tau = \{\tau_k\}$ is the collection of all the τ_k's. The penalized log likelihood for the hierarchical model is redefined as

$$\psi(\theta) = \sum_{j=1}^{M} \left[y_j \ln \mu_j + (1 - y_j) \ln(1 - \mu_j) \right]$$

$$+ \sum_{k=1}^{N} \ln N(\gamma_k | 0, \tau_k)$$

$$+ \sum_{k=1}^{N} \ln \text{Inv} - \chi^2(\tau_k | a, b) \qquad (22.16)$$

where $(a, b) = (0, 0)$ are the hyperparameter values. Of course, alternative parameter values can also be selected. The first part of the above penalized log likelihood function is the usual log likelihood function $L(\theta)$; the second and third parts represent the penalty and denoted by $P(\theta)$. Therefore, the penalized log likelihood is expressed as

$$\psi(\theta) = L(\theta) + P_1(\theta) + P_2(\theta) \qquad (22.17)$$

The solution of the penalized likelihood can be obtained by maximizing $\psi(\theta)$ with respect to θ. Theoretically, any algorithm can be adopted for the maximization. Practically, it is difficult due to the large number of parameters involved in the log likelihood function. The dimensionality of the parameters for the hierarchical model is $1 + 2N$, including one intercept, N gene effects, and N variance components. We now introduce the following coordinate descent algorithm for parameter estimation.

22.3   The Coordinate Descent Algorithm

This algorithm starts with some initial values of the parameters and then maximizes the penalized log likelihood function with one parameter at a time conditional on the values of other parameters at their previous values. Let $\psi(\beta | \cdots) = L(\beta | \cdots)$ be the conditional penalized log likelihood function for β conditional on the values of other parameters. Note that the intercept is not penalized and thus the conditional likelihood does not include the penalty. The conditional solution for β is obtained using the Newton–Raphson algorithm as described in the previous section,

$$\beta^{(t+1)} = \left[\sum_{j=1}^{M} X_j^T W_j(\theta^{(t)}) X_j \right]^{-1} \left[\sum_{j=1}^{M} X_j^T W_j(\theta^{(t)}) y_j^* \right] \qquad (22.18)$$

where

$$y_j^* = X_j \beta^{(t)} + W_j^{-1}(\theta^{(t)})(y_j - \mu_j) \qquad (22.19)$$

is the pseudodata point. Let

$$\psi(\gamma_k | \cdots) = L(\gamma_k | \cdots) + P_1(\gamma_k | \cdots) \qquad (22.20)$$

be the penalized log likelihood function for γ_k conditional on the values of other parameters. The Newton–Raphson iteration is

$$\gamma_k^{(t+1)} = \left[\sum_{j=1}^{M} Z_{jk}^T W_j(\theta^{(t)}) Z_{jk} + \frac{1}{\tau_k} \right]^{-1} \left[\sum_{j=1}^{M} Z_{jk}^T W_j(\theta^{(t)}) y_j^* \right] \qquad (22.21)$$

where

$$y_j^* = Z_{jk} \gamma_k^{(t)} + W_j^{-1}(\theta^{(t)})(y_j - \mu_j) \qquad (22.22)$$

is the pseudodata point. Let

$$\psi(\tau_k | \cdots) = P_1(\tau_k | \cdots) + P_2(\tau_k | \cdots) \qquad (22.23)$$

be the conditional log likelihood of τ_k given all other parameters. This variance parameter only appears in the penalty. The partial derivative is

$$\frac{\partial \psi(\tau_k | \cdots)}{\partial \tau_k} = -\frac{a+3}{2} \ln(\tau_k) - \frac{\gamma_k^2 + b}{2\tau_k} \qquad (22.24)$$

Setting the partial derivative to zero, we can solve for τ_k, and the solution happens to have the following explicit form:

$$\tau_k = \frac{\gamma_k^2 + b}{a+3} \qquad (22.25)$$

After each component of θ is updated, we complete one iteration. The process repeats for many iterations until a certain criterion of convergence is satisfied. The final solution of the parameter vector, denoted by $\hat{\theta}$, is the penalized estimate. When $(a, b) = (0, 0)$, the above solution becomes $\tau_k = \gamma_k^2 / 3$, and the shrinkage is very strong. Choosing $(a, b) = (-2, 0)$ leads to $\tau_k = \gamma_k^2$, which corresponds to the uniform prior for τ_k, and the shrinkage is weak.

22.4 Cross Validation

The penalized maximum likelihood method requires hyperparameters a and b in the scaled inverse chi-square distribution. The $(a, b) = (0, 0)$ choice is called the Jeffreys' prior for a variance component. This prior is not necessarily optimal. We need some criteria to evaluate the optimality of the hyperparameters. The most reliable measure of the model fit is the average prediction error (PE) defined as

$$\Omega = \frac{1}{M} \sum_{j=1}^{M} (y_j - \hat{\mu}_j)^2 \tag{22.26}$$

where

$$\hat{\mu}_j = \frac{e^{X_j \hat{\beta} + Z_j \hat{\gamma}}}{1 + e^{X_j \hat{\beta} + Z_j \hat{\gamma}}} \tag{22.27}$$

is the predicted outcome of sample j and $(\hat{\beta}, \hat{\gamma})$ are estimated parameters from the population that excludes the jth sample. This type of cross validation is called the leave-one-out cross validation or n-fold cross validation where the n means the sample size. In the contest of microarray data analysis in this chapter, the sample size is denoted by M, and thus, it should be called the M-fold cross validation. In general, one can choose any K-fold cross validation to select the optimal hyperparameter values, where $K = 2, \ldots, M$. The basic principle of cross validation is that individuals predicted do not contribute to parameter estimation. Under each selection of (a, b), we can calculate the PE, denoted by $\Omega(a, b)$, as a function of (a, b). The value of (a, b) that minimizes $\Omega(a, b)$ is the optimal choice of the hyperparameters.

22.5 Prediction of Disease Outcome

The ultimate purpose of the logistic regression is to predict the disease outcome of new subjects using the microarray data. Let X_j^* and Z_j^* be the design matrices for the tissue sample from a new subject. Based on the estimated parameters $(\hat{\beta}, \hat{\gamma})$, we can calculate the probability of the unknown disease outcome of individual j using

$$\Pr(y_j^* = 1) = \mu_j^* = \frac{e^{X_j^* \hat{\beta} + Z_j^* \hat{\gamma}}}{1 + e^{X_j^* \hat{\beta} + Z_j^* \hat{\gamma}}}$$

where y_j^* is the unknown disease outcome. This probability can help physicians to make recommendation to patients regarding how to prevent the disease.

22.6 Multiple-Category Classification

Multiple-category classifications using microarray data have been extensively studied (Lee and Lee 2003; Yeung and Bumgarner 2003; Dagliyan et al. 2011). We now extend binary classification into multiple-category classification using the penalized logistic regression analysis. Assume that there are C categories of the tissue samples. Let $y_j = \begin{bmatrix} y_{j1} & \cdots & y_{jC} \end{bmatrix}^T$ be the outcome of the disease category for the jth tissue sample, where

$$y_{jc} = \begin{cases} 1 & \text{if } j \text{ belongs to category } c \\ 0 & \text{if } j \text{ belongs to category } c' \end{cases} \tag{22.28}$$

where $c' \neq c$. Let

$$\mathrm{E}(y_j) = \mu_j = \begin{bmatrix} \mu_{j1} & \cdots & \mu_{jC} \end{bmatrix}^T \tag{22.29}$$

be the expectation of vector y_j, where

$$\mu_{jc} = \frac{e^{\beta_c + \sum_{k=1}^{N} z_{jk}\, \gamma_{kc}}}{\sum_{c'=1}^{C} e^{\beta_{c'} + \sum_{k=1}^{N} z_{jk}\, \gamma_{kc'}}} \tag{22.30}$$

For each category, there is an intercept and a set of regression coefficients. We now define

$$\beta = \begin{bmatrix} \beta_1 & \cdots & \beta_C \end{bmatrix}^T \tag{22.31}$$

and

$$\gamma_k = \begin{bmatrix} \gamma_{k1} & \cdots & \gamma_{kC} \end{bmatrix}^T \tag{22.32}$$

The penalized likelihood function is defined as

$$\psi(\theta) = \sum_{j=1}^{M} (y_{j1} \ln \mu_{j1} + \cdots + y_{jC} \ln \mu_{jC})$$

$$+ \sum_{k=1}^{N} \ln N(\gamma_k | 0, I\, \tau_k)$$

$$+ \sum_{k=1}^{N} \ln \mathrm{Inv} - \chi^2(\tau_k | a, b) \tag{22.33}$$

A simple extension of the binary penalized logistic regression to the multiple-category case can be made. The sequential estimate of each parameter set is shown below. Let

$$W_j(\theta) = \mathrm{diag}\begin{bmatrix} \mu_{j1} & \cdots & \mu_{jC} \end{bmatrix} - \begin{bmatrix} \mu_{j1} & \cdots & \mu_{jC} \end{bmatrix}^T \begin{bmatrix} \mu_{j1} & \cdots & \mu_{jC} \end{bmatrix} \tag{22.34}$$

be the variance–covariance matrix of y_j, and define the generalized inverse of $W_j(\theta)$ as

$$W_j^{-1}(\theta) = \text{diag}\left[\mu_{j1}^{-1} \cdots \mu_{jC}^{-1} \right] \tag{22.35}$$

Conditional on all other parameters, the intercept is updated using

$$\beta^{(t+1)} = \left[\sum_{j=1}^{M} W_j(\theta^{(t)}) \right]^{-1} \left[\sum_{j=1}^{M} W_j(\theta^{(t)}) y_j^* \right] \tag{22.36}$$

where

$$y_j^* = \beta^{(t)} + W_j^{-1}(\theta^{(t)})(y_j - \mu_j) \tag{22.37}$$

is the pseudodata point. The regression coefficients for gene k are updated using

$$\gamma_k^{(t+1)} = \left[\sum_{j=1}^{M} Z_{jk}^T W_j(\theta^{(t)}) Z_{jk} + \frac{1}{\tau_k} I \right]^{-1} \left[\sum_{j=1}^{M} Z_{jk}^T W_j(\theta^{(t)}) y_j^* \right] \tag{22.38}$$

where

$$y_j^* = Z_{jk}\gamma_k^{(t)} + W_j^{-1}(\theta^{(t)})(y_j - \mu_j) \tag{22.39}$$

is the pseudodata point. Given γ_k, the updated τ_k happens to have the following explicit form:

$$\tau_k = \frac{\gamma_k^T \gamma_k + b}{a + C + 2} \tag{22.40}$$

Iterations are required, and the penalized logistic estimates of the parameters take the converged values of the parameters.

To predict the outcomes of a new tissue sample, we use the estimated parameters and the new gene expression Z_{jk}^* data to calculate

$$\Pr(y_{jc}^* = 1) = \mu_{jc}^* = \frac{e^{\hat{\beta}_c + \sum_{k=1}^{N} Z_{jk}^* \hat{\gamma}_{kc}}}{\sum_{c'=1}^{C} e^{\hat{\beta}_{c'} + \sum_{k=1}^{N} Z_{jk}^* \hat{\gamma}_{kc'}}}, \forall c = 1, \ldots, C \tag{22.41}$$

The jth sample will be classified into the class which has the maximum probability.

Chapter 23
Time-Course Microarray Data Analysis

In a time-course microarray experiment, each condition or sample represents a time point after a particular treatment on tissue samples. Because mRNA abundance of different time points are correlated, we may express the mRNA abundance as a function of time. This function is usually nonlinear, and the visual plot of this function against time is called gene expression profile. Similar to time-course microarray experiment, each condition or sample may represent a particular dose of a drug injection. We may also express the mRNA abundance as a function of the dosage. This kind of experiment is called dose-response microarray experiment.

23.1 Gene Expression Profiles

Cluster analysis of time-course or dose-response microarray data may be performed based on the functional relationships of gene expression and the time points, rather than based on the original expression levels of different time points. Because a curve is governed by a few parameters, cluster analysis is actually performed based on the parameters that generate the curves. Gene expression profiles can have many different shapes; each different shape may be described by a different function, e.g., linear, quadratic, exponential, or logistic. However, orthogonal polynomial is a method of curve fitting that can be applied to more complicated functional relationships (Seber 1977; Narula 1979).

B-spline is another commonly used curve fitting approach (de Boor 1978; Welham et al. 2007). Unlike the orthogonal polynomial curve fitting that a particular degree of polynomial is fit to the entire range of the time course, the B-spline fits a particular degree of polynomial to multiple ranges of the time course. A B-spline function is a piecewise polynomial function, i.e., the entire curve is divided into many different pieces and the curve in each piece is a polynomial. All the piecewise polynomials have exactly the same degree (order) and are connected smoothly at the joint points whose abscissa values, referred to as knots, are prespecified. We can use B-spline functions to fit curves to a wide variety of data, not limited to time

S. Xu, *Principles of Statistical Genomics*, DOI 10.1007/978-0-387-70807-2_23,
© Springer Science+Business Media, LLC 2013

course. Like the orthogonal polynomial where we use the polynomial function to describe the relationship between the time point and the response variable, the B-spline uses the B-spline function to describe the relationship. Both B-spline and orthogonal polynomial will be discussed in this chapter, but more detail will be given to the polynomial cure fitting with a very brief description on the B-spline.

Recent methodology developments and applications in time-course microarray data can be found in Cullinan et al. (1995), Peddada et al. (2003), Park et al. (2003), Schliep et al. (2003), Luan and Li (2003), Glonek and Solomon (2004), Ma et al. (2006), and Storey et al. (2005). The method we will introduce in this chapter follows that of Luan and Li (2003). Although Luan and Li (2003) used B-splines, we will adopt their method of clustering in both B-splines and orthogonal polynomials.

23.2   Orthogonal Polynomial

Let t be the time point and y be the gene expression. The polynomial relationship between y and t is expressed as

$$y = f(t) + \varepsilon = \gamma_0 + t\gamma_1 + t^2\gamma_2 + ,\ldots, +t^d\gamma_d + \varepsilon \qquad (23.1)$$

where $\gamma_r, \forall r = 0,\ldots,d$ is the polynomial coefficient and d is the degree of the polynomial. We can choose d to reflect the complexity of the curve. For example, $d = 1$ and $d = 2$ represent linear and quadratic curves, respectively. The polynomial coefficients can be estimated from the data. One desirable property of the polynomial function is that the function is linear on the parameters, leading to an easy way to estimate the parameters.

Let $t_k, \forall k = 1,\ldots,M$, be the actual time point for the kth condition and y_{kj} be the expression level of gene j at time t_k. For example, if gene expressions are measured at five time points with equal time interval, say, 0 min, 20 min, 40 min, 60 min, and 80 min, then $t_1 = 0$, $t_2 = 20$, $t_3 = 40$, $t_4 = 60$, and $t_5 = 80$. Note that the expression data are stored in an $M \times N$ matrix with the jth column representing the expressions of gene j across M time points. To fit the orthogonal polynomial, we need to rescale the actual time points into time points in a standardized scale so that the range of the standardized time points runs from -1 to $+1$. The standardize time point is defined as

$$\tau_k = 2\frac{(t_k - t_{\min})}{(t_{\max} - t_{\min})} - 1 \qquad (23.2)$$

where $(t_{\min}, t_{\max})$ is the actual range of the time points. The standardized time points need to be transformed again before used to fit the orthogonal polynomial. The transformed time point is denoted by $\psi_r(\tau_k), \forall r = 0, 1,\ldots,d$.

Note that we are dealing with thousands of genes, each of which is fitted by a polynomial model. For the jth gene, the polynomial model is

$$y_{kj} = \psi_0(\tau_k)\gamma_{0j} + \psi_1(\tau_k)\gamma_{1j} + \cdots + \psi_d(\tau_k)\gamma_{dj} + \varepsilon_{kj} \qquad (23.3)$$

where $\psi_0(\tau_k) = 1, \forall k = 1, \ldots, M$. The orthogonal property is reflected by

$$\sum_{k=1}^{M} \psi_r(\tau_k)\psi_s(\tau_k) = 0 \qquad (23.4)$$

for all $r \neq s$ and

$$\sum_{k=1}^{M} \psi_r(\tau_k) = 0 \qquad (23.5)$$

for $r = 1, \ldots, d$. The coefficients of orthogonal polynomial can be obtained in a number of ways. We used the Hayes' (1974) three-term recurrence relationship,

$$\psi_{r+1}(\tau_k) = 2(\tau_k - a_r)\psi_r(\tau_k) - b_r\psi_{r-1}(\tau_k) \qquad (23.6)$$

Beginning with initial polynomials $\psi_0(\tau_k) = 1$, $\psi_1(\tau_k) = 2(\tau_k - a_0)$, a_r and b_r are chosen to make the orthogonal relations hold, namely,

$$a_r = \frac{\sum_{k=1}^{N} \tau_k \psi_r^2(\tau_k)}{\sum_{k=1}^{n} \psi_r^2(\tau_k)} \text{ and } b_r = \frac{\sum_{k=1}^{n} \psi_r^2(\tau_k)}{\sum_{k=1}^{N} \psi_{r-1}^2(\tau_k)} \qquad (23.7)$$

where $r = 0, \ldots, d-1$, $b_0 = 0$ and $a_0 = \frac{1}{M}\sum_{k=1}^{M} \tau_k$. The transformed time points can be used as the independent variables for data analysis. The a and b vector, each with dimension $d + 1$, will be used later when we predict the gene expression for any time point within the range of the time course. After all the $\psi_r(\tau_k)$ are found, we may estimate the regression coefficients γ_j, classify the genes based on their γ_j values, and predict the curves for each gene in each cluster. Now let us define the following variables using a matrix notation so that the model can be rewritten in a compact matrix form. Let

$$y_j = \begin{bmatrix} y_{1j} & y_{2j} & \cdots & y_{Mj} \end{bmatrix}^T \qquad (23.8)$$

be an $M \times 1$ vector of expressions for the jth gene,

$$\gamma_j = \begin{bmatrix} \gamma_{0j} & \gamma_{1j} & \cdots & \gamma_{dj} \end{bmatrix}^T \qquad (23.9)$$

be an $(d + 1) \times 1$ vector of the regression coefficients,

$$\varepsilon_j = \begin{bmatrix} \varepsilon_{1j} & \varepsilon_{2j} & \cdots & \varepsilon_{Mj} \end{bmatrix}^T \qquad (23.10)$$

be an $M \times 1$ vector of the residual errors, and

$$\psi(\tau) = \begin{bmatrix} \psi_0(\tau_1) & \psi_1(\tau_1) & \cdots & \psi_d(\tau_1) \\ \psi_0(\tau_2) & \psi_1(\tau_2) & \cdots & \psi_d(\tau_2) \\ \cdots & \cdots & \cdots & \cdots \\ \psi_0(\tau_M) & \psi_1(\tau_M) & \cdots & \psi_d(\tau_M) \end{bmatrix} \qquad (23.11)$$

be an $M \times (d + 1)$ matrix for the orthogonal polynomial coefficients. The linear model of expression for gene j in matrix notation is

$$y_j = \psi(\tau)\gamma_j + \varepsilon_j \qquad (23.12)$$

The residual error vector is assumed to be normally distributed $\varepsilon_j \sim N(0, I\sigma^2)$ where σ^2 is a common residual variance shared by all time points. We now use an example to demonstrate the polynomial coefficient matrix $\psi(\tau)$. Let $M = 6$ be the number of time points and

$$t = \begin{bmatrix} 0 & 20 & 40 & 50 & 60 & 80 \end{bmatrix} \qquad (23.13)$$

be the original time point vector. The standardized time points are

$$\tau = \begin{bmatrix} -1.00 & -0.50 & 0.00 & +0.25 & +0.50 & +1.00 \end{bmatrix} \qquad (23.14)$$

Let $d = 3$ be the degree of the polynomial to be fit. The polynomial coefficient matrix is

$$\psi(\tau) = \begin{bmatrix} 1 & -2.083333 & 2.1428571 & -1.363636 \\ 1 & -1.083333 & -0.785714 & 2.7000000 \\ 1 & -0.083333 & -1.714286 & 0.4909091 \\ 1 & 0.4166667 & -1.428571 & -1.090909 \\ 1 & 0.9166667 & -0.642857 & -1.990909 \\ 1 & 1.9166667 & 2.4285714 & 1.2545455 \end{bmatrix} \qquad (23.15)$$

The values of vector a and b are

$$a = \begin{bmatrix} 0.0416667 \\ 0.0000000 \\ -0.077381 \\ 0.1038961 \end{bmatrix} \quad \text{and} \quad b = \begin{bmatrix} 0.0000000 \\ 1.7013889 \\ 1.6163265 \\ 0.0000000 \end{bmatrix} \qquad (23.16)$$

which will be used later to calculate the orthogonal polynomial coefficients for new time points. For example, suppose that we want to predict gene expressions for three new time points in addition to the existing six time points. The new time sequence will be

$$t = \begin{bmatrix} 0 & 10 & 20 & 30 & 40 & 50 & 60 & 70 & 80 \end{bmatrix}$$

After standardization, the time points in the standardized scale are

$$\tau = \begin{bmatrix} -1.00 & -0.75 & -0.50 & -0.25 & 0.00 & 0.25 & 0.50 & 0.75 & 1.00 \end{bmatrix}$$

The new time points that are not observed are 10, 30, and 70. Using the updated list of time points with the new time points and the existing a and b vectors calculated before, we get the new $\psi(\tau)$ matrix,

$$\psi(\tau) = \begin{bmatrix} 1 & -2.083333 & 2.1428571 & -1.363636 \\ 1 & -1.583333 & 0.4285714 & 1.8272727 \\ 1 & -1.083333 & -0.785714 & 2.7000000 \\ 1 & -0.583333 & -1.500000 & 2.0045455 \\ 1 & -0.083333 & -1.714286 & 0.4909091 \\ 1 & 0.4166667 & -1.428571 & -1.090909 \\ 1 & 0.9166667 & -0.642857 & -1.990909 \\ 1 & 1.4166667 & 0.6428571 & -1.459091 \\ 1 & 1.9166667 & 2.4285714 & 1.2545455 \end{bmatrix} \qquad (23.17)$$

Some software packages, e.g., the SAS/IML program (SAS Institute 2008a), have an intrinsic function to calculate the orthogonal polynomial coefficients. Users provide the observed time points and the degree of the polynomial and call the orthogonal polynomial function $\psi(\tau) = \text{OrPol}(\tau, d)$. The returned value of $\psi(\tau)$, however, will be different from what is calculated using the Hayes' three-term recurrent algorithm. For example, calling the orthogonal polynomial function using the six-time-point sequence with degree of three will return a $\psi(\tau)$ matrix

$$\psi(\tau) = \begin{bmatrix} 0.4082483 & -0.652051 & 0.527535 & -0.339657 \\ 0.4082483 & -0.339066 & -0.193429 & 0.6725208 \\ 0.4082483 & -0.026082 & -0.422028 & 0.1222765 \\ 0.4082483 & 0.1304101 & -0.35169 & -0.271726 \\ 0.4082483 & 0.2869023 & -0.15826 & -0.495899 \\ 0.4082483 & 0.5998866 & 0.597873 & 0.3124844 \end{bmatrix} \qquad (23.18)$$

The difference between matrix (23.18) and matrix (23.15) is due to the normalization of the OrPol function in SAS. The normalization process is expressed as

$$\psi_r(\tau_k) = \frac{\psi_r(\tau_k)}{\sqrt{\sum_{k=1}^{N} \psi_r^2(\tau_k)}}, \forall r = 0, 1, \ldots, d \qquad (23.19)$$

Fig. 23.1 Polynomial curve
with order three given
in (23.20)

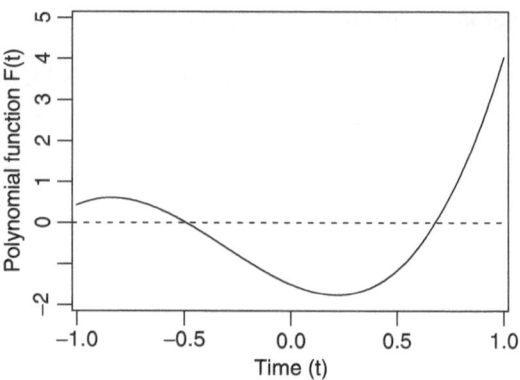

where the $\psi_r(\tau_k)$ values in the right-hand side of (23.19) are those calculated from
the Hayes' algorithm and the $\psi_r(\tau)$ in the left-hand side of (23.19) is the normalized
value. After the normalization, the sum of squares of each column of $\psi(\tau)$ equals
unity, i.e., $\sum_{k=1}^{M} \psi_r^2(\tau_k) = 1$. Rescaling of the time points for the polynomial
analysis is necessary, but normalization of the polynomial coefficients should be
avoided because we cannot calculate the polynomial coefficients correctly for new
time points after the normalization.

We now use the time points (rescaled) given in (23.14) as an example to
demonstrate the smooth polynomial function for a given set of parameters (γ).
Using the τ vector and $d = 3$, we already calculated vectors a and b as shown
in (23.16). Once the a and b vectors are calculated, the time point vector τ is no
longer useful for drawing a smooth curve. This original time point vector is only
used to find polynomial coefficient matrix $\psi(\tau)$ for parameter estimation (discussed
later). Let us create a new vector of time points ranging from -1 to $+1$ with 0.01
as a constant step of increment. This new vector of time points has 201 elements.
For each element (time point), say τ_k, we used (23.6) to calculate

$$\psi(\tau_k) = \begin{bmatrix} \psi_0(\tau_k) & \psi_1(\tau_k) & \psi_2(\tau_k) & \psi_3(\tau_k) \end{bmatrix}$$

Let

$$\gamma = \begin{bmatrix} \gamma_0 & \gamma_1 & \gamma_2 & \gamma_3 \end{bmatrix}^T = \begin{bmatrix} 0.01 & 0.50 & 1.00 & 0.50 \end{bmatrix}^T$$

be the parameter values (either given or estimated from data). The predicted gene
expression at time τ_k is then obtained through

$$\tilde{y}(\tau_k) = \gamma_0\psi_0(\tau_k) + \gamma_1\psi_1(\tau_k) + \gamma_2\psi_2(\tau_k) + \gamma_3\psi_3(\tau_k) \tag{23.20}$$

We then calculated the predicted gene expression for all $k = 1,\ldots,201$ to form a
vector $\tilde{y}$. The plot of $\tilde{y}$ against the new τ vector (with 201 elements) forms a smooth
curve as shown in Fig. 23.1.

23.3   B-spline

The B-spline bases are equivalent to the orthogonal polynomial coefficients, and they are derived before a B-spline function is fit. For a B-spline with d degrees and s internal knots, there will be $p = d + s + 1$ B-spline bases, excluding the intercept. An internal knot value must be taken between (but not including) the minimum and maximum values of the time points. Using the same notation as the orthogonal polynomial, let $\psi_r(\tau)$, for $r = 0, 1, \ldots, p$, be the rth B-spline base at time τ that is defined either in the standardized scale or in the original scale, the B-spline function is defined as

$$y_j(\tau) = \psi_0(\tau)\gamma_{0j} + \psi_1(\tau)\gamma_{1j} + \cdots + \psi_p(\tau)\gamma_{pj} \qquad (23.21)$$

where $\psi_0(\tau) = 1$, γ_{0j} is the intercept and $\gamma_{1j}, \cdots, \gamma_{pj}$ are the regression coefficients. The B-spline bases

$$\psi(\tau) = \left[\psi_1(\tau) \cdots \psi_p(\tau) \right] \qquad (23.22)$$

are calculated using the algorithm given by de Boor (1978). The algorithm is not too difficult to code, but it is hard to express in the current notation system within the text. Some software package has a function to create the B-spline bases given the degree and the knots, such as the B-spline function in the SAS/IML language (SAS Institute 2008a). What we need to know is to define the knots and call the B-spline function to create the p vectors of the B-spline bases. There are two ways to define the knots given the range of the time points; one is to select the knots manually, and the other is to select equally spaced knots. We now use an example to show how to select the knots. Assume that the time points are

$$\tau = \left[\tau_1 \ \tau_2 \ \tau_3 \ \tau_4 \right] = \left[2.5 \ 3.0 \ 4.5 \ 5.1 \right]$$

The knot sequence must contain s interval knots, d external knots below the minimum time point, and d external knots above the maximum time point for $d \geq 1$. The minimum and maximum time points cannot be included in the selected knot sequence. The total number of knots in the sequence is $s + 2d$. Let us assume that $d = 3$ and we want to choose $s = 3$ internal knots. The following knot sequence is legal:

$$\phi = \left[0 \ 1 \ 2 \ 3 \ 4 \ 5 \ 6 \ 7 \ 8 \right]$$

where the three internal knots are 3, 4, and 5. The three external knots below 2.5 (the minimum time point) are 0, 1, and 2, and the three external knots above 5.1 (the maximum time point) are 6, 7, and 8. When we call the B-spline function in SAS/IML $\psi(\tau) = \mathrm{Bspline}(\tau, d, \phi)$, we will get

$$\psi(\tau) = \begin{bmatrix} 0.02083 & 0.47917 & 0.47917 & 0.02083 & 0 & 0 & 0 \\ 0 & 0.16667 & 0.66667 & 0.16667 & 0 & 0 & 0 \\ 0 & 0 & 0.02083 & 0.47917 & 0.47917 & 0.02083 & 0 \\ 0 & 0 & 0 & 0.1215 & 0.65717 & 0.22117 & 0.00017 \end{bmatrix}$$

(23.23)

To define equally spaced knot sequence, we first need to specify the number of internal knots, say $s = 3$, and find the three interval knots, which are

$$[\min(\tau) + \text{space}, \min(\tau) + 2 \times \text{space}, \min(\tau) + 3 \times \text{space}]$$

where

$$\text{space} = \frac{[\max(\tau) - \min(\tau)]}{s + 1} = \frac{5.1 - 2.5}{3 + 1} = 0.65 \qquad (23.24)$$

Let $\omega = 10^{-12}$ be a small positive number. The d external knots below $\min(\tau)$ are

$$[\min(\tau) - \omega - 2 \times \text{space}, \min(\tau) - \omega - \text{space}, \min(\tau) - \omega]$$

and the d external knots above $\max(\tau)$ are

$$[\min(\tau) + \omega, \min(\tau) + \omega + \text{space}, \min(\tau) + \omega + 2 \times \text{space}]$$

The small number ω added or subtracted in the external knots is to make sure that the external knots do not include the maximum and minimum values of the time points. Adding all the external and internal knots together, we get the overall knot sequence

$$\phi = \begin{bmatrix} 1.20 - \omega & 1.85 - \omega & 2.50 - \omega & 3.15 & 3.80 & 4.45 & 5.10 + \omega & 5.75 + \omega & 6.40 + \omega \end{bmatrix}$$

We then call the B-spline function in SAS $\psi(\tau) = \text{Bspline}(\tau, d, \phi)$; we will get

$$\psi(\tau) = \begin{bmatrix} 0.16667 & 0.66667 & 0.16667 & 0 & 0 & 0 & 0 \\ 0.00205 & 0.30253 & 0.61956 & 0.07586 & 0 & 0 & 0 \\ 0 & 0 & 0 & 0.13109 & 0.66098 & 0.20786 & 0.00008 \\ 0 & 0 & 0 & 0 & 0.16667 & 0.66667 & 0.16667 \end{bmatrix}$$

(23.25)

We can also rely on the B-spline function to compute a knot vector. For example, the above B-spline bases (matrix (23.25)) can also be generated using the $\psi(\tau) = \text{Bspline}(\tau, d, ., s)$ function in SAS, where the knot sequence ϕ has been replaced by a missing value (represented by a period) and the number of equally spaced internal knots $s = 3$. The B-spline function actually generates the same knot sequence ϕ and then uses this sequence to create the B-spline bases. Note that the knot sequence is only determined by the minimum and maximum values of the actual time points in the τ vector, i.e., the interval values of vector τ are irrelevant to the knot sequence.

Special properties of the B-spline bases include the following: (1) The sum of each row equals one, and (2) each row has at most $d + 1 = 4$ nonzero values. Although B-spline can fit data to more complicated curves, the B-spline bases do not have a meaningful statistical property such as linear or quadratic interpretation of the curve. This property, however, holds for the orthogonal polynomial coefficients where the first column of $\psi(\tau)$ corresponds to linear, the second column of $\psi(\tau)$ corresponds quadratic, etc. On the other hand, when predicting gene expression for a new time point, we can use exactly the same B-spline function to call the new $\psi(\tau)$ matrix. Consider the time points

$$\tau = \begin{bmatrix} 2.5 \ 3.0 \ 4.5 \ 5.1 \end{bmatrix}$$

and the knot sequence

$$\phi = \begin{bmatrix} 0 \ 1 \ 2 \ 3 \ 4 \ 5 \ 6 \ 7 \ 8 \end{bmatrix}$$

We now want to calculate the $\psi(\tau)$ matrix by adding two new time points so that

$$\tau = \begin{bmatrix} 2.5 \ 3.0 \ 3.5 \ 4.5 \ 5.0 \ 5.1 \end{bmatrix}$$

Using the same knot sequence and the same degree of the polynomial to call the B-spline function, we get

$$\psi(\tau) = \begin{bmatrix} 0.02083 & 0.47917 & 0.47917 & 0.02083 & 0 & 0 & 0 \\ 0 & 0.16667 & 0.66667 & 0.16667 & 0 & 0 & 0 \\ 0 & 0.02083 & 0.47917 & 0.47917 & 0.02083 & 0 & 0 \\ 0 & 0 & 0.02083 & 0.47917 & 0.47917 & 0.02083 & 0 \\ 0 & 0 & 0 & 0.16667 & 0.66667 & 0.16667 & 0 \\ 0 & 0 & 0 & 0.12150 & 0.65717 & 0.22117 & 0.00017 \end{bmatrix}$$

$$(23.26)$$

Comparing this matrix (matrix (23.26)) with the one without the new points (matrix (23.23)), we can see that the values corresponding to the observed time points remain the same.

Note that the $\psi(\tau)$ matrix returned from the B-spline function call does not have the intercept. We now have to add an additional column of unity $\psi_0(\tau) = 1$ to the left of matrix $\psi(\tau)$ to make it an $M \times (p + 1)$ matrix so that the intercept γ_{0j} can also be estimated along with the regression coefficients.

Let τ be a vector of time points ranging from -1 to $+1$ with 0.01 as a constant step of increment. This vector of time points has 201 elements, the same τ used

Fig. 23.2 B-spline curve
with $s = 2$ equally spaced
internal knots and $d = 3$
polynomial degree for each
interval. The B-spline
function is given in (23.27)

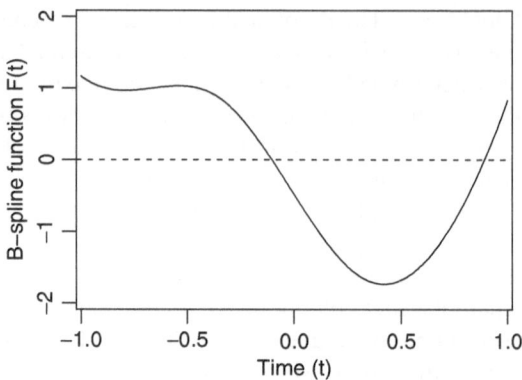

to demonstrate the polynomial curve (Fig. 23.1). We want to create B-spline bases using this τ with $s = 2$ equally spaced internal knots between -1 and $+1$ and $d = 3$ degrees of each polynomial segment. The number of B-spline bases is $p = d + s + 1 = 6$. Including the intercept, we should have seven $\psi(\tau_k)$ coefficients for each time point τ_k for $k = 1, \ldots, 201$. The ψ corresponding to the intercept is denoted by $\psi_0(\tau_k) = 1$. The six B-spline bases are calculated using the B-spline function in SAS as $\psi(\tau) = \text{Bspline}(\tau, d, ., s)$ where $d = 3$ and $s = 2$. Let

$$\gamma = [\gamma_0, \gamma_1, \gamma_2, \gamma_3, \gamma_4, \gamma_5, \gamma_6] = [0, 5, 0, 2, -3, 0, 8]$$

be the parameters (regression coefficients) either given or estimated from data. The predicted gene expression at time τ_k is obtained through

$$\tilde{y}(\tau_k) = \sum_{r=0}^{6} \gamma_r \psi_r(\tau_k) \tag{23.27}$$

We then calculated the predicted gene expression for all $k = 1, \ldots, 201$ to form a vector $\tilde{y}$. The plot of $\tilde{y}$ against τ that forms a smooth B-spline curve is shown in Fig. 23.2.

23.4   Mixed Effect Model

This section only presents some models for parameter estimation. The actual computational algorithms will be given in later sections. Instead of clustering genes based on y_j (with dimension M), we now try to cluster genes based on γ_j. The dimension of γ_j is $d + 1$ for polynomial and $p + 1$ for B-spline. Both $d + 1$ and $p + 1$ can be substantially smaller than M. Assume that these γ_j's are sampled from C multivariate normal distributions (clusters). Let $\eta_j = 1, \ldots, C$ be the cluster index, i.e., if the jth gene belongs to the κth cluster, then $\eta_j = \kappa$. If we know that γ_j belongs to cluster κ, we can use the following model to describe γ_j:

$$\gamma_j|_{\eta_j=\kappa} = \mu_\kappa + \alpha_j \tag{23.28}$$

where

$$\mu_\kappa = \begin{bmatrix} \mu_{0\kappa} & \mu_{1\kappa} & \dots & \mu_{d\kappa} \end{bmatrix}^T. \tag{23.29}$$

is the mean vector for cluster κ and

$$\alpha_j = \begin{bmatrix} \alpha_{0j} & \alpha_{1j} & \dots & \alpha_{dj} \end{bmatrix}^T \tag{23.30}$$

is a vector of random deviations of γ_j from the mean, called the random effects. As a vector of random effects, we assume $\alpha_j \sim N(0, \Sigma_\kappa)$, where Σ_κ is a $(d+1) \times (d+1)$ covariance matrix. We now rewrite (23.12) as

$$y_j|_{\eta_j=\kappa} = \psi(\tau)\mu_\kappa + \psi(\tau)\alpha_j + \varepsilon_j \tag{23.31}$$

This is a mixed effect model because μ_κ now represents the fixed effect and α_j represents the random effect. A model containing both the fixed and random effects is called a mixed effect model, or simply mixed model. Note that mixed model and mixture model are completely different concepts. Conditional on the cluster label $\eta_j = \kappa$, the model is a mixed model.

23.5 Mixture Mixed Model

When the cluster label is unknown, the model becomes a mixture model. Overall, we have to deal with a situation with a mixture of several mixed models. To estimate parameters using the original gene expression data, we must examine the distribution of $y_j|_{\eta_j=\kappa}$, which turns out to be multivariate normal with mean

$$E(y_j|\eta_j = \kappa) = \psi(\tau)\mu_\kappa \tag{23.32}$$

and variance

$$\text{var}(y_j|\eta_j = \kappa) = V_\kappa = \psi(\tau)\Sigma_\kappa\psi^T(\tau) + I\sigma^2 \tag{23.33}$$

The multivariate normal density can be expressed as

$$p(y_j|\eta_j = \kappa) = N(y_j|\psi(\tau)\mu_\kappa, V_\kappa)$$

$$= \frac{1}{|V_\kappa|^{1/2}} \exp\left\{-\frac{1}{2}[y_j - \psi(\tau)\mu_\kappa]^T V_\kappa^{-1}[y_j - \psi(\tau)\mu_\kappa]\right\}$$

$$\tag{23.34}$$

The mixture distribution can be written as

$$p(y_j|\theta) = \sum_{\kappa=1}^{C} \pi_\kappa N(y_j|\psi(\tau)\mu_\kappa, V_\kappa) \tag{23.35}$$

where $\pi_\kappa, \forall \kappa = 1, \ldots, C$ is the mixing proportion. The overall log likelihood function is

$$L(\theta) = \sum_{j=1}^{M} \ln p(y_j|\theta) \tag{23.36}$$

where

$$\theta = \{\pi_1, \ldots, \pi_C, \mu_1, \ldots, \mu_C, \Sigma_1, \ldots, \Sigma_C, \sigma^2\} \tag{23.37}$$

is a vector containing all the parameters.

Numerous algorithms are available to estimate the parameters. We will introduce two algorithms in this chapter, which are the expectation-maximization (EM) algorithm and the stochastic expectation-maximization (SEM) algorithm. The Bayesian method can also be applied here, but we defer it to Chap. 24 when we deal with quantitative trait-associated microarray data analysis.

23.6   EM Algorithm

The EM algorithm cluster analysis developed by Luan and Li (2003) is adopted here. The algorithm is summarized by the following steps:

1. Set $t = 0$ and initialize the mixing proportions by

$$\pi_\kappa^{(t)} = \frac{1}{C}, \forall \kappa = 1, \ldots, C \tag{23.38}$$

   and the probabilities of cluster assignments by

$$\rho_{j\kappa}^{(t)} = \pi_\kappa^{(t)}, \forall j = 1, \ldots, N \ \& \ \kappa = 1, \ldots, C \tag{23.39}$$

   where N is the number of genes.
2. Update the fixed effects (mean vectors) by

$$\mu_\kappa^{(t)} = \left[N\pi_\kappa^{(t-1)}\psi^T(\tau)V_\kappa^{-1}\psi(\tau) \right]^{-1} \psi^T(\tau)V_\kappa^{-1} \sum_{j=1}^{M} \pi_{j\kappa}^{(t-1)} y_j \tag{23.40}$$

   where

$$V_\kappa = \psi(\tau)\Sigma_\kappa^{(t-1)}\psi^T(\tau) + I\sigma^{2(t-1)} \tag{23.41}$$

   for all $\kappa = 1, \ldots, C$.

3. Update the random effects by

$$\alpha_j^{(t)} = \Sigma_\kappa^{(t-1)} \psi^T(\tau) V_\kappa^{-1} \left(y_j - \psi(\tau) \mu_\kappa^{(t)} \right) \tag{23.42}$$

and compute the conditional covariance matrix of α_j

$$S_j^{(t)} = \Sigma_\kappa^{(t-1)} - \Sigma_\kappa^{(t-1)} \psi^T(\tau) V_\kappa^{-1} \psi(\tau) \Sigma_\kappa^{(t-1)} \tag{23.43}$$

for all $j = 1, \ldots, N$ & $\kappa = 1, \ldots, C$.
4. Update the covariance matrices by

$$\begin{aligned}
\Sigma_\kappa^{(t)} &= \frac{1}{N \pi_\kappa^{(t-1)}} \left[\sum_{j=1}^N \rho_{j\kappa}^{(t-1)} E(\alpha_j \alpha_j^T) \right] \\
&= \frac{1}{N \pi_\kappa^{(t-1)}} \left[\sum_{j=1}^M \rho_{j\kappa}^{(t-1)} \left(S_j^{(t)} + \alpha_j^{(t)} (\alpha_j^{(t)})^T \right) \right]
\end{aligned} \tag{23.44}$$

5. Update the residual variance by

$$\sigma^{2(t)} = \frac{1}{NM} \sum_{j=1}^N y_j^T \left[y_j - \sum_{\kappa=1}^C \rho_{j\kappa}^{(t-1)} \psi(\tau) \left(\mu_\kappa^{(t)} + \alpha_j^{(t)} \right) \right] \tag{23.45}$$

6. Update the posterior probabilities of cluster assignments,

$$\rho_{j\kappa}^{(t)} = \frac{\pi_\kappa^{(t-1)} N(y_j | \psi(\tau) \mu_\kappa^{(t)}, \psi(\tau) \Sigma_\kappa^{(t)} \psi^T(\tau) + I \sigma^{2(t)})}{\sum_{\kappa'}^C \pi_{\kappa'}^{(t-1)} N(y_j | \psi(\tau) \mu_{\kappa'}^{(t)}, \psi(\tau) \Sigma_\kappa^{(t)} \psi^T(\tau) + I \sigma^{2(t)})} \tag{23.46}$$

for all $j = 1, \ldots, N$ & $\kappa = 1, \ldots, C$.
7. Update the mixing proportions,

$$\pi_\kappa^{(t)} = \frac{1}{N} \sum_{j=1}^N \rho_{j\kappa}^{(t)}, \forall \kappa = 1, \ldots, C \tag{23.47}$$

8. Increment t by 1, and repeat Steps 2–7 until a certain criterion of convergence is reached.

The most important information from this analysis is the cluster label $\rho_{\kappa j}$. We can use the

$$\rho_{j\kappa} = \max\{\rho_{j1}, \ldots, \rho_{jC}\} \tag{23.48}$$

rule to decide that gene j should be classified into cluster κ. Alternatively, we can set a cutoff point, say 0.9. When $\rho_{j\kappa} > 0.9$, gene j is classified into cluster κ; otherwise, it is declared as not resolved.

23.7 Best Linear Unbiased Prediction

Once the parameters are estimated, they can be used to predict the gene expression profiles using a method called the best linear unbiased prediction (Henderson 1975; Robinson 1991). The observed expression for gene j at time point τ_k is $y_j(\tau_k) = y_{jk}$; we can plot y_{jk} against τ_k to see the profile (the change of gene expression across the time points). This plot is called the observed expression profile. Using the BLUP method, we can draw a predicted profile that is a smooth curve covering observed time points as well as time points between any two observed time points. Assume that gene j has been classified into cluster κ, and let

$$\hat{\gamma}_{rj} = \hat{\mu}_{r\kappa} + \hat{\alpha}_{rj}, \forall r = 0, \ldots, d \tag{23.49}$$

be the estimated effects for gene j. The BLUP profile for gene j is

$$\hat{y}_j(\tau)|_{\eta_j=\kappa} = \psi_0(\tau)\hat{\gamma}_{0j} + \psi_1(\tau)\hat{\gamma}_{1j} + \cdots + \psi_d(\tau)\hat{\gamma}_{jd} \tag{23.50}$$

where the standardize time point can take any value within the range $-1 \leq \tau \leq +1$. An example of the BLUP profile is illustrated in Fig. 23.3, showing both the observed and the predicted gene expression profiles.

Figure 23.4 illustrates the mean profiles of four clusters and the expression profiles of individual genes within each cluster.

In practice, most genes will be classified into a "neutral cluster," which represents all genes whose expressions do not change during the time course. Such a neutral cluster contains all genes with a constant expression profile. Mathematically, it is possible to exclusively define such a neutral cluster, say cluster 1, by forcing

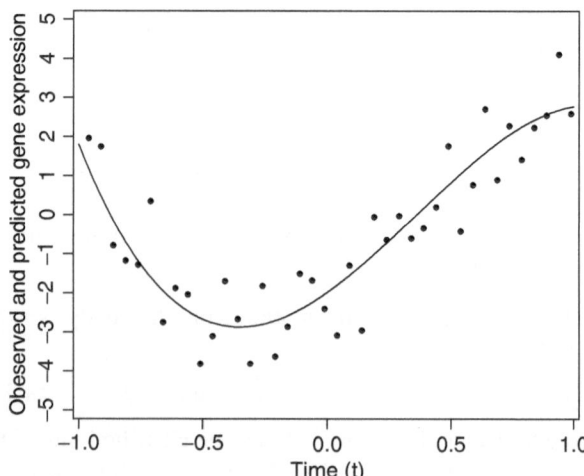

Fig. 23.3 Observed (*dots*) and predicted (*smooth curve*) time-course gene expression profiles

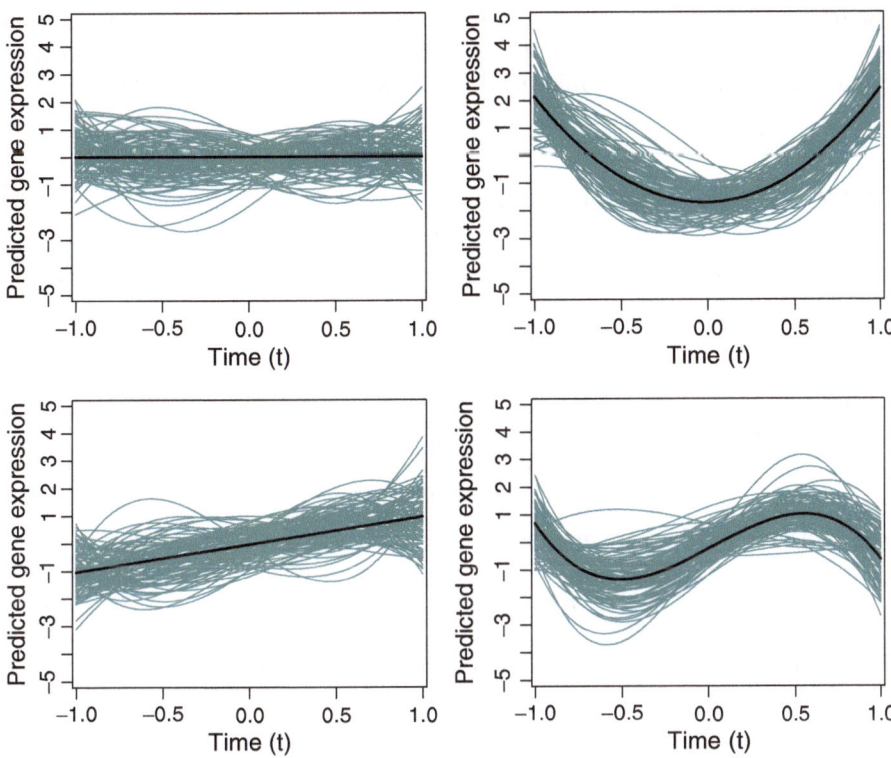

Fig. 23.4 Mean expression profiles of four clusters (*black*) and the expression profiles of 100 genes within each cluster (*green*)

$$\mu_1 = \begin{bmatrix} \mu_{01} & \mu_{11} & \cdots & \mu_{d1} \end{bmatrix}^T = \begin{bmatrix} \mu_{01} & 0 & \cdots & 0 \end{bmatrix}^T \qquad (23.51)$$

where only μ_{01}, the intercept, is estimated from the data. None of the genes classified into this cluster are differentially expressed. Therefore, the time-course microarray gene expression analysis can also be used to detect differentially expressed genes.

23.8 SEM Algorithm

The traditional EM algorithm converges to a local solution of the parameters. For a problem as complicated as gene clustering under the Gaussian mixture, the number of local solutions for the parameters is unknown. We often noticed that some clusters identified using the EM algorithm have cluster means very close to each other. This implies that there might be multiple solutions for the parameters and the one we obtained may be just one of many local solutions. To separate two clusters with similar mean vectors, we can adopt the SEM algorithm, which is a hybrid method between the Markov chain Monte Carlo (MCMC) and the EM algorithm.

23.8.1 Monte Carlo Sampling

Let us define a vector of indicator variables of cluster assignment for gene j as

$$\delta_j = \begin{bmatrix} \delta_{j1} & \delta_{j2} & \cdots & \delta_{jC} \end{bmatrix} \tag{23.52}$$

where

$$\delta_{j\kappa} = \begin{cases} 1 & \text{for } \eta_j = \kappa \\ 0 & \text{for } \eta_j \neq \kappa \end{cases} \tag{23.53}$$

This indicator variables are multinomial of sample size 1 with probabilities

$$\rho_j = \begin{bmatrix} \rho_{j1} & \rho_{j2} & \cdots & \rho_{jC} \end{bmatrix} \tag{23.54}$$

These probabilities are calculated as the posterior probabilities described in the EM algorithm. Rather than using the multinomial probabilities to update all other parameters, we use them to sample the unknown δ_j vector. Once sampled, the δ_j vector is used in place of the posterior probabilities for updating other parameters. The SEM steps are similar to the EM steps described in the previous section but with an additional step of sampling the class label for each gene.

23.8.2 SEM Steps

1. Initialize the mixing proportions by

$$\pi_\kappa^{(0)} = \frac{1}{C}, \forall \kappa = 1, \ldots, C \tag{23.55}$$

and the probabilities of cluster assignments by

$$\rho_{\kappa j}^{(0)} = \pi_\kappa^{(0)}, \forall j = 1, \ldots, N \ \& \ \kappa = 1, \ldots, C \tag{23.56}$$

where N is the number of genes and C is the number of clusters.
2. Sample δ_j from

$$\delta_j \sim \text{Multinomial}(1, \rho_{j1} \cdots \rho_{jC}) \tag{23.57}$$

3. Update the fixed effects (mean vectors) by

$$\mu_{\kappa}^{(t)} = \left[N \pi_{\kappa}^{(t-1)} \psi^T(\tau) V_{\kappa}^{-1} \psi(\tau) \right]^{-1} \psi^T(\tau) V_{\kappa}^{-1} \sum_{j=1}^{N} \delta_{j\kappa} y_j \qquad (23.58)$$

where

$$V_{\kappa} = \psi(\tau) \Sigma_{\kappa}^{(t-1)} \psi^T(\tau) + I \sigma^{2(t-1)} \qquad (23.59)$$

for all $\kappa = 1, \ldots, C$.

4. Update the random effects by

$$\alpha_j^{(t)} = \mathrm{E}(\alpha_j | \eta_j = \kappa) = \Sigma_{\kappa}^{(t-1)} \psi^T(\tau) V_{\kappa}^{-1} \left(y_j - \psi(\tau) \mu_{\kappa}^{(t)} \right) \qquad (23.60)$$

and compute the conditional covariance matrix of α_j

$$S_j^{(t)} = \mathrm{var}(\alpha_j | \eta_j = \kappa) = \Sigma_{\kappa}^{(t-1)} - \Sigma_{\kappa}^{(t-1)} \psi^T(\tau) V_{\kappa}^{-1} \psi(\tau) \Sigma_{\kappa}^{(t-1)} \qquad (23.61)$$

for all $j = 1, \ldots, N$ & $\kappa = 1, \ldots, C$.

5. Update the covariance matrices by

$$\begin{aligned} \Sigma_{\kappa}^{(t)} &= \frac{1}{N \pi_{\kappa}^{(t-1)}} \left[\sum_{j=1}^{N} \delta_{j\kappa} E(\alpha_j \alpha_j^T) \right] \\ &= \frac{1}{N \pi_{\kappa}^{(t-1)}} \left[\sum_{j=1}^{M} \delta_{j\kappa} \left(S_j^{(t)} + \alpha_j^{(t)} (\alpha_j^{(t)})^T \right) \right] \end{aligned} \qquad (23.62)$$

6. Update the residual variance by

$$\sigma^{2(t)} = \frac{1}{NM} \sum_{j=1}^{N} y_j^T \left[y_j - \sum_{\kappa=1}^{C} \delta_{j\kappa} \psi(\tau) (\mu_{\kappa}^{(t)} + \alpha_j^{(t)}) \right] \qquad (23.63)$$

7. Update the posterior probabilities of cluster assignments,

$$\rho_{j\kappa}^{(t)} = \frac{\pi_{\kappa}^{(t-1)} N(y_j | \psi(\tau) \mu_{\kappa}^{(t)}, \psi(\tau) \Sigma_{\kappa}^{(t)} \psi^T(\tau) + I \sigma^{2(t)})}{\sum_{\kappa'}^{C} \pi_{\kappa'}^{(t-1)} N(y_j | \psi(\tau) \mu_{\kappa'}^{(t)}, \psi(\tau) \Sigma_{\kappa}^{(t)} \psi^T(\tau) + I \sigma^{2(t)})} \qquad (23.64)$$

for all $j = 1, \ldots, N$ & $\kappa = 1, \ldots, C$.

8. Update the mixing proportions,

$$\pi_{\kappa}^{(t)} = \frac{1}{N} \sum_{j=1}^{N} \delta_{j\kappa}, \forall \kappa = 1, \ldots, C \qquad (23.65)$$

9. Increment t by 1, and repeat Steps 2–8 until a certain criterion of convergence is reached.

Recall that the SEM algorithm does not converge to a constant; rather, each parameter converges to a stationary distribution. We need to collect a sample of values for each parameter to obtain an empirical mean. This empirical mean value is the SEM estimate of that parameter.

Chapter 24
Quantitative Trait-Associated Microarray Data Analysis

Differential expression analysis often applies to discrete phenotypes (primarily dichotomous phenotypes). The phenotype is often defined as "normal" or "affected." If a phenotype is measured quantitatively, it is often converted into two or a few discrete phenotype groups so that a differential expression analysis or an ANOVA method for multiple comparisons can be applied. It is obvious that such discretization will result in loss of information. The current microarray data analysis technique has not been able to efficiently analyze the association of gene expression with a continuous phenotype. Pearson correlation between gene expression and a continuous phenotype has been proposed. Blalock et al. (2004) ranked genes according to the correlation coefficients of gene expression with MMSE, a quantitative measurement of the severity of Alzheimer disease, and detected many genes that are associated with Alzheimer disease. Pearson correlation is intuitive and easy to calculate. However, it may not be optimal because (1) the correlation coefficient may not be the best indicator of the association, (2) higher order association cannot be detected, (3) data are analyzed individually with one gene at a time, and (4) the method cannot be extended to association study of gene expression with multiple continuous phenotypes. Potokina et al. (2004) investigated the association of gene expression with six malting quality phenotypes (quantitative traits) of ten barley cultivars. They compared the distance matrix of each gene expression among the ten cultivars with each of the distance matrix calculated from the phenotypes using the G-test statistic. The distance matrix comparison approach may have the same flaws as the correlation analysis. Recently, we proposed to use the regression coefficient of the expression on a continuous phenotype as the indicator of the strength of association (Jia and Xu 2005). Instead of analyzing one gene at a time, we took a model-based clustering approach to studying all genes simultaneously. Qu and Xu (2006) extended the model-based clustering algorithm to capture genes with higher order association with the phenotype.

24.1   Linear Association

24.1.1   Linear Model

Let $Z = \begin{bmatrix} Z_1 \cdots Z_M \end{bmatrix}^T$ be the phenotypic values of a quantitative trait (a continuous variable) for M individuals who are also microarrayed for N genes. Let $y_j = \begin{bmatrix} y_{1j} \cdots y_{Mj} \end{bmatrix}^T$ be the expressions of the jth genes on all the M individuals for $j = 1, \ldots, N$ where N is the total number of genes. The linear model for gene expression associated with the phenotype is

$$y_j = 1\beta_j + Z\gamma_j + \varepsilon_j \tag{24.1}$$

where 1 is a vector of unity with dimension $M \times 1$, β_j is the intercept, and γ_j is the regression coefficient representing the association of gene j with the phenotype under investigation. The residual error ε_j is an $M \times 1$ vector with an assumed $N(0, I\sigma^2)$ distribution. Since β_j is irrelevant to the association study, it can be eliminated from the model. The simplest way to eliminate β_j is to centralize the expression by $y_j = y_j - \bar{y}_j$, where

$$\bar{y}_j = \frac{1}{M} \sum_{k=1}^{M} y_{kj} \tag{24.2}$$

The phenotypic value should also be centralized using $Z = Z - \bar{Z}$ where

$$\bar{Z} = \frac{1}{M} \sum_{k=1}^{M} Z_k \tag{24.3}$$

The linear model for the centralized gene expression becomes

$$y_j = Z\gamma_j + \varepsilon_j \tag{24.4}$$

Through centralization, we have eliminated β_j from the model. We can now focus on γ_j because it represents the strength of the association of y_j with Z.

24.1.2   Cluster Analysis

Clustering genes based on the regression coefficients of gene expressions on a quantitative trait was first proposed by Jia and Xu (2005). We now use a Gaussian mixture with C components to describe the regression coefficients,

$$p(\gamma_j) = \sum_{\kappa=1}^{C} \pi_\kappa N(\gamma_j | \mu_\kappa, \Sigma) \tag{24.5}$$

where μ_κ is the mean of cluster κ for $\kappa = 1, \ldots, C$ and Σ is a common variance (a single value, not a matrix). Again, the same EM algorithm described in the time-course microarray data analysis can be applied here, except that $\psi(\tau)$ in the time-course microarray is now replaced by Z in the quantitative trait-associated microarray data analysis.

The number of clusters C can be inferred using the BIC analysis (Schwarz 1978). Most genes will be classified into the "neutral cluster" in which the cluster mean is close to zero.

24.1.3 Three-Cluster Analysis

We now use a Gaussian mixture with $C = 3$ components to describe the regression coefficients,

$$p(\gamma_j) = \sum_{\kappa=1}^{3} \pi_\kappa N(\gamma_j | \mu_\kappa, \Sigma) \tag{24.6}$$

where μ_κ is the mean of cluster κ for $\kappa = 1, 2, 3$ and Σ is a common variance. The means of the three clusters are restricted with $\mu_1 > 0$, $\mu_2 = 0$, and $\mu_3 < 0$. Under these restrictions, the neutral cluster is cluster 2 because $\mu_2 = 0$. All genes classified into this neutral cluster are neutral genes (not associated with the trait), while all other genes are differentially expressed or associated with the trait.

The usual EM algorithm is incapable of dealing with such a cluster analysis with constrained cluster means. Therefore, we adopted the stochastic EM algorithm. The SEM steps are similar to those described in the time-course microarray data analysis with the step of updating the cluster means modified to constrain the means within their defined domains. Let

$$\delta_j = \begin{bmatrix} \delta_{j1} & \delta_{j2} & \delta_{j3} \end{bmatrix} \tag{24.7}$$

be the cluster indicator variables for the jth gene and

$$\rho_j = \begin{bmatrix} \rho_{j1} & \rho_{j2} & \rho_{j3} \end{bmatrix} \tag{24.8}$$

be the corresponding posterior probabilities of δ_j. Note that the proportions of genes belonging to cluster κ are denoted by π_κ for all $\kappa = 1, 2, 3$. Let

$$\mathrm{var}(y_j) = V = Z\Sigma^{-1}Z^T + I\sigma^2 \tag{24.9}$$

be the variance–covariance matrix of y_j. The cluster label δ_j is missing, and thus, it is sampled from its posterior distribution, a multinomial distribution of sample size one and a probability vector ρ_j. If we ignore the constraints, the posterior mean and posterior variance of μ_κ would be

$$\xi_\kappa = (\pi_\kappa N Z^T V^{-1} Z)^{-1} \sum_{j=1}^{N} \delta_{j\kappa} Z^T V^{-1} y_j \tag{24.10}$$

and

$$\varphi_\kappa^2 = (\pi_\kappa N Z^T V^{-1} Z)^{-1} \tag{24.11}$$

respectively, for $\kappa = 1, 3$. The constrained estimate of $\mu_1 > 0$ is

$$\mu_1 = \xi_1 + \varphi_1 \frac{\phi\left(\frac{0-\xi_1}{\varphi_1}\right) - \phi\left(\frac{\infty-\xi_1}{\varphi_1}\right)}{\Phi\left(\frac{\infty-\xi_1}{\varphi_1}\right) - \Phi\left(\frac{0-\xi_1}{\varphi_1}\right)} = \xi_1 + \varphi_1 \frac{\phi\left(\frac{\xi_1}{\varphi_1}\right)}{\Phi\left(\frac{\xi_1}{\varphi_1}\right)} \tag{24.12}$$

where $\phi(x)$ and $\Phi(x)$ are the standardized normal density and the standardized normal function, respectively. The constrained estimate for $\mu_3 < 0$ is obtained through

$$\mu_3 = \xi_3 + \varphi_3 \frac{\phi\left(\frac{-\infty-\xi_3}{\varphi_3}\right) - \phi\left(\frac{0-\xi_3}{\varphi_3}\right)}{\Phi\left(\frac{0-\xi_3}{\varphi_3}\right) - \Phi\left(\frac{-\infty-\xi_3}{\varphi_3}\right)} = \xi_3 - \varphi_3 \frac{\phi\left(\frac{\xi_3}{\varphi_3}\right)}{1 - \Phi\left(\frac{\xi_3}{\varphi_3}\right)} \tag{24.13}$$

Equations (24.12) and (24.13) were derived following the theory of truncated normal distribution given by Cohen (1991). By definition, the mean of the neutral cluster is always $\mu_2 = 0$, and no estimation is required for μ_2. Originally, the class label for the jth gene was denoted by η_j, which is assigned $\eta_j = \kappa$ if gene j belongs to cluster κ, for $\kappa = 1, 2, 3$. The η_j variable was eventually converted into a 1×3 vector δ_j, a multinomial variable with sample size one. Given $\eta_j = \kappa$, we now rewrite the linear model in the form of a mixed model

$$y_j|_{\eta_j = \kappa} = Z\mu_\kappa + Z\alpha_j + \varepsilon_j \tag{24.14}$$

and perform the SEM iterations by sampling δ_j and updating the parameters. The SEM steps are summarized as follows:

1. Initialize all parameters within their legal domains.
2. Calculate the posterior probability that gene j belonging to cluster κ using

$$\rho_{j\kappa} = \frac{\pi_\kappa N(y_j|Z\mu_\kappa, V)}{\sum_{\kappa'=1}^{3} \pi_{\kappa'} N(y_j|Z\mu_{\kappa'}, V)} \tag{24.15}$$

3. Sample δ_j from

$$p(\delta_j) = \text{Multinomial}(\delta_j|1, \rho_j) \tag{24.16}$$

4. Update the cluster means μ_κ using the means of the truncated normal distributions given in (24.12) for μ_1 and (24.13) for μ_3 while forcing $\mu_2 = 0$.
5. Calculate the posterior mean and posterior variance for α_j using

$$\hat{\alpha}_j = \text{E}(\alpha_j|\eta_j = \kappa) = \Sigma Z^T V^{-1} \delta_{j\kappa}(y_j - Z\mu_\kappa) \tag{24.17}$$

and

$$\hat{S}_j = \text{var}(\alpha_j | \eta_j = \kappa) = \Sigma - \Sigma Z^T V^{-1} Z \Sigma \tag{24.18}$$

6. Update the common variance of all clusters using

$$\Sigma = \frac{1}{N} \sum_{j=1}^{N} \sum_{\kappa=1}^{3} \delta_{j\kappa} \left(\hat{\alpha}_j^2 + \hat{S}_j \right) \tag{24.19}$$

7. Update the residual error variance using

$$\sigma^2 = \frac{1}{MN} \sum_{j=1}^{N} \sum_{\kappa=1}^{3} \delta_{j\kappa} y_j^T (y_j - Z\mu_\kappa - Z\hat{\alpha}_j) \tag{24.20}$$

8. Update the proportion of genes for each cluster using

$$\pi_\kappa = \frac{1}{N} \sum_{j=1}^{N} \delta_{j\kappa} \tag{24.21}$$

9. Repeat Steps 2–8 until all parameters converge to their stationary distributions.

After the SEM analysis, genes will be classified based on their posterior distributions of the clusters, i.e., gene j will be classified into cluster κ if

$$\rho_{j\kappa} = \max(\rho_{j1}, \rho_{j2}, \rho_{j3}) \tag{24.22}$$

Genes classified into the neutral cluster, $\kappa = 2$, will be excluded from the list of associated genes. Assume that gene j is classified into cluster κ; the predicted expression for gene j is calculated via

$$\hat{y}_j |_{\eta_j = \kappa} = Z(\hat{\mu}_\kappa + \hat{\alpha}_j) \tag{24.23}$$

where $\hat{\mu}_\kappa$ and $\hat{\alpha}_j$ are the estimated values obtained from the SEM analysis. Figure 24.1 illustrates the predicted gene expressions for the three designated clusters with $\mu_1 = 0.5$, $\mu_2 = 0$, $\mu_3 = -0.5$, and $\Sigma = 0.01$.

24.1.4 Differential Expression Analysis

An alternative method to detect genes associated with quantitative trait is through the differential expression analysis. This time, we use a Gaussian mixture with two components to model the distribution of the regression coefficients,

$$p(\gamma_j) = \pi N(\gamma_j | 0, \Sigma_1) + (1 - \pi) N(\gamma_j | 0, \Sigma_0) \tag{24.24}$$

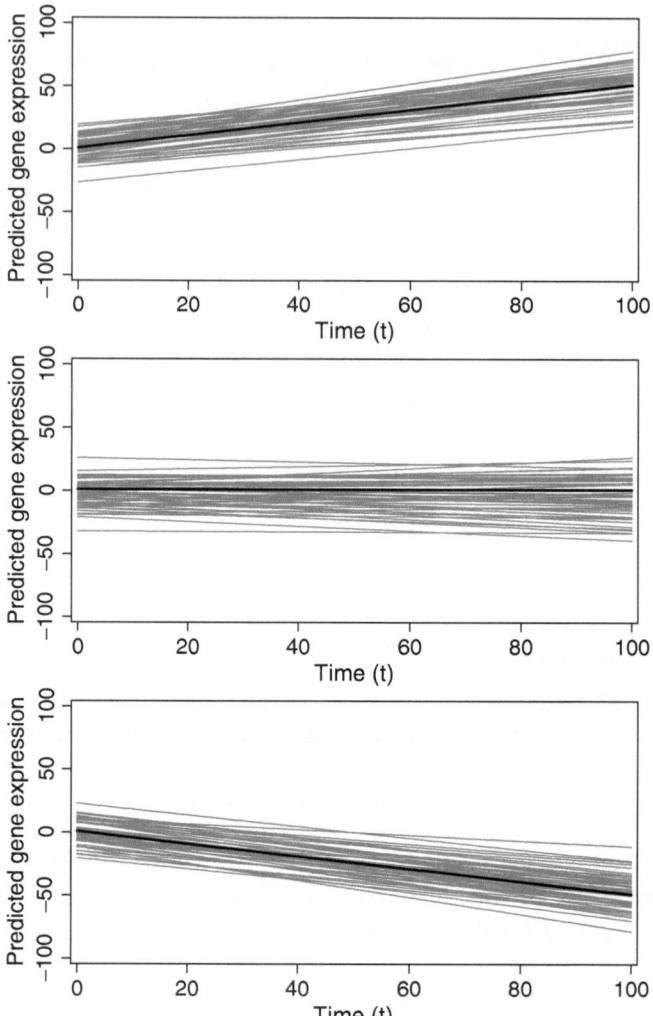

Fig. 24.1 Predicted expressions of 50 genes from each of the three designated clusters with $\mu_1 = 0.5$, $\mu_2 = 0$, $\mu_3 = -0.5$, and $\Sigma = 0.01$. The *lines in green* color represent the predicted expressions for individual genes, and the *line in black* within each cluster represents the mean profile of the cluster

where both clusters have the same mean (zero value) but the two clusters have different cluster variances, Σ_0 and Σ_1. Let cluster 0 be the neutral cluster and cluster 1 be the differentially expressed cluster where Σ_1 is treated as a parameter and Σ_0 is set to a small constant, say $\Sigma_0 = 10^{-5}$. The proportion of genes coming from cluster 1 is denoted by π. The distributions of the two components of the Gaussian mixture are illustrated in Fig. 24.2. All genes with γ classified into the

Fig. 24.2 Gaussian mixture with two components. The *sharp curve* represents $N(\gamma|0, \Sigma_0)$, the density of cluster 0, and the *flat curve* represents $N(\gamma|0, \Sigma_1)$, the density of cluster 1

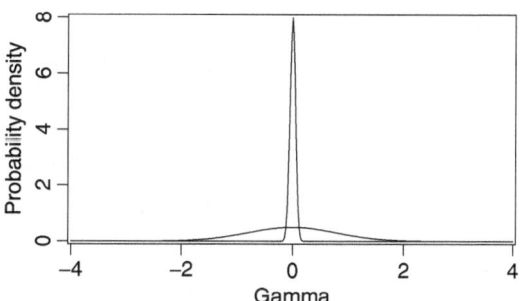

distribution represented by the flat curve are associated with the trait, while the remaining genes (with γ classified into the distribution represented by the sharp curve) are neutral genes.

Under this model, the usual EM algorithm works well for parameter estimation. However, the SEM algorithm is preferred because of the high chance of finding the global maximum likelihood estimates of the parameters. Let η_j be an indicator variable with a value of one if gene j belongs to cluster 1 and zero if the gene belongs to cluster 0. Let ρ_j be the posterior probability of $\eta_j = 1$. Define

$$\text{var}(y_j|\eta_j) = V_j = Z\left[\eta_j \Sigma_1 + (1 - \eta_j)\Sigma_0\right]Z^T + I\sigma^2 \qquad (24.25)$$

as the variance–covariance matrix of y_j. The SEM steps are summarized as follows:

1. Initialize all parameters within their legal domains.
2. Calculate the posterior probability that gene j belonging to cluster 1 using

$$\rho_j = \frac{\pi N(y_j|0, \Gamma_1)}{\pi N(y_j|0, \Gamma_1) + (1 - \pi)N(y_j|0, \Gamma_0)} \qquad (24.26)$$

where

$$\Gamma_1 = Z\Sigma_1^{-1}Z^T + I\sigma^2 \qquad (24.27)$$

and

$$\Gamma_0 = Z\Sigma_0^{-1}Z^T + I\sigma^2 \qquad (24.28)$$

3. Sample η_j from the following Bernoulli distribution:

$$p(\eta_j) = \text{Bernoulli}(\eta_j|\rho_j) \qquad (24.29)$$

4. Calculate the posterior mean and posterior variance for γ_j using

$$\hat{\gamma}_j = \text{E}(\gamma_j|\cdots) = \Theta_j Z^T V_j^{-1} y_j \qquad (24.30)$$

and

$$\hat{S}_j = \mathrm{var}(\gamma_j | \cdots) = \Theta_j - \Theta_j Z^T V_j^{-1} Z \Theta_j \qquad (24.31)$$

where

$$\Theta_j = \eta_j \Sigma_1 + (1 - \eta_j) \Sigma_0 \qquad (24.32)$$

5. Update the variance of cluster one using

$$\Sigma_1 = \frac{1}{\pi N} \sum_{j=1}^{N} \eta_j \mathrm{E}(\gamma_j^2) = \frac{1}{\pi N} \sum_{j=1}^{N} \eta_j \left(\hat{\gamma}_j^2 + \hat{S}_j \right) \qquad (24.33)$$

6. Update the residual error variance

$$\sigma^2 = \frac{1}{MN} \sum_{j=1}^{N} y_j^T (y_j - \eta_j Z \hat{\gamma}_j) \qquad (24.34)$$

7. Update the proportion of genes for cluster 1 using

$$\pi = \frac{1}{N} \sum_{j=1}^{N} \eta_j \qquad (24.35)$$

8. Repeat Steps 2–7 until all parameters converge to their stationary distributions.

After the SEM analysis, genes will be classified based on their posterior distributions of the clusters, i.e., gene j will be classified into cluster 1 if $\rho_j \geq 0.9$. All genes classified into cluster 1 will be declared as being associated with the phenotype.

Before we proceed to the next section, a comment on the second step of the SEM algorithm is helpful to inexperienced students. This step is used to calculate the posterior probability of gene j belonging to cluster 1 (the differentially expressed cluster). One would have thought to use the following equation:

$$\rho_j = \frac{\pi N(\gamma_j | 0, \Sigma_1)}{\pi N(\gamma_j | 0, \Sigma_1) + (1 - \pi) N(\gamma_j | 0, \Sigma_0)} \qquad (24.36)$$

because it is simpler than the one used in Step 2. However, the densities of the two distributions for γ_j require value of γ_j, which is a missing quantity. We cannot simply replace the missing γ_j by the conditional expectation like we did for the other quantities that involve the missing γ_j. Therefore, we cannot use any probability densities containing missing parameters to calculate the conditional posterior probabilities of clustering assignment. In the next section, we will discuss the MCMC-implemented Bayesian method, in which γ_j will be sampled. With the sampled γ_j, we can use the distributions of γ_j to calculate ρ_j.

24.2 Polynomial and B-spline

The linear association analysis cannot detect genes that are associated with the trait in higher orders. Although most associated genes may show linear relationship with the trait, some genes may show nonlinear association with the trait. Figure 24.3 shows various forms of associations of genes with a quantitative trait. The polynomial and B-spline analyses can be used for detecting these genes. The procedure is identical to that described in the time-course microarray data analysis by replacing the time points with the phenotypic values of the quantitative trait, and thus, no further discussion will be given here for the EM and SEM algorithm. The intercepts can be included in the analysis or excluded from the analysis via centralization. This section will focus on the Bayesian analysis implemented via the MCMC algorithm. In addition, we will deal with differential expression analysis, in which only two clusters are considered, one is the neutral cluster and the other is the differentially expressed cluster.

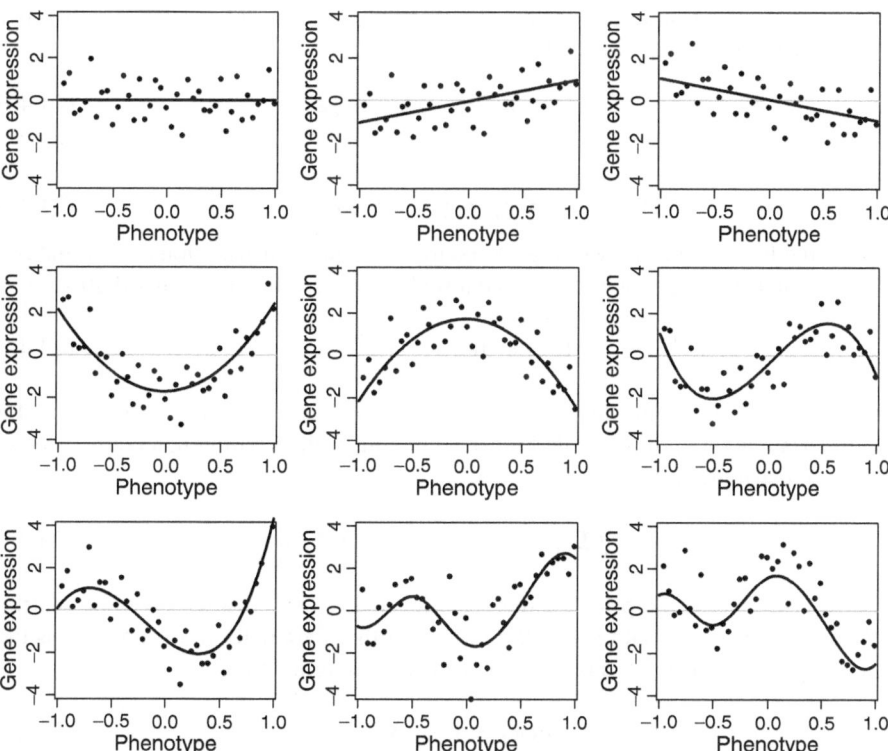

Fig. 24.3 Various forms of associations of gene expressions with a quantitative trait

Let ζ be the scaled phenotypic value ranging from -1 to $+1$ and d be the degree of the polynomial to be fit. The linear model is

$$y_j = \psi(\zeta)\gamma_j + \varepsilon_j \tag{24.37}$$

where $\psi(\zeta)$ is an $M \times d$ orthogonal polynomial coefficient matrix and

$$\gamma_j = \left[\gamma_{1j} \cdots \gamma_{dj}\right]^T \tag{24.38}$$

are the regression coefficients. Again, we use a Gaussian mixture with two components to describe the regression coefficients

$$p(\gamma_j) = \pi N(\gamma_j|0, \Sigma_1) + (1 - \pi)N(\gamma_j|0, \Sigma_0) \tag{24.39}$$

where both clusters have a mean zero, but with cluster-specific variance–covariance matrices, Σ_1 and Σ_0, each is a $d \times d$ matrix. Let cluster 0 be the neutral cluster and cluster 1 be the differentially expressed cluster where Σ_1 is treated as a parameter matrix and $\Sigma_0 = 10^{-5}I_{d \times d}$ be a constant matrix with diagonal values taking a value of virtually zero. Again, let η_j be the cluster label and ρ_j be the posterior probability of gene j coming from cluster 1. Define

$$\Theta_j = \eta_j \Sigma_1 + (1 - \eta_j)\Sigma_0 \tag{24.40}$$

and

$$\mathrm{var}(y_j|\eta_j) = V_j = \psi(\zeta)\Theta_j\psi(\zeta)^T + I\sigma^2 \tag{24.41}$$

Under the Bayesian framework, we need to assign prior distributions to Σ_1 and σ^2. The residual error variance is assigned a scaled inverse chi-square distribution,

$$p(\sigma^2) = \mathrm{Inv} - \chi^2(\sigma^2|\tau, \omega) \tag{24.42}$$

where $\omega = \tau = 0$ (the Jeffreys' prior). The covariance matrix Σ_1 is assigned a multivariate version of the scaled inverse chi-square distribution named the inverse Wishart distribution,

$$p(\Sigma_1) = \mathrm{Inv} - \mathrm{Wishart}(\Sigma_1|\tau, \omega) \tag{24.43}$$

where $\tau > d - 1$ is a prior degree of belief and $\omega > 0$ is a $d \times d$ scale matrix. We simply set $\tau = d$ and $\omega = 10^{-5}I_{d \times d}$. Another parameter in the analysis is the mixing proportion denoted by π. A beta prior is assigned to π,

$$p(\pi) = \mathrm{Beta}(\pi|1, 1) \tag{24.44}$$

These priors are conjugate, and thus, the posterior distributions of these parameters have the same forms of distributions as the priors.

The MCMC sampling process is summarized as follows:

1. Sample all unknown variables and parameters from their prior distributions.
2. Sample η_j from Bernoulli distribution,

$$p(\eta_j | \cdots) = \text{Bernoulli}(\eta_j | 1, \rho_j) \qquad (24.45)$$

where ρ_j is a posterior probability calculated using

$$\rho_j = \frac{\pi N(\gamma_j | 0, \Sigma_1)}{\pi N(\gamma_j | 0, \Sigma_1) + (1 - \pi) N(\gamma_j | 0, \Sigma_0)} \qquad (24.46)$$

3. Sample γ_j from its posterior distribution, which is multivariate normal

$$p(\gamma_j | \cdots) = N \left[\gamma_j | \Theta_j \psi(\zeta)^T V_j^{-1} y_j, \Theta_j - \Theta_j \psi(\zeta)^T V_j^{-1} \psi(\zeta) \Theta_j \right] \qquad (24.47)$$

4. Sample the variance matrix from

$$p(\Sigma_1 | \cdots) = \text{Inv} - \text{Wishart} \left(\Sigma_1 \,\middle|\, \tau + \sum_{j=1}^{N} \eta_j, \omega + \sum_{j=1}^{N} \eta_j \gamma_j \gamma_j^T \right) \qquad (24.48)$$

5. Sample the residual error variance from

$$p(\sigma^2 | \cdots) = \text{Inv} - \chi^2 \left[\sigma^2 \,\middle|\, \tau + NM, \omega + \sum_{j=1}^{N} y_j^T (y_j - \eta_j \psi(\zeta) \gamma_j) \right] \qquad (24.49)$$

6. Sample the mixture proportion from

$$p(\pi | \cdots) = \text{Beta} \left(\pi \,\middle|\, 1 + \sum_{j=1}^{N} \eta_j, 1 + N - \sum_{j=1}^{N} \eta_j \right) \qquad (24.50)$$

7. Repeat Steps 2–6 until a desired length of the Markov chain is reached.

After the post-MCMC analysis (burn-in deletion and autocorrelation thinning), genes will be classified based on their posterior distributions of the clusters, i.e., gene j will be classified into cluster 1 if $\rho_j > 0.9$. All genes classified into cluster 1 will be declared as being associated with the phenotype. Further analysis may be conducted on the Bayesian estimate of γ_j for all the differentially expressed genes. For example, we can use the K-means method to cluster all the γ_j into a few clusters. Different clusters represent different patterns (curves) of the expression profiles.

The MCMC algorithm developed for polynomial analysis also applies to B-spline analysis. The only difference is the dimension of the model. In the polynomial analysis, the dimension of γ_j is $d \times 1$, and the dimension for $\psi(\zeta)$ is $M \times d$. In the B-spline analysis, the dimension for γ_j is $p \times 1$, and the dimension for $\psi(\zeta)$ is $M \times p$, where $p = d + s + 1$ and d is the degree of the piecewise polynomial and s is the number of internal knots.

24.3   Multiple-Trait Association

It is possible to study the association of genes with multiple quantitative traits. Let

$$Z_r = \left[Z_{r1} \cdots Z_{rM} \right]^T \tag{24.51}$$

be the phenotypic values of trait r collected from all the M individuals microarrayed. Denote the regression coefficient of gene j on the rth trait by γ_{rj} for $r = 1, \ldots, T$, where T is the number of traits. Assume that all gene expressions have been centralized and the phenotypic values of all the T traits have be standardized (centralized and normalized). The linear model for y_j can be expressed as

$$y_j = \sum_{r=1}^{T} Z_r \gamma_{jr} + \varepsilon_j = Z\gamma_j + \varepsilon_j \tag{24.52}$$

where

$$Z = \left[Z_1 \cdots Z_T \right] \tag{24.53}$$

is an $M \times T$ matrix and

$$\gamma_j = \left[\gamma_{j1} \cdots \gamma_{jT} \right]^T \tag{24.54}$$

be a $T \times 1$ vector.

Cluster analysis can be used to classify all genes into different clusters, depending on their associations with the multiple traits. Methods for the time-course microarray analysis and the single-trait polynomial and/or B-spline association study apply to the multiple-trait association study. The only difference is in the notation where $\psi(\zeta)$ in the polynomial analysis is replaced by matrix Z in the multiple-trait analysis. Note that all the associations are linear in the multiple-trait analysis. Extension to nonlinear association of multiple traits is possible, but it is difficult to implement. One has to define the degree of polynomial for each trait, and different traits may be modeled using different degrees. If the degrees of polynomials are the same for all traits, the total number of effects in the γ_j vector is $T \times d$ for polynomial analysis and $T \times p$ for B-spline analysis where $p = d + s + 1$.

Chapter 25
Mapping Expression Quantitative Trait Loci

The transcript abundance can be treated as a classical quantitative trait, and thus, mapping can be done on the transcript (Brem et al. 2002; Cheung and Spielman 2002; Schadt et al. 2003). Mendelian loci in the genome that control the expression levels of transcripts are called expression quantitative trait loci (eQTL). In eQTL analysis, an expression trait is mapped to genomic locations represented by *cis*- or *trans*-loci. The *cis*-eQTL represent sequence variants that encode transcriptional differences. The *trans*-eQTL, however, represent remote genes that regulate the expression of the gene being transcribed. The purpose of a linkage study is to identify the *cis*- and *trans*-eQTL for each transcript. Results from the eQTL analysis may provide more detailed information about the biological processes of the gene network than the classical quantitative trait analysis (Emilsson et al. 2008; Cookson et al. 2009). The usual quantitative traits are often gross clinical measurements and may be far remote from the biological processes giving rise to clinical traits.

In eQTL analysis, we often deal with thousands of expression traits simultaneously. Methods developed for multiple quantitative trait QTL mapping may not apply here because of the high dimensionality of the model. Two naive approaches may be taken in eQTL analysis: (1) individual transcript analysis in which a single expression trait is mapped at a time and the entire eQTL mapping involves separate analysis of thousands of traits and (2) individual marker analysis where differentially expressed transcripts are detected based on their association with the segregation pattern of an individual marker and the entire analysis requires scanning markers of the entire genome. The first naive approach requires only a single-trait QTL mapping procedure, e.g., the interval mapping (Lander and Botstein 1989), the composite interval mapping (Zeng 1994), or the multiple QTL mapping (Kao et al. 1999). A common practice for handling thousands of transcript traits is to select a small number of target transcripts based on some criterion of preselection and map QTL only for these prescreened transcripts. The second naive approach requires only a method for differential expression analysis, e.g., the regularized *t*-test, the hierarchical mixture model of Newton et al. (2004), or the model-based cluster analysis (Pan et al. 2002). In differential expression analysis, one requires samples from at least two conditions, the control and the treatment. When applied to

S. Xu, *Principles of Statistical Genomics*, DOI 10.1007/978-0-387-70807-2_25,
© Springer Science+Business Media, LLC 2013

expression-marker association studies, the conditions become the marker genotypes, e.g., individuals carrying one genotype are arbitrarily designated to the control and those carrying the other genotype to the treatment.

It appears that eQTL mapping has been treated as either a QTL mapping problem for multiple traits or a microarray differential expression problem for multiple treatment comparisons. Prior to the method described in Kendziorski et al. (2006) and Kendziorski and Wang (2006), there has been no unique method particularly designed for eQTL analysis. Neither the first nor the second aforementioned naive approach is optimal because data are not analyzed jointly. The mixture over marker (MOM) approach developed by Kendziorski et al. (2006) is the first attempt to analyze transcripts and markers jointly. The method is called MOM because the expression level of a transcript is described by a mixture model over markers. A transcript is either associated with a marker or not associated with any markers at all. Given that the transcript is associated with a marker, it is associated with one and only one of the markers. We believe that the assumption of a transcript associated with at most one marker is too stringent and needs to be relaxed. The MOM approach is able to detect either the *cis*-locus or one of the *trans*-loci but not both. This will seriously limit the application of the MOM method. Only until recently, more advanced statistical methods have been developed. All these methods are related to Bayesian shrinkage statistics. These methods include the Bayesian method of Jia and Xu (2007), the sparse partial least-squares regression of Chun and Keles (2009), and the Bayesian method of Bottolo et al. (2011).

The method of Bottolo et al. (2011) was developed based on Jia and Xu's (2007) model but with improved prior distributions. Their method is more powerful than the original Bayesian method. The Bayesian method combines the two naive approaches into a single step of analysis so that parameters are inferred using multiple transcripts and multiple markers simultaneously. This joint approach will capture the maximum information from the microarray experiment. Like a regular quantitative trait, a transcript can be mapped to many different locations, including *cis*- and *trans*-loci. In multiple QTL mapping, we face a variable selection problem. To avoid variable selection, we have adopted the Bayesian shrinkage analysis, in which marker loci of the entire genome are evaluated simultaneously (Wang et al. 2005b). Markers with small effects are forced to shrink their effects to zero, and markers with large effects are subject to no shrinkage. The shrinkage estimation is made possible through the Bayesian hierarchical modeling (Gelman 2005, 2006; Gelman et al. 2008). In this chapter, we will first review the second naive method and then describe the Bayesian method of Jia and Xu (2007). The chapter ends with a brief description of the HESS algorithm of Bottolo et al. (2011).

25.1 Individual Marker Analysis

Let N be the number of transcripts and y_j for $j = 1, \ldots, N$ be an $M \times 1$ vector for the expressions of the jth transcript measured from M subjects of a mapping population. We use an F_2 mapping population as an example to demonstrate the method, which, of course, can be applied to other mapping populations, such as BC, RIL, DH, and FW, with very little modification. Assume that there are p markers available for the analysis and each is genotyped for all the M individuals. The analysis is focused on one marker at a time. Let Z_k be an $M \times 1$ vector for the genotypes of all individuals at marker k for $k = 1, \ldots, p$. For the individual marker analysis, one analyzes each marker at a time, and the entire analysis is conducted p times, one for each marker. The following description focuses on one marker only, and thus, the subscript k is dropped for simplicity. The linear model for y_j is

$$y_j = 1\beta_j + Z\gamma_j + \varepsilon_j \qquad (25.1)$$

where 1 is an $M \times 1$ vector of unity, β_j is the intercept, γ_j is the eQTL effect for the particular marker under study, and ε_j is an $M \times 1$ vector for the residual errors with an assumed multivariate $N(0, I\sigma^2)$ distribution. Let us assign a normal distribution to β_j so that

$$p(\beta_j) = N(\beta_j | \mu_\beta, \Sigma_\beta) \qquad (25.2)$$

where μ_β and Σ_β are the unknown mean and unknown variance of β_j for $j = 1, \ldots, N$. Describe γ_j by a Gaussian mixture distribution with two components,

$$p(\gamma_j) = (1 - \pi)N(\gamma_j | 0, \Sigma_0) + \pi N(\gamma_j | 0, \Sigma_1) \qquad (25.3)$$

where π is the proportion of genes belonging to cluster one (the associated cluster), Σ_1 is an unknown variance representing the associated cluster, and $\Sigma_0 = 10^{-5}$ is a small positive number (constant) representing the neutral cluster. Gene detection and parameter estimation are conducted using one of two approaches, the SEM algorithm and the MCMC algorithm.

25.1.1 SEM Algorithm

The SEM algorithm is the same as that described in the quantitative trait-associated microarray data analysis except that we now fit an intercept to the model. Let η_j be the class label for gene j, i.e., $\eta_j = 0$ if gene j belongs to the neutral cluster and $\eta_j = 1$ otherwise. Conditional on η_j, we have

$$E(y_j | \eta_j) = 1\mu_\beta \qquad (25.4)$$

and

$$\text{var}(y_j | \eta_j) = V_j = 1\Sigma_\beta 1^T + Z\Theta_j Z^T + I\sigma^2 \qquad (25.5)$$

where

$$\Theta_j = \eta_j \Sigma_1 + (1 - \eta_j)\Sigma_0 \qquad (25.6)$$

This variance matrix is gene specific because it depends on η_j. Let us define

$$\Gamma_1 = 1\Sigma_\beta 1^T + Z\Sigma_1 Z^T + I\sigma^2 \tag{25.7}$$

and

$$\Gamma_0 = 1\Sigma_\beta 1^T + Z\Sigma_0 Z^T + I\sigma^2 \tag{25.8}$$

The stochastic sampling step involves sampling the cluster label η_j from its conditional posterior distribution,

$$\rho_j = \frac{\pi N(y_j|1\mu_\beta, \Gamma_1)}{\pi N(y_j|1\mu_\beta, \Gamma_1) + (1-\pi)N(y_j|1\mu_\beta, \Gamma_0)} \tag{25.9}$$

Combining this step with the EM algorithm leads to the following SEM algorithm:

1. Initialize all parameters.
2. Calculate ρ_j using (25.9), and sample η_j from

$$p(\eta_j|\cdots) = \text{Bernoulli}(\eta_j|\rho_j) \tag{25.10}$$

3. Update μ_β using

$$\mu_\beta = \left(\sum_{j=1}^N 1^T V_j^{-1} 1\right)^{-1} \left(\sum_{j=1}^N 1^T V_j^{-1} y_j\right) \tag{25.11}$$

4. Calculate the posterior mean and posterior variance for β_j using

$$\hat{\beta}_j = \text{E}(\beta_j|\cdots) = \Sigma_\beta 1^T V_j^{-1}(y_j - 1\mu_\beta) \tag{25.12}$$

and

$$\hat{W}_j = \text{var}(\beta_j|\cdots) = \Sigma_\beta - \Sigma_\beta 1^T V_j^{-1} 1\Sigma_\beta \tag{25.13}$$

5. Update Σ_β using

$$\Sigma_\beta = \frac{1}{N}\sum_{j=1}^N \text{E}\left[(\beta_j - \mu_\beta)(\beta_j - \mu_\beta)^T\right]$$

$$= \frac{1}{N}\sum_{j=1}^N \left[(\hat{\beta}_j - \mu_\beta)(\hat{\beta}_j - \mu_\beta)^T + \hat{W}_j\right] \tag{25.14}$$

6. Calculate the posterior mean and posterior variance for γ_j using

$$\hat{\gamma}_j = \text{E}(\gamma_j|\cdots) = \Theta_j Z^T V_j^{-1}(y_j - 1\mu_\beta) \tag{25.15}$$

and

$$\hat{S}_j = \mathrm{var}(\gamma_j | \cdots) = \Theta_j - \Theta_j Z^T V_j^{-1} Z \Theta_j \qquad (25.16)$$

where V_j and Θ_j are defined in (25.5) and (25.6), respectively, prior to the SEM steps.

7. Update the variance of the associated cluster Σ_1 using

$$\Sigma_1 = \frac{1}{\pi N} \sum_{j=1}^{N} \eta_j \, \mathrm{E}(\gamma_j^2) = \frac{1}{\pi N} \sum_{j=1}^{N} \eta_j (\hat{\gamma}_j^2 + \hat{S}_j) \qquad (25.17)$$

8. Update the residual error variance using

$$\sigma^2 = \frac{1}{MN} \sum_{j=1}^{N} y_j^T (y_j - 1\hat{\beta}_j - \eta_j Z \hat{\gamma}_j) \qquad (25.18)$$

9. Update the mixing proportion of gene clustering using

$$\pi = \frac{1}{N} \sum_{j=1}^{N} \eta_j \qquad (25.19)$$

10. Repeat Steps 2–9 until all parameters have converged to their stationary distributions.

After the analysis, genes with $\rho_j \geq 0.9$ will be claimed as being associated with the current marker under investigation. The 0.9 criterion is arbitrary, and other criteria can be used, depending on the preference of the investigator. To scan markers of the entire genome, the above analysis is repeated for each marker. This process mimics the individual marker analysis or interval mapping for QTL of a quantitative trait. The only difference is that here the quantitative trait is represented by the expressions of N microarrayed transcripts. When the marker density is sufficiently high, the genotype indicator variable Z will be observed. For sparse marker maps, the conditional probabilities of marker genotypes will be calculated using the multipoint method described in Chap. 4. These probabilities will then be used to calculate the conditional expectations of $U = \mathrm{E}(Z)$ so that Z will be replaced by U in the eQTL analysis.

25.1.2 MCMC Algorithm

The MCMC-implemented Bayesian method requires prior distribution for each parameter. The parameters that have not been described with a prior distribution in the SEM algorithm include π, μ_β, Σ_β, Σ_1, and σ^2. A beta prior can be assigned to π, i.e., $p(\pi) = \mathrm{Beta}(\pi | 1, 1)$. The overall population mean μ_β can be assigned a flat prior distribution $p(\mu_\beta) = 1$. Each of the variance parameters can be assigned a scaled inverse chi-square distribution, e.g.,

$$p(\Sigma_\beta) = \text{Inv} - \chi^2(\Sigma_\beta | \tau, \omega) \tag{25.20}$$

The fully conditional posterior distributions of all the variables (including the parameters) are conjugate, and thus, the Gibbs sampler algorithm is used to simulate every variable involved in the process. We now describe the fully conditional posterior distributions in the order in which they will be sampled in the MCMC process.

1. The fully conditional posterior distribution for π is beta with updated parameters given below:

$$p(\pi | \cdots) = \text{Beta}\left(\pi \left| 1 + \sum_{j=1}^{N} \eta_j, 1 + N - \sum_{j=1}^{N} \eta_j \right. \right) \tag{25.21}$$

2. Once π is sampled, we need to calculate ρ_j using the Bayes' theorem,

$$\rho_j = \frac{\pi N(\gamma_j | 0, \Sigma_1)}{\pi N(\gamma_j | 0, \Sigma_1) + (1 - \pi) N(\gamma_j | 0, \Sigma_0)} \tag{25.22}$$

3. The fully conditional posterior of η_j is Bernoulli with parameter ρ_j

$$p(\eta_j | \cdots) = \text{Bernoulli}(\eta_j | \rho_j) \tag{25.23}$$

4. The overall mean μ_β has a normal fully conditional posterior distribution,

$$p(\mu_\beta | \cdots) = N\left(\mu_\beta \left| \frac{1}{MN} \sum_{j=1}^{N} 1^T y_j^*, \frac{1}{MN} \sigma^2 \right. \right) \tag{25.24}$$

where

$$y_j^* = y_j - 1(\beta_j - \mu_\beta) - \eta_j Z \gamma_j \tag{25.25}$$

is an adjusted gene expression vector by subtraction of all other model effects except μ_β from the original observed gene expression.

5. The gene-specific intercept β_j has a normal fully conditional posterior distribution,

$$p(\beta_j | \cdots) = N(\beta_j | \hat{\beta}_j, \hat{W}_j) \tag{25.26}$$

where

$$\hat{\beta}_j = \left(\frac{N}{\sigma^2} + \frac{1}{\Sigma_\beta} \right)^{-1} \left(\frac{1^T y_j^*}{\sigma^2} + \frac{\mu_\beta}{\Sigma_\beta} \right) \tag{25.27}$$

and

$$\hat{W}_j = \text{var}(\beta_j | \cdots) = \left(\frac{N}{\sigma^2} + \frac{1}{\Sigma_\beta} \right)^{-1} \tag{25.28}$$

The offset for y_j, denoted by y_j^*, is redefined as

$$y_j^* = y_j - \eta_j Z \gamma_j \tag{25.29}$$

6. The fully conditional posterior distribution of γ_j is normal as given below:

$$p(\gamma_j | \cdots) = N(\gamma_j | \hat{\gamma}_j, \hat{S}_j) \tag{25.30}$$

where

$$\hat{\gamma}_j = \left(Z^T Z + \sigma^2 \Theta_j^{-1} \right)^{-1} Z^T y_j^* \tag{25.31}$$

and

$$\hat{S}_j = \left(Z^T Z + \sigma^2 \Theta_j^{-1} \right)^{-1} \sigma^2 \tag{25.32}$$

The offset y_j^* is redefined as

$$y_j^* = y_j - 1\beta_j \tag{25.33}$$

7. The fully conditional posterior of Σ_β remais a scaled inverse chi-square distribution,

$$p(\Sigma_\beta | \cdots) = \text{Inv} - \chi^2 \left[\Sigma_\beta \middle| \tau + N, \omega + \sum_{j=1}^{N} (\beta_j - \mu_\beta)^2 \right] \tag{25.34}$$

8. The variance of the associated cluster also has a fully conditional posterior distribution of scaled inverse chi-square,

$$p(\Sigma_1 | \cdots) = \text{Inv} - \chi^2 \left(\Sigma_1 \middle| \tau + \sum_{j=1}^{N} \eta_j, \omega + \sum_{j=1}^{N} \eta_j \gamma_j^2 \right) \tag{25.35}$$

9. Finally, the residual variance has the following scaled inverse chi-square distribution:

$$p(\sigma^2 | \cdots) = \text{Inv} - \chi^2 \left(\sigma^2 | \tau + MN, \omega + SS \right) \tag{25.36}$$

where

$$SS = \sum_{j=1}^{N} (y_j - 1\beta_j - \eta_j Z \gamma_j)^T (y_j - 1\beta_j - \eta_j Z \gamma_j) \tag{25.37}$$

is the sum of squares of the residuals.

Post-MCMC analysis is performed using the MCMC-generated posterior sample of all the variables.

25.2 Joint Analysis of All Markers

Joint analysis of all markers refers to an analysis that the marker effects of the entire genome are estimated simultaneously in a single model. With the current molecular technology, genomes of many important species have been saturated by markers, especially, SNP markers. The optimal approach of eQTL mapping is the joint analysis using markers of the entire genome. There is a limitation with the joint analysis, that is, the analysis cannot handle markers that are cosegregating, i.e., perfectly correlated marker genotypes across the mapping population. Therefore, a preliminary screening of the markers is required prior to eQTL mapping. The conventional strategy for marker screening is to select one putative position in every d centiMorgan, where d can be 2 cM, 5 cM, or any other values, depending on the sample size of the experiment. Larger samples sizes allow us to handle markers with higher density. If a putative position happens to be located in an existing marker, the genotypes of the putative position are observed; otherwise, the genotypes of the putative position are inferred from neighboring markers. The conditional expectations of the genotype indicator variables are then used instead.

25.2.1 Multiple eQTL Model

Let Z_k be an $M \times 1$ vector for the genotype indicator variables of marker k for $k = 1, \ldots, p$, where p is the total number of markers or putative positions included in the model. The following linear model is used to describe the expressions of gene j measured from all M individuals:

$$y_j = 1\beta_j + \sum_{k=1}^{p} Z_k \gamma_{jk} + \varepsilon_j \tag{25.38}$$

where y_j is an $M \times 1$ vector for the expressions of all individuals, 1 is an $M \times 1$ vector of unity, β_j is the intercept, γ_{jk} is the eQTL effect for gene j at marker k for $k = 1, \ldots, p$, and ε_j is an $M \times 1$ vector for the residual errors with an assumed multivariate $N(0, I\sigma_j^2)$ distribution. Note that the residual error variance is gene specific, i.e., each gene has its own residual variance. In the individual marker analysis, we have assumed that all genes share the same error variance. That assumption is not necessary, and it is introduced simply for convenience of presentation. We now relax this assumption and introduce this gene-specific error variance. Let us assign a normal distribution to β_j so that

$$p(\beta_j) = N(\beta_j | \mu_\beta, \Sigma_\beta) \tag{25.39}$$

where μ_β and Σ_β are the unknown mean and unknown variance of these β_j's. Describe γ_{kj} by a Gaussian mixture distribution with two components,

$$p(\gamma_{jk}) = (1 - \pi_k)N(\gamma_{jk}|0, \Sigma_0) + \pi_k N(\gamma_{jk}|0, \Sigma_k) \tag{25.40}$$

where π_k is the proportion of genes belonging to cluster one (the cluster containing all associated genes), Σ_k is an unknown variance for the associated cluster, and $\Sigma_0 = 10^{-5}$ is a small positive number (constant) representing the neutral cluster. Again, we will take one of two approaches for the analysis, the SEM algorithm and the MCMC algorithm.

25.2.2 SEM Algorithm

Let η_{jk} be the class label for gene j at locus k, i.e., $\eta_{jk} = 0$ if gene j belongs to the neutral cluster and $\eta_{jk} = 1$ otherwise. Let

$$\eta_j = \begin{bmatrix} \eta_{j1} \cdots \eta_{jp} \end{bmatrix}^T \tag{25.41}$$

be a $p \times 1$ vector of the class label indicators for gene j. Conditional on η_j, we can write the variance of γ_{jk} as

$$\text{var}(\gamma_{jk}|\eta_{jk}) = \Theta_{jk} = \eta_{jk}\Sigma_k + (1 - \eta_{jk})\Sigma_0 \tag{25.42}$$

The expectation and variance of y_j given η_j are

$$E(y_j|\eta_j) = 1\mu_\beta \tag{25.43}$$

and

$$\text{var}(y_j|\eta_j) = V_j = \sum_{k=1}^{p} Z_k \Theta_{jk} Z_k^T + I\sigma_j^2 \tag{25.44}$$

respectively. The stochastic sampling step involves sampling the cluster label η_{jk} from its conditional posterior distribution,

$$\rho_{jk} = \frac{\pi_k N(y_j|1\mu_\beta, \Gamma_{jk})}{\pi_k N(y_j|1\mu_\beta, \Gamma_{jk}) + (1 - \pi_k)N(y_j|1\mu_\beta, \Gamma_{j0})} \tag{25.45}$$

where

$$\Gamma_{jk} = \sum_{k' \neq k}^{p} Z_{k'} \Theta_{jk'} Z_{k'}^T + Z_k \Sigma_k Z_k^T + I\sigma_j^2 \tag{25.46}$$

and

$$\Gamma_{j0} = \sum_{k' \neq k}^{p} Z_{k'} \Theta_{jk'} Z_{k'}^T + Z_k \Sigma_0 Z_k^T + I\sigma_j^2 \tag{25.47}$$

Combining this step with the EM algorithm leads to the following SEM algorithm:

1. Initialize all parameters.
2. Calculate ρ_{jk} using (25.45), and sample η_{jk} from

$$p(\eta_{jk}|\cdots) = \text{Bernoulli}(\eta_{jk}|\rho_{jk}) \qquad (25.48)$$

3. Update μ_β using

$$\mu_\beta = \left(\sum_{j=1}^{N} 1^T V_j^{-1} 1^T\right)^{-1} \left(\sum_{j=1}^{N} 1^T V_j^{-1} y_j\right) \qquad (25.49)$$

4. Calculate the posterior mean and posterior variance for β_j using

$$\hat{\beta}_j = \text{E}(\beta_j|\cdots) = \Sigma_\beta 1^T V_j^{-1}(y_j - 1\mu_\beta) \qquad (25.50)$$

and

$$\hat{W}_j = \text{var}(\beta_j|\cdots) = \Sigma_\beta - \Sigma_\beta 1^T V_j^{-1} 1 \Sigma_\beta \qquad (25.51)$$

5. Update Σ_β using

$$\Sigma_\beta = \frac{1}{N} \sum_{j=1}^{N} \text{E}\left[(\beta_j - \mu_\beta)(\beta_j - \mu_\beta)^T\right]$$

$$= \frac{1}{N} \sum_{j=1}^{N} \left[(\hat{\beta}_j - \mu_\beta)(\hat{\beta}_j - \mu_\beta)^T + \hat{W}_j\right] \qquad (25.52)$$

6. Calculate the posterior mean and posterior variance for γ_{jk} using

$$\hat{\gamma}_{jk} = \text{E}(\gamma_{jk}|\cdots) = \Theta_{jk} Z_k^T V_j^{-1}(y_j - 1\mu_\beta) \qquad (25.53)$$

and

$$\hat{S}_{jk} = \text{var}(\gamma_{jk}|\cdots) = \Theta_{jk} - \Theta_{jk} Z_k^T V_j^{-1} Z_k \Theta_{jk} \qquad (25.54)$$

where Θ_{jk} and V_j are defined in (25.42) and (25.44), respectively, prior to the SEM steps.

7. Update the variance of the associated cluster Σ_k using

$$\Sigma_k = \frac{1}{\pi_k N} \sum_{j=1}^{N} \eta_{jk} \text{E}\left(\gamma_{jk}^2\right) = \frac{1}{\pi_k N} \sum_{j=1}^{N} \eta_{jk}\left(\hat{\gamma}_{jk}^2 + \hat{S}_{jk}\right) \qquad (25.55)$$

8. Update the residual error variance for gene j using

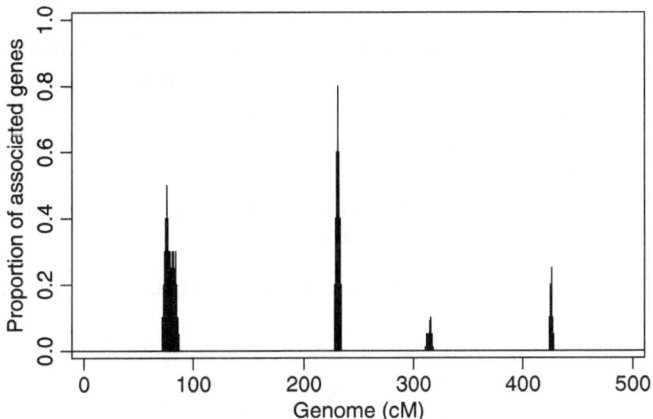

Fig. 25.1 Proportion of associated genes $\pi(\lambda)$ against genome location (λ)

$$\sigma_j^2 = \frac{1}{M}\left(y_j - 1\hat{\beta}_j\right)^T \left(y_j - 1\hat{\beta}_j - \sum_{k=1}^{p} \eta_{jk} Z_k \hat{\gamma}_{jk}\right) \tag{25.56}$$

9. Update the proportion of genes in the associated cluster using

$$\pi_k = \frac{1}{N}\sum_{j=1}^{N} \eta_{jk} \tag{25.57}$$

10. Repeat Steps 2–9 until all parameters have converged to their stationary distributions.

After the analysis, genes with $\rho_{jk} \geq 0.9$ will be claimed to be associated with marker k. The 0.9 criterion is arbitrary, and other criterion can be used, depending on the preference of the investigator. An important parameter in eQTL mapping is π_k, the proportion of genes associated with marker k. If k is replaced by the genome position of marker k, say λ, we can plot $\pi(\lambda)$ against λ to form a π profile. This profile can show all the "hot" spots of the genomes, where a "hot" spot represents a location of the genome containing many associated genes (transcripts). Figure 25.1 illustrates such a π profile of a hypothetical genome. There are four "hot" spots in the genome. About 30% of the genes are associated with a wide range of genome around 75 cM. Almost 80% of the genes are associated with a narrow region around 230 cM of the genome.

25.2.3 MCMC Algorithm

The parameters that have not been assigned a prior distribution include π_k, μ_β, Σ_β, Σ_k, and σ_j^2. A beta prior is assigned to the π_k, i.e., $p(\pi_k) = \text{Beta}(\pi_k | 1, 1)$.

The overall population mean μ_β is assigned a flat prior distribution $p(\mu_\beta) = 1$. Each of the variance parameters is assigned a scaled inverse chi-square distribution, e.g., $p(\sigma_j^2) = \text{Inv} - \chi^2(\sigma_j^2|\tau, \omega)$. Any variance–covariance matrix is assigned an inverse Wishart distribution, e.g., $p(\Sigma_\beta) = \text{Inv} - \text{Wishart}(\Sigma_\beta|\tau, \omega)$. The fully conditional posterior distributions of all the variables (including the parameters) are conjugate, and thus, the Gibbs sampler is used to simulate every variable involved in the process. We now describe the fully conditional posterior distributions in the order in which they are sampled in the MCMC process.

1. The fully conditional posterior distribution for π_k is beta with updated parameters given below:

$$p(\pi_k|\cdots) = \text{Beta}\left(\pi_k \left| 1 + \sum_{j=1}^N \eta_{jk}, 1 + N - \sum_{j=1}^N \eta_{jk}\right.\right) \tag{25.58}$$

2. Once π_k is sampled, we calculate ρ_{jk} using the Bayes' theorem,

$$\rho_{jk} = \frac{\pi_k N(\gamma_{jk}|0, \Sigma_k)}{\pi_k N(\gamma_{kj}|0, \Sigma_k) + (1 - \pi_k)N(\gamma_{jk}|0, \Sigma_0)} \tag{25.59}$$

3. The fully conditional posterior of η_{jk} is Bernoulli with parameter ρ_{jk}

$$p(\eta_{jk}|\cdots) = \text{Bernoulli}(\eta_{jk}|\rho_{jk}) \tag{25.60}$$

4. The overall mean μ_β has a normal conditional posterior distribution,

$$p(\mu_\beta|\cdots) = N\left[\mu_\beta \left| A^{-1}\left(\sum_{j=1}^N \frac{1}{\sigma_j^2} 1^T y_j^*\right), A^{-1}\right.\right] \tag{25.61}$$

where

$$A = M \sum_{j=1}^N \frac{1}{\sigma_j^2} \tag{25.62}$$

and

$$y_j^* = y_j - 1(\beta_j - \mu_\beta) - \sum_{k=1}^p \eta_{jk} Z_k \gamma_{jk} \tag{25.63}$$

is an adjusted gene expression vector by subtraction of all other model effects except μ_β from the original observed vector of gene expressions.

5. The gene-specific intercept β_j has a normal fully conditional posterior distribution as given below:

$$p(\beta_j|\cdots) = N(\beta_j|\hat{\beta}_j, \hat{W}_j) \tag{25.64}$$

where

$$\hat{\beta}_j = \left(\frac{N}{\sigma^2} + \frac{1}{\Sigma_\beta}\right)^{-1} \left(\frac{1}{\sigma^2} 1^T y_j^* + \frac{1}{\Sigma_\beta} \mu_\beta\right) \tag{25.65}$$

and

$$\hat{W}_j = \mathrm{var}(\beta_j | \cdots) = \left(\frac{N}{\sigma^2} + \frac{1}{\Sigma_\beta}\right)^{-1} \tag{25.66}$$

where the offset is redefined as

$$y_j^* = y_j - \sum_{k=1}^{p} \eta_{kj} Z_k \gamma_{jk} \tag{25.67}$$

6. Each of the γ_{jk} effects has a normal distribution,

$$p(\gamma_{jk} | \cdots) = N(\gamma_{jk} | \hat{\gamma}_{jk}, \hat{S}_{jk}) \tag{25.68}$$

where

$$\hat{\gamma}_{jk} = \left(Z_k^T Z_k + \sigma_j^2 \Theta_{jk}^{-1}\right)^{-1} Z_k^T y_j^* \tag{25.69}$$

and

$$\hat{S}_{kj} = \left(Z_k^T Z_k + \sigma_j^2 \Theta_{kj}^{-1}\right)^{-1} \sigma_j^2 \tag{25.70}$$

The adjusted vector of gene expression y_j^* is redefined as

$$y_j^* = y_j - 1\beta_j - \sum_{k' \neq k}^{p} \eta_{jk'} Z_{k'} \gamma_{jk'} \tag{25.71}$$

7. The fully conditional posterior of Σ_β is an inverse Wishart distribution,

$$p(\Sigma_\beta | \cdots) = \mathrm{Inv} - \mathrm{Wishart} \left[\Sigma_\beta | \tau + N, \omega + SS_\beta\right] \tag{25.72}$$

where

$$SS_\beta = \sum_{j=1}^{N} (\beta_j - \mu_\beta)(\beta_j - \mu_\beta)^T \tag{25.73}$$

is the sum of squares of all the β_j's.
8. The variance of the associated cluster for marker k has a fully conditional posterior distribution of scaled inverse chi-square,

$$p(\Sigma_k | \cdots) = \mathrm{Inv} - \chi^2 \left(\Sigma_k \left| \tau + \sum_{j=1}^{N} \eta_{jk}, \omega + \sum_{j=1}^{N} \eta_{jk} \gamma_{jk}^2\right.\right) \tag{25.74}$$

9. Finally, the fully conditional posterior distribution for the residual variance is the following scaled inverse chi-square:

$$p(\sigma_j^2 | \cdots) = \text{Inv} - \chi^2 \left(\sigma_j^2 | \tau + M, \omega + SS_j \right) \qquad (25.75)$$

where

$$SS_j = \left(y_j - 1\beta_j - \sum_{k=1}^{p} \eta_{jk} Z_k \gamma_{jk} \right)^T \left(y_j - 1\beta_j - \sum_{k=1}^{p} \eta_{jk} Z_k \gamma_{jk} \right) \qquad (25.76)$$

is the residual sum of squares for gene j.

The MCMC sampling process generates posterior samples for all variables. Burn-in deletion and chain thinning apply to the samples prior to the post-MCMC analysis. An important parameter is the proportion of genes associated with each marker π_k. Replacing k by the genome location λ, we can draw a profile of associated genes $\pi(\lambda)$ against λ.

25.2.4   Hierarchical Evolutionary Stochastic Search

Most recently, Bottolo et al. (2011) developed a hierarchical model for eQTL mapping. The model is similar to that of Jia and Xu (2007) but with a set of new hierarchical priors. Their method is more powerful than the original method due to the use of improved prior distributions. Because their prior is complicated, the usual MCMC sampling process is not efficient. The authors then used an improved sampling strategy called hierarchical evolutionary stochastic search (HESS) algorithm. This algorithm is an extension of the evolutionary stochastic search (ESS) algorithm developed by Bottolo and Richardson (2010). We will briefly introduce this new method and state the differences of it from the Bayesian method of Jia and Xu (2007).

The multiple eQTL model remains the same as that of Jia and Xu (2007) given in (25.38), but it is rewritten in a more compact form as given below:

$$y_j = 1\beta_j + Z\gamma_j + \varepsilon_j \qquad (25.77)$$

where p is the number of markers, $\gamma_j = [\gamma_{j1}, \ldots, \gamma_{jp}]^T$ is a $p \times 1$ vector of eQTL effects for gene j, and $Z = [Z_1, \ldots, Z_p]$ is an $M \times p$ matrix for the marker genotype indicator variables. The residual error is assumed to be $N(0, \sigma_j^2)$ distributed. The difference between Bottolo et al. (2011) hierarchical model and that of Jia and Xu (2007) comes from the prior setup. Recall that $\eta_j = [\eta_{j1}, \ldots, \eta_{jp}]$ is a vector of indicator variables for inclusion of $\gamma_j = [\gamma_{j1}, \ldots, \gamma_{jp}]^T$ in the model. Given η_j, the prior distribution for γ_j is

$$p\left(\gamma_j|\eta_j,\alpha,\sigma_j^2\right) = N\left[\gamma_j \left|0,\alpha\left(Z_\eta^T Z_\eta\right)^{-1}\sigma_j^2\right.\right] \qquad (25.78)$$

where α is a shrinkage parameter and σ_j^2 is the residual variance for gene j. The new design matrix Z_η is a subset of matrix Z that contains only the columns corresponding to nonzero η_j. Accordingly, γ_j is a subset of γ_j corresponding the nonzero elements indicated by vector η_j. This particular prior will capture the covariance structure of the marker genotypes.

The hierarchical model completes by specifying priors for σ_j^2 and α. The prior for σ_j^2 is

$$p\left(\sigma_j^2\right) = \text{Inv-Gamma}\left(\sigma_j^2|a,b\right) \qquad (25.79)$$

where $a > 0$ and $b > 0$ are hyperparameters. This prior is different from the inverse chi-square prior used by Jia and Xu (2007). The prior for α is

$$p(\alpha) = \text{Inv-Gamma}\left(\alpha|\tfrac{1}{2},\tfrac{M}{2}\right) \qquad (25.80)$$

This prior leads to a heavy-tailed normal distribution for γ_j, called the Cauchy distribution,

$$p\left(\gamma_j|\eta_j,\sigma_j^2\right) = \text{Cauchy}\left[\gamma_j \left|0,n\left(Z_\eta^T Z_\eta\right)^{-1}\sigma_j^2\right.\right] \qquad (25.81)$$

For the indicator variable η_{jk}, they assign a Bernoulli prior

$$p(\eta_{jk}|\rho_{jk}) = \text{Bernoulli}(\eta_{jk}|\rho_{jk}) \qquad (25.82)$$

which is the same as that of Jia and Xu (2007). However, Bottolo et al. (2011) decomposed ρ_{jk} into

$$\rho_{jk} = \pi_j \times \varphi_k, \ \forall j = 1,\ldots,N \ \& \ k = 1,\ldots,p \qquad (25.83)$$

where π_j and φ_k represent "row" and "column" effects, respectively, and $0 \leq \pi_j \leq 1$ and $\varphi_k \geq 0$. The idea of this decomposition is to control the level of sparsity for gene j through a suitable choice of the hyperparameters (a_j,b_j) of the beta distribution for π_j,

$$p(\pi_j) = \text{Beta}(\pi_j|a_j,b_j) \qquad (25.84)$$

Note that π_j defined here is different from π_k used in (25.40). Here, π_j represents the proportion of markers that affect gene j, while π_k defined before represents the proportion of genes that are affected by marker k. The newly defined φ_k here more or less serves the same role as π_k in the MCMC sampling algorithms because a large φ_k indicates that marker k is most likely a hot spot (affecting the expressions of many genes). Bottolo et al. (2011) concluded their prior setup by specifying a prior for φ_k as

$$p(\varphi_k) = \text{Gamma}(\varphi_k|c,d) \tag{25.85}$$

The hierarchical model eventually ends up with hyperparameters (a_j, b_j) for π_j where $j = 1, \ldots, N$ and hyperparameters (c, d) for φ_k where $k = 1, \ldots, p$. Selection of the hyperparameters remains a challenging task, and some suggestions can be found in Bottolo et al. (2011).

The prior setup of Bottolo et al. (2011) has a nice property of allowing explicit integrations for many parameters, e.g., β_j, γ_{jk}, and σ_j^2. After integration of these parameters, the parameters left are α, η, and ρ where

$$\eta = \{\eta_{jk}, \forall j = 1, \ldots, N \ \& \ k = 1, \ldots, p\}$$

and

$$\rho = \{\rho_{jk}, \forall j = 1, \ldots, N \ \& \ k = 1, \ldots, p\}$$

The joint density of parameters and data is

$$p(y, Z, \alpha, \eta, \rho) = p(y|Z, \eta, \alpha)p(\eta|\rho)p(\rho)p(\alpha) \tag{25.86}$$

where

$$p(y|Z, \eta, \alpha) = \prod_{j=1}^{N} p(y_j|Z, \eta_j, \alpha), \tag{25.87}$$

$$p(\eta|\rho) = \prod_{j=1}^{N} \prod_{k=1}^{p} p(\eta_{jk}|\rho_{jk}) \tag{25.88}$$

and

$$p(\rho) = \prod_{j=1}^{N} p(\pi_j) \prod_{k=1}^{p} p(\varphi_k) \tag{25.89}$$

As described earlier, $p(\alpha)$ has an inverse gamma prior (see (25.80)). The posterior distribution is proportional to the above joint distribution, and thus, MCMC sampling can be conducted using (25.86) as the target distribution.

Sampling η is very challenging since the complex structures in Z create problems of multimodality of the model space even for a single gene. Here the computational challenge is even higher because we are dealing with a huge model space of dimension $(2^p)^N$. Bottolo and Richardson (2010) proposed the ESS sampling algorithm, which has been adopted to the hierarchical model, named HESS, by Bottolo et al. (2011) for eQTL mapping. The idea of HESS is that for each gene (response), HESS relies on running multiple chains with different "temperature" in parallel. These chains exchange information for some covariates. Since chains with high temperatures flatten the posterior density, global moves (between chains) allow the algorithm to jump from one local mode to another. Local moves (within chains) permit fine exploration of alternative models, resulting in a combined algorithm ensuring that the chains mix efficiently and do not become trapped locally. The entire MCMC sampling process involves the ESS step and the Metropolis steps. A brief summary of the iteration process is given below:

1. Given ρ and α, update η according to the ESS procedure using global and local moves.
2. Given η and α, sample π_j and φ_k using a random walk Metropolis step with adaptive proposals.
3. Given η and ρ, sample α using a random walk Metropolis step with a fixed proposal.
4. Go back to Step (1) to Step (3) until the chains reach a desired length.

The HESS algorithm concludes with the post-MCMC analysis. The most important information in the analysis is the posterior propensity of each marker to be a "hot spot." We can declare marker k as a hot spot if

$$\Pr(\varphi_k | y, Z) \geq t \tag{25.90}$$

where $t = 0.8$ or another number defined by the investigator. Another interesting quantity is the "linkage" for the (j, k) pair. One may define the marginal probability of $\eta_{jk} = 1$ and declare a linkage if

$$\Pr(\eta_{jk} = 1 | y, Z) \geq t \tag{25.91}$$

The HESS algorithm is much too complicated to describe fully here. We only introduced the prior setup and the concept of HESS. The description provided here is certainly not sufficient for readers to code the HESS algorithm. Interested readers should read the original studies (Bottolo and Richardson 2010; Bottolo et al. 2011) for detailed information.

References

Abdi H (2007) Bonferroni and Šidák corrections for multiple comparisons. Encyclopedia of Measurement and Statistics. Sage, Thousand Oaks, California

Almasy L, Blangero J (1998) Multipoint quantitative-trait linkage analysis in general pedigrees. Am J Human Genet 62(5):1198–1211

Amos CI (1994) Robust variance-components approach for assessing genetic linkage in pedigrees. Am J Human Genet 54(3):535–543

Baldi P, Long AD (2001) A Bayesian framework for the analysis of microarray expression data: regularized t-test and statistical inferences of gene changes. Bioinformatics 17(6):509–519

Banerjee S, Yandell BS, Yi N (2008) Bayesian quantitative trait loci mapping for multiple traits. Genetics 179(4):2275–2289

Benjamini Y, Hochberg Y (1995) Controlling the false discovery rate – a practical and powerful approach to multiple testing. J Roy Stat Soc Ser B (Stat Methodol) 57(1):289–300

Blalock EM, Geddes JW, Chen KC, Porter NM, Markesbery WR, Landfield PW (2004) Incipient Alzheimer's disease: microarray correlation analyses reveal major transcriptional and tumor suppressor responses. Proc Nat Acad Sci USA 101(7):2173–2178

Bottolo L, Petretto E, Blankenberg S, Cambien F, Cook SA, Tiret L, Richardson S (2011) Bayesian detection of expression quantitative trait loci hot-spots. Genetics 189(4):1449–1459

Bottolo L, Richardson S (2010) Evolutionary stochastic search for Bayesian model exploration. Bayesian Anal 5(3):583–618

Box GEP, Cox DR (1964) An analysis of transformations. J Roy Stat Soc Ser B (Stat Methodol) 26(2):211–252

Box GEP, Tiao GC (1973) Bayesian inference in statistical analysis. Wiley, New York

Brem RB, Yvert G, Clinton R, Kruglyak L (2002) Genetic dissection of transcriptional regulation in budding yeast. Science 296(5568):752–755

Broman KW, Speed TP (2002) A model selection approach for the identification of quantitative trait loci in experimental crosses. J Roy Stat Soc Ser B (Stat Methodol) 64(4):641–656

Cai X, Huang A, Xu S (2011) Fast empirical Bayesian LASSO for multiple quantitative trait locus mapping. BMC Bioinformatics 12(1):211

Che X, Xu S (2010) Significance test and genome selection in Bayesian shrinkage analysis. Int J Plant Genomics 2010:doi:10.1155/2010/893206

Chen M, Presting G, Barbazuk WB, Goicoechea JL, Blackmon, B, Fang G, Ki H, Frisch D, Yu Y, Sun S, Higingbottom S, Phimphilai J, Phimphilai D, Thurmond S, Gaudette B, Li P, Liu J, Hatfield J, Main D, Farrar K, Henderson C, Barnett L, Costa R, Williams B, Walser S, Atkins M, Hall C, Budiman MA, Tomkins JP, Luo M, Bancroft I, Salse J, Regad F, Mohapatra T, Singh NK, Tyagi AK, Soderlund C, Dean RA, Wing RA (2002) An integrated physical and genetic map of the rice genome. Plant Cell 14(3):537–545

S. Xu, *Principles of Statistical Genomics*, DOI 10.1007/978-0-387-70807-2,
© Springer Science+Business Media, LLC 2013

Cheung VG, Spielman RS (2002) The genetics of variation in gene expression. Nat Genet 32(Supp):522–525

Chun H, Keles S (2009) Expression quantitative trait loci mapping with multivariate sparse partial least squares regression. Genetics 182(1):79–90

Churchill GA, Doerge RW (1994) Empirical threshold values for quantitative trait mapping. Genetics 138(3):963–971

Civardi L, Xia Y, Edwards EJ, Schnable PS, Nikolau BJ (1994) The relationship between genetic and physical distances in the cloned al-h2 interval of the Zea mays L. genome. Proc Nat Acad Sci USA 91(17):8268–8272

Cohen AC (1991) Truncated and censored samples:theory and applications, vol 119 of Statistics: textbooks and monographs, 1st edn. Marcel Dekker Inc., New York

Cookson W, Liang L, Abecasis G, Moffatt M, Lathrop M (2009) Mapping complex disease traits with global gene expression. Nat Rev Genet 10(3):184–194

Cullinan WE, Herman JP, Battaglia DF, Akil H, Watson SJ (1995) Pattern and time course of immediate early gene expression in rat brain following acute stress. Neuroscience 64(2): 477–505

Dagliyan O, Uney-Yuksektepe F, Kavakli IH, Turkay M (2011) Optimization based tumor classification from microarray gene expression data. Publ Libr Sci One 6(2):e14579

de Boor C (1978) A practical guide to splines. Springer, New York

Dempster AP, Laird NM, Rubin DB (1977) Maximum likelihood from incomplete data via the EM algorithm. J Roy Stat Soc Ser B (Stat Methodol) 39(1):1–38

Dou B, Hou B, Xu H, Lou X, Chi X, Yang J, Wang F, Ni Z, Sun Q (2009) Efficient mapping of a female sterile gene in wheat (Triticum aestivum l.). Genet Res 91(05):337–343

Dunn OJ (1961) Multiple comparisons among means. J Am Stat Assoc 56(293):52–64

Efron B (1979) Bootstrap methods: another look at the jackknife. Ann Stat 7(1):1–26

Efron B, Hastie T, Johnstone I, Tibshirani R (2004) Least angle regression. Ann Stat 32(2):407–499

Efron B, Tibshirani R, Storey JD, Tusher V (2001) Empirical Bayes analysis of a microarray experiment. J Am Stat Assoc 96(456):1151–1160

Eisen MB, Spellman PT, Brown PO, Botstein D (1998) Cluster analysis and display of genome-wide expression patterns. Proc Nat Acad Sci USA 95(25):14863–14868

Elston RC, Steward J (1971) A general model for the genetic analysis of pedigree data. Hum Hered 21(6):523–542

Emilsson V, Thorleifsson G, Zhang B, Leonardson AS, Zink F, Zhu J, Carlson S, Helgason A, Walters GB, Gunnarsdottir S, Mouy M, Steinthorsdottir V, Eiriksdottir GH, Bjornsdottir G, Reynisdottir I, Gudbjartsson D, Helgadottir A, Jonasdottir A, Styrkarsdottir U, Gretarsdottir S, Magnusson KP, Stefansson H, Fossdal R, Kristjansson K, Gislason HG, Stefansson T, Leifsson BG, Thorsteinsdottir U, Lamb JR, Gulcher JR, Reitman ML, Kong A, Schadt EE, Stefansson K (2008) Genetics of gene expression and its effect on disease. Nature 452:423–428

Falconer DS, Mackay TFC (1996) Introduction to quantitative genetics, 4th edn. Longman Group Ltd., London

Feenstra B, Skovgaard IM, Broman KW (2006) Mapping quantitative trait loci by an extension of the Haley-Knott regression method using estimating equations. Genetics 173(4):2269–2282

Felsenstein J (1981a) Evolutionary trees from DNA sequences: a maximum likelihood approach. J Mol Evol 17(6):368–376

Felsenstein J (1981b) Evolutionary trees from gene frequencies and quantitative characters: finding maximum likelihood estimates. Evolution 35(6):1229–1242

Felsenstein J (1985) Confidence limits on phylogenies: an approach using the bootstrap. Evolution 39(4):783–791

Fisher RA (1946) A system of scoring linkage data, with special reference to the pied factors in mice. Am Nat 80(794):568–578

Fraley C, Raftery AE (2002) Model-based clustering, discriminant analysis, and density estimation. J Am Stat Assoc 97(458):611–631

Friedman J, Hastie T, Tibshirani R (2010) Regularization paths for generalized linear models via coordinate descent. J Stat Software 33(1):1–22

Fu YB, Ritland K (1994) On estimating the linkage of marker genes to viability genes controlling inbreeding depression. Theoret Appl Genet 88(8):925–932

Fulker DW, Cardon LR (1994) A sib-pair approach to interval mapping of quantitative trait loci. Am J Hum Genet 54(6):1092–1103

Gelfand AE, Hills SE, Racine-Poon A, Smith AFM (1990) Illustration of Bayesian inference in normal data models using Gibbs sampling. J Am Stat Assoc 85(412):972–985

Gelman A (2005) Analysis of variance – why it is more important than ever. Ann Stat 33(1):1–53

Gelman A (2006) Prior distributions for variance parameters in hierarchical models (Comment on article by Browne and Draper). Bayesian Anal 1(3):515–533

Gelman A, Jakulin A, Pittau MG, Su YS (2008) A weakly informative default prior distribution for logistic and other regression models. Ann Appl Stat 2(4):1360–1383

Geman S, Geman D (1984) Stochastic relaxation, Gibbs distribution, and the Bayesian restoration of images. IEEE Trans Pattern Anal Mach Intell PAMI-6(6):721–741

George EI, McCulloch RE (1993) Variable selection via Gibbs sampling. J Am Stat Assoc 88(423):881–889

George EI, McCulloch RE (1997) Approaches for Bayesian variable selection. Statistica Sinica 7:339–373

Ghosh D, Chinnaiyan AM (2002) Mixture modelling of gene expression data from microarray experiments. Bioinformatics 18(2):275–286

Gilks WR, Richardson S, Spiegelhalter DJ (1996) Markov chain Monte Carlo in practice. Chapman and Hall/CRC, London

Glonek G, Solomon P (2004) Factorial and time course designs for cDNA microarray experiments. Biostatistics 5(1):89–111

Goldgar DE (1990) Multipoint analysis of human quantitative genetic variation. Am J Hum Genet 47(6):957–967

Golub GH, Van Loan CF (1996) Matrix computations, 3rd edn. The Johns Hopkins University Press, Baltimore

Golub TR, Slonim DK, Tamayo P, Huard C, Gaasenbeek M, Mesirov JP, Coller H, Loh ML, Downing JR, Caligiuri MA, Bloomfield CD, Lander ES (1999) Molecular classification of cancer: class discovery and class prediction by gene expression monitoring. Science 286(5439):531–537

Green PJ (1995) Reversible jump Markov chain Monte Carlo computation and Bayesian model determination. Biometrika 82(4):711–732

Hackett CA, Meyer RC, Thomas WTB (2001) Multi-trait QTL mapping in barley using multivariate regression. Genet Res 77(1):95–106

Hackett CA, Weller JI (1995) Genetic mapping of quantitative trait loci for traits with ordinal distributions. Biometrics 51(4):1252–1263

Haldane JBS (1919) The combination of linkage values and the calculation of distances between the loci of linked factors. J Genet 8(29):299–309

Haldane JBS, Waddington CH (1931) Inbreeding and linkage. Genetics 16(4):357–374

Haley CS, Knott SA (1992) A simple regression method for mapping quantitative trait loci in line crosses using flanking markers. Heredity 69(4):315–324

Haley CS, Knott SA, Elsen JM (1994) Mapping quantitative trait loci in crosses between outbred lines using least squares. Genetics 136(3):1195–1207

Han L, Xu S (2008) A Fisher scoring algorithm for the weighted regression method of QTL mapping. Heredity 101(5):453–464

Han L, Xu S (2010) Genome-wide evaluation for quantitative trait loci under the variance component model. Genetica 138(9–10):1099–1109

Hardy GH (1908) Mendelian proportions in a mixed population. Science 28(706):49–50

Hartigan J, Wong MA (1979) Algorithm AS 136: a K-means clustering algorithm. J Roy Stat Soc Ser C (Appl Stat) 28(1):100–108

Hartigan JA (1975) Clustering algorithms. Wiley, New York

Hartl DL, Clark AG (1997) Principles of population genetics, 3rd edn. Sinauer Associates Inc., Sunderland, Massachusetts

Haseman JK, Elston RC (1972) The investigation of linkage between a quantitative trait and a marker locus. Behav Genet 2(1):3–19

Hastings WK (1970) Monte Carlo sampling methods using Markov chains and their applications. Biometrika 57(1):97–109

Hayes JG (1974) Numerical methods for curve and surface fitting. Bull Inst Math Appl 10(5/6):144–152

Hayes PM, Liu BH, Knapp SJ, Chen F, Jones B, Blake T, Franckowiak J, Rasmusson D, Sorrells M, Ullrich SE, Wesenberg D, Kleinhofs A (1993) Quantitative trait locus effects and environmental interaction in a sample of North American barley germ plasm. Theor Appl Genet 87(3):392–401

Heath SC (1997) Markov chain Monte Carlo segregation and linkage analysis of oligogenic models. Am J Hum Genet 61(3):748–760

Henderson CR (1950) Estimation of genetic parameters (abstract). Ann Math Stat 21(2):309–310

Henderson CR (1975) Best linear unbiased estimation and prediction under a selection model. Biometrics 31(2):423–447

Henshall JM, Goddard ME (1999) Multiple-trait mapping of quantitative trait loci after selective genotyping using logistic regression. Genetics 151(2):885–894

Hoerl AE, Kennard RW (1970) Ridge regression: Biased estimation for nonorthogonal problems. Technometrics 12(2):55–67

Horton NJ, Laird NM (1999) Maximum likelihood analysis of generalized linear models with missing covariates. Stat Methods Med Res 8(1):37–50

Hu Z, Xu S (2009) PROC QTL – a SAS procedure for mapping quantitative trait loci. Int J Plant Genom 2009:1–3, doi:10.1155/2009/141234

Huelsenbeck JP Ronquist F, Nielsen R, Bollback JP (2001) Bayesian inference of phylogeny and its impact on evolutionary biology. Science 294(5550):2310–2314

Ibrahim JG (1990) Incomplete data in generalized linear models. J Am Stat Assoc 85(411): 765–769

Ibrahim JG, Chen MH, Lipsitz SR (2002) Bayesian methods for generalized linear models with covariates missing at random. Can J Stat 30(1):55–78

Ibrahim JG, Chen MH, Lipsitz SR, Herring AH (2005) Missing-data methods for generalized linear models. J Am Stat Assoc 100(469):332–346

Jia Z, Xu S (2005) Clustering expressed genes on the basis of their association with a quantitative phenotype. Genet Res 86(3):193–207

Jia Z, Xu S (2007) Mapping quantitative trait loci for expression abundance. Genetics 176(1): 611–623

Jiang C, Zeng ZB (1995) Multiple trait analysis of genetic mapping for quantitative trait loci. Genetics 140(3):1111–1127

Jiang C, Zeng ZB (1997) Mapping quantitative trait loci with dominance and missing markers in various crosses from two inbred lines. Genetica 101(1):47–58

Jirapech-Umpai T, Aitken S (2005) Feature selection and classification for microarray data analysis: evolutionary methods for identifying predictive genes. BMC Bioinform 6:148

Kao CH (2000) On the differences between the maximum likelihood and the regression interval mapping in the analysis of quantitative trait loci. Genetics 156(2):855–865

Kao CH, Zeng ZB, Teasdale RD (1999) Multiple interval mapping for quantitative trait loci. Genetics 152(3):1203–1216

Kendziorski C, Wang P (2006) A review of statistical methods for expression quantitative trait loci mapping. Mamm Genome 17(6):509–517

Kendziorski CM, Chen M, Yuan M, Lan H, Attie AD (2006) Statistical methods for expression quantitative trait loci (eQTL) mapping. Biometrics 62(1):19–27

Knott SA, Haley CS (2000) Multitrait least squares for quantitative trait loci detection. Genetics 156(2):899–911

Korol AB, Ronin YI, Itskovich AM, Peng J, Nevo E (2001) Enhanced efficiency of quantitative trait loci mapping analysis based on multivariate complexes of quantitative traits. Genetics 157(4):1789–1803

Korol AB, Ronin YI, Kirzhner VM (1995) Interval mapping of quantitative trait loci employing correlated trait complexes. Genetics 140(3):1137–1147

Kosambi DD (1943) The estimation of map distances from recombination values. Ann Hum Genet 12(1):172–175

Lan H, Chen M, Flowers JB, Yandell BS, Stapleton DS, Mata CM, Mui ET, Flowers MT, Schueler KL, Manly KF, Williams RW, Kendziorski C, Attie AD (2006) Combined expression trait correlations and expression quantitative trait locus mapping. Pub Lib Sci Genet 2(1):e6

Land AH, Doig AG (1960) An automatic method of solving discrete programming problems. Econometrica 28(3):497–520

Lander ES, Botstein D (1989) Mapping Mendelian factors underlying quantitative traits using RFLP linkage maps. Genetics 121(1):185–199

Lee Y, Lee C (2003) Classification of multiple cancer types by multicategory support vector machines using gene expression data. Bioinformatics 19(9):1132–1139

Li CC (1955) Population genetics. University of Chicago Press, Chicago

Liao J, Chin K (2007) Logistic regression for disease classification using microarray data: model selection in a large p and small n case. Bioinformatics 23(15):1945–1951

Liu BH (1998) Statistical genomics: linkage, mapping and qtl analysis, 1st edn. CRC, Boca Raton

Lorieux M, Goffinet B, Perrier X, Leon DG, Lanaud C (1995a) Maximum-likelihood models for mapping genetic markers showing segregation distortion. 1. Backcross populations. Theor Appl Genet 90(1):73–80

Lorieux M, Perrier X, Goffinet B, Lanaud C, Leon DG (1995b) Maximum-likelihood models for mapping genetic markers showing segregation distortion. 2. F2 populations. Theor Appl Genet 90(1):81–89

Loudet O, Chaillou S, Camilleri C, Bouchez D, Daniel-Vedele F (2002) Bay-0 × Shahdara recombinant inbred line population: a powerful tool for the genetic dissection of complex traits in Arabidopsis. Theor Appl Genet 104(6):1173–1184

Louis T (1982) Finding the observed information matrix when using the EM algorithm. J Roy Stat Soc Ser B (Stat Methodol) 44(2):226–233

Luan Y, Li H (2003) Clustering of time-course gene expression data using a mixed-effects model with B-splines. Bioinformatics 19(4):474–482

Luo L, Xu S (2003) Mapping viability loci using molecular markers. Heredity 90(6):459–467

Luo L, Zhang YM, Xu S (2005) A quantitative genetics model for viability selection. Heredity 94(3):347–355

Luo ZW, Zhang RM, Kearsey MJ (2004) Theoretical basis for genetic linkage analysis in autotetraploid species. Proc Nat Acad Sci USA 101(18):7040–7045

Luo ZW, Zhang Z, Leach L, Zhang RM, Bradshaw JE, Kearsey MJ (2006) Constructing genetic linkage maps under a tetrasomic model. Genetics 172(4):2635–2645

Lynch M, Walsh B (1998) Genetics and analysis of quantitative traits, 1st edn. Sinauer Associates Inc., Sunderland

Ma P, Castillo-Davis CI, Zhong W, Liu JS (2006) A data-driven clustering method for time course gene expression data. Nucleic Acids Res 34(4):1261–1269

MacQueen JB (1967) Some methods for classification and analysis of multivariate observations. Proceedings of the 5th Berkeley symposium on mathematical statistics and probability, vol 1, pp 281–297, Berkeley, California

Mangin B, Thoquet P, Grimsley N (1998) Pleiotropic QTL analysis. Biometrics 54(1):88–99

McCullagh P, Nelder JA (1999) Generalized linear models. Monograph on statistics and applied probability. Chapman and Hall/CRC, London

McCulloch CE, Searle SR (2001) Generalized linear and mixed models. Wiley, New York

McLachlan G, Peel D (2000) Finite mixture models. Wiley, New York

McLachlan GJ, Bean RW, Peel D (2002) A mixture model-based approach to the clustering of microarray expression data. Bioinformatics 18(3):413–422

McNicholas PD, Murphy TB (2010) Model-based clustering of microarray expression data via latent Gaussian mixture models. Bioinformatics 26(21):2705–2712

Metropolis N, Rosenbluth AW, Rosenbluth MN, Teller AH, Teller E (1953) Equation of state calculations by fast computing machines. J Chem Phys 21(6):1087–1092

Mitchell-Olds T (1995) Interval mapping of viability loci causing heterosis in Arabidopsis. Genetics 140(3):1105–1109

Morgan TH (1928) The theory of the gene. Yale University Press, New Haven

Morgan TH, Bridges CB (1916) Sex-linked inheritance in drosophila. Carniegie Institute of Washington, Washington DC

Narula SC (1979) Orthogonal polynomial regression. Int Stat Rev 47(1):31–36

Nelder JA, Mead R (1965) A simplex method for function minimization. Comput J 7(4):308–313

Nelder JA, Wedderburn RWM (1972) Generalized linear models. J Roy Stat Soc Ser A (General) 135(3):370–384

Nettleton D, Doerge RW (2000) Accounting for variability in the use of permutation testing to detect quantitative trait loci. Biometrics 56(1):52–58

Newton MA, Noueiry A, Sarkar D, Ahlquist P (2004) Detecting differential gene expression with a semiparametric hierarchical mixture method. Biostatistics 5(2):155–176

Ouyang M, Welsh WJ, Georgopoulos P (2004) Gaussian mixture clustering and imputation of microarray data. Bioinformatics 20(6):917–923

Pan W, Lin J, Le CT (2002) Model-based cluster analysis of microarray gene expression data. Genome Biol 3(2):research0009.1–0009.8

Park T, Casella G (2008) The Bayesian Lasso. J Am Stat Assoc 103(482):681–686

Park T, Yi SG, Lee S, Lee SY, Yoo DH, Ahn JI, Lee YS (2003) Statistical tests for identifying differentially expressed genes in time-course microarray experiments. Bioinformatics 19(6):694–703

Peddada SD, Lobenhofer EK, Li L, Afshari CA, Weinberg CR, Umbach DM (2003) Gene selection and clustering for time-course and dose-response microarray experiments using order-restricted inference. Bioinformatics 19(7):834–841

Piepho HP (2001) A quick method for computing approximate thresholds for quantitative trait loci detection. Genetics 157(1):425–432

Potokina E, Caspers M, Prasad M, Kota R, Zhang H, Sreenivasulu N, Wang M, Graner A (2004) Functional association between malting quality trait components and cDNA array based expression patterns in barley. Mol Breed 14(2):153–170

Qu Y, Xu S (2004) Supervised cluster analysis for microarray data based on multivariate Gaussian mixture. Bioinformatics 20(12):1905–1913

Qu Y, Xu S (2006) Quantitative trait associated microarray gene expression data analysis. Mol Biol Evol 23(8):1558–1573

Robinson GK (1991) That BLUP is a good thing: the estimation of random effects. Stat Sci 6(1):15–32

Rubin NB (1987) Multiple imputation for nonresponse in survey. Wiley, New York

Rubinstein R (1981) Simulation and the Monte Carlo method. Wiley, New York

Saitou N, Nei M (1987) The neighbor-joining method: a new method for reconstructing phylogenetic trees. Mol Biol Evol 4(4):406–425

SAS Institute (2008a). SAS/IML 9.2 user's guide. SAS Institute Inc, Cary, North Carolina

SAS Institute (2008b) SAS/STAT 9.2 user's guide. SAS Institute Inc., Cary, North Carolina

Satagopan JM, Yandell BS, Newton MA, Osborn TC (1996) A Bayesian approach to detect quantitative trait loci using Markov chain Monte Carlo. Genetics 144(2):805–816

Schadt EE, Monks SA, Drake TA, Lusis AJ, Che N, Colinayo V, Ruff TG, Milligan SB, Lamb JR, Cavet G, Linsley PS, Mao M, Stoughton RB, Friend SH (2003) Genetics of gene expression surveyed in maize, mouse and man. Nature 422:297–302

Schena M, Shalon D, Davis RW, Brown PO (1995) Quantitative monitoring of gene expression patterns with a complementary DNA microarray. Science 270(5235):467–470

Schliep A, Schnhuth A, Steinhoff C (2003) Using hidden Markov models to analyze gene expression time course data. Bioinformatics 19(supp 1):i255–i263

Schork NJ (1993) Extended multipoint identity-by-descent analysis of human quantitative traits: efficiency, power, and modeling considerations. Am J Hum Genet 53(6):1306–1319

Schwarz GE (1978) Estimating the dimension of a model. Ann Stat 6(2):461–464

Searle SR, Casella G, McCulloch CE (1992) Variance components. Wiley, New Yok

Seber GAF (1977) Linear regression analysis, 1st edn. Wiley, New York

Sillanpaa MJ, Arjas E (1998) Bayesian mapping of multiple quantitative trait loci from incomplete inbred line cross data. Genetics 148(3):1373–1388

Sillanpää MJ, Arjas E (1999) Bayesian mapping of multiple quantitative trait loci from incomplete outbred offspring data. Genetics 151(4):1605–1619

Smyth GK (2004) Linear models and empirical Bayes methods for assessing differential expression in microarray experiments. Stat Appl Genet Mol Biol 3(2):Article 3

Sobel E, Sengul H, Weeks DE (2001) Multipoint estimation of identity-by-descent probabilities at arbitrary positions among marker loci on general pedigrees. Hum Hered 52(3):121–131

Sober E (1983) Parsimony in systematics: philosophical issues. Ann Rev Ecol Systemat 14: 335–357

Sokal R, Michener C (1958) A statistical method for evaluating systematic relationships. Univ Kansas Sci Bull 38:1409–1438

Sorensen D, Gianola D (2002) Likelihood, Bayesian, and MCMC methods in quantitative genetics. Springer, New York

Statnikov A, Aliferis CF, Tsamardinos I, Hardin D, Levy S (2005) A comprehensive evaluation of multicategory classification methods for microarray gene expression cancer diagnosis. Bioinformatics 21(5):631–643

Steeb W, Hardy Y (2011) Matrix calculus and Kronecker product: a practical approach to linear and multilinear algebra. World Scientific Publishing Company, Singapore

Storey JD, Xiao W, Leek JT, Tompkins RG, Davis RW (2005) Significance analysis of time course microarray experiments. Proc Nat Acad Sci USA 102(36):12837–12842

Studier JA, Keppler KJ (1988) A note on the neighbor-joining algorithm of Saitou and Nei. Mol Biol Evol 5(6):729–731

Swofford DL, Olsen GJ, Waddell PJ, Hillis DM (1996) Phylogenetic inference. In: Hillis DM, Moritz C, Mable BK (eds) Molecular systematics, 2nd edn. Sinauer Associates, Sunderland, Mass., pp 407–514

ter Braak CJF, Boer MP, Bink MCAM (2005) Extending Xu's Bayesian model for estimating polygenic effects using markers of the entire genome. Genetics 170(3):1435–1438

Tibshirani R (1996) Regression shrinkage and selection via the Lasso. J Roy Stat Soc Ser B (Stat Methodol) 58(1):267–288

Tipping ME (2001) Sparse Bayesian learning and the relevance vector machine. J Mach Learn Res 1:211–244

Tusher VG, Tibshirani R, Chu G (2001) Significance analysis of microarrays applied to the ionizing radiation response. Proc Nat Acad Sci USA 98(9):5116–5121

Visscher PM, Haley CS, Knott SA (1996) Mapping QTLs for binary traits in backcross and F2 populations. Genet Res 68(01):55–63

Vogl C, Xu S (2000) Multipoint mapping of viability and segregation distorting loci using molecular markers. Genetics 155(3):1439–1447

Wald A (1943) Tests of statistical hypotheses concerning several parameters when the number of observations is large. Trans Am Math Soc 54(3):426–482

Wang C, Zhu C, Zhai H, Wan J (2005a) Mapping segregation distortion loci and quantitative trait loci for spikelet sterility in rice (Oryza sativa l.). Genet Res 86(2):97–106

Wang H, Zhang Y, Li X, Masinde GL, Mohan S, Baylink DJ, Xu S (2005b) Bayesian shrinkage estimation of quantitative trait loci parameters. Genetics 170(1):465–480

Wedderburn RWM (1974) Quasi-likelihood functions, generalized linear models, and the Gauss-Newton method. Biometrika 61(3):439–447

Weinberg W (1908) Über den nachweis der vererbung beim menschen. Jahreshefte des Vereins für vaterländische Naturkunde in Württemberg 64:368–382

Welham S, Cullis B, Kenward M, Thompson R (2007) A comparison of mixed model splines for curve fitting. Aust New Zeal J Stat 49(1):1–23

Williams JT, Van Eerdewegh P, Almasy L, Blangero J (1999) Joint multipoint linkage analysis of multivariate qualitative and quantitative traits. I. Likelihood formulation and simulation results. Am J Hum Genet 65(4):1134–1147

Wolfinger RD, Gibson C, Wolfinger ED, Bennet L, Hamadeh H, Rishel P, Afshari C, Paules RS (2001) Assessing gene significance from cDNA microarray expression data via mixed models. J Comput Biol 8(6):625–637

Xie C, Xu S (1999) Mapping quantitative trait loci with dominant markers in four-way crosses. Theor Appl Genet 98(6):1014–1021

Xu C, Li Z, Xu S (2005) Joint mapping of quantitative trait loci for multiple binary characters. Genetics 169(2):1045–1059

Xu C, Wang X, Li Z, Xu S (2009) Mapping QTL for multiple traits using Bayesian statistics. Genet Res 91(1):23–37

Xu C, Xu S (2003) A SAS/IML program for mapping QTL in line crosses. Proceedings of the twenty-eighth annual SAS users group international conference (SUGI), Cary, NC. SAS Institute

Xu S (1995) A comment on the simple regression method for interval mapping. Genetics 141(4):1657–1659

Xu S (1996) Mapping quantitative trait loci using four-way crosses. Genet Res 68(02):175–181

Xu S (1998a) Further investigation on the regression method of mapping quantitative trait loci. Heredity 80(3):364–373

Xu S (1998b) Iteratively reweighted least squares mapping of quantitative trait loci. Behav Genet 28(5):341–355

Xu S (2003) Estimating polygenic effects using markers of the entire genome. Genetics 163(2):789–801

Xu S (2007) An empirical Bayes method for estimating epistatic effects of quantitative trait loci. Biometrics 63(2):513–521

Xu S (2008) Quantitative trait locus mapping can benefit from segregation distortion. Genetics 180(4):2201–2208

Xu S, Atchley WR (1995) A random model approach to interval mapping of quantitative trait loci. Genetics 141(3):1189–1197

Xu S, Hu Z (2009) Mapping quantitative trait loci using distorted markers. Int J Plant Genom 2009, doi:10.1155/2009/410825

Xu S, Hu Z (2010) Generalized linear model for interval mapping of quantitative trait loci. Theor Appl Genet 121(1):47–63

Xu S, Xu C (2006) A multivariate model for ordinal trait analysis. Heredity 97(6):409–417

Xu S, Yi N (2000) Mixed model analysis of quantitative trait loci. Proc Nat Acad Sci USA 97(26):14542–14547

Xu S, Yi N, Burke D, Galecki A, Miller RA (2003) An EM algorithm for mapping binary disease loci: application to fibrosarcoma in a four-way cross mouse family. Genet Res 82(2):127–138

Yeung KY, Bumgarner RE (2003) Multiclass classification of microarray data with repeated measurements: application to cancer. Genome Biol 4(12):R83

Yi N (2004) A unified Markov chain Monte Carlo framework for mapping multiple quantitative trait loci. Genetics 167(2):967–975

Yi N, George V, Allison DB (2003) Stochastic search variable selection for identifying multiple quantitative trait loci. Genetics 164(3):1129–1138

Yi N, Shriner D (2008) Advances in Bayesian multiple QTL mapping in experimental designs. Heredity 100(3):240–252

Yi N, Xu S (1999) A random model approach to mapping quantitative trait loci for complex binary traits in outbred populations. Genetics 153(2):1029–1040

Yi N, Xu S (2000) Bayesian mapping of quantitative trait loci for complex binary traits. Genetics 155(3):1391–1403

Yi N, Xu S (2001) Bayesian mapping of quantitative trait loci under complicated mating designs. Genetics 157(4):1759–1771

Yi N, Xu S (2008) Bayesian LASSO for quantitative trait loci mapping. Genetics 179(2): 1045–1055

Yuan M, Lin Y (2005) Efficient empirical Bayes variable selection and estimation in linear models. J Am Stat Assoc 100(472):1215–1225

Zeng ZB (1994) Precision mapping of quantitative trait loci. Genetics 136(4):1457–1468

Zhan H, Chen X, Xu S (2011) A stochastic expectation and maximization algorithm for detecting quantitative trait-associated genes. Bioinformatics 27(1):63–69

Zhao H, Speed TP (1996) On genetic map functions. Genetics 142(4):1369–1377

Zhu J, Hastie T (2004) Classification of gene microarrays by penalized logistic regression. Biostatistics 5(3):427–443

Index

S. Xu, *Principles of Statistical Genomics*, DOI 10.1007/978-0-387-70807-2,
© Springer Science+Business Media, LLC 2013